SAFE MOBILITY: CHALLENGES, METHODOLOGY AND SOLUTIONS

TRANSPORT AND SUSTAINABILITY

Series Editors: Stephen Ison, John Shaw and Maria Attard

Recent Volumes:

Volume 1:	Cycling and Sustainability
Volume 2:	Transport and Climate Change
Volume 3:	Sustainable Transport for Chinese Cities
Volume 4:	Sustainable Aviation Futures
Volume 5:	Parking: Issues and Policies
Volume 6:	Sustainable Logistics
Volume 7:	Sustainable Urban Transport
Volume 8:	Paratransit: Shaping the Flexible Transport Future
Volume 9:	Walking: Connecting Sustainable Transport with Health
Volume 10:	Transport, Travel and Later Life

TRANSPORT AND SUSTAINABILITY
VOLUME 11

SAFE MOBILITY: CHALLENGES, METHODOLOGY AND SOLUTIONS

EDITED BY

DOMINIQUE LORD
Texas A&M University, College Station, TX, USA

SIMON WASHINGTON
The University of Queensland, Brisbane, Australia

emerald PUBLISHING

United Kingdom – North America – Japan
India – Malaysia – China

Emerald Publishing Limited
Howard House, Wagon Lane, Bingley BD16 1WA, UK

First edition 2018

British Library Cataloguing in Publication Data
A catalogue record for this book is available from the British Library

ISBN: 978-1-78635-224-8 (Print)
ISBN: 978-1-78635-223-1 (Online)
ISBN: 978-1-78714-892-5 (Epub)

ISSN: 2044-9941 (Series)

ISOQAR certified
Management System,
awarded to Emerald
for adherence to
Environmental
standard
ISO 14001:2004.

ISOQAR
REGISTERED
Certificate Number 1985
ISO 14001

INVESTOR IN PEOPLE

For Dominique Lord:

"For my wife Leah and our son Javier, and to my mother Diane and my brother Sébastien and his family.

And especially for my father Laurent, who sadly passed away shortly before this book was finished."

For Simon Washington:

"I'd like to dedicate this book to my sister Karen, who is the most kind, compassionate, and loving sister–thank you for your unwavering support and encouragement over my career and life."

CONTENTS

LIST OF TABLES

LIST OF FIGURES

EDITORIAL BOARD

LIST OF CONTRIBUTORS

Dr. Dominique Lord is a Professor in the Zachry Department of Civil Engineering at Texas A&M University. Over the last 20 years, Dr. Lord has conducted several research studies in the United States, Canada and across the world. His work has led to the identification of important issues in highway safety research and the development of new and innovative methodologies for analysing crash data. The results of his research have been used by researchers across the world and several areas, including medicine, accounting, mathematics, statistics, biology and most engineering disciplines among others. Dr. Lord's primary interests are conducting fundamental research on accident analysis methodology, new and innovative statistical methods for analysing motor vehicle collisions and before–after evaluation techniques. He has published more than 110 papers published in peer-reviewed journals and presented more than 100 papers at international conferences. He is also the recipient of numerous and prestigious awards.

Simon Washington is currently Head of School and Professor, School of Civil Engineering at the University of Queensland. He is recognised internationally for his contributions in the fields of behavioural econometrics applied in the areas of transport safety and risk across travel modes, transport and urban planning and travel behaviour. He is associate editor or Editorial Advisory Board Member for six leading international transport journals. At the time of this book's completion, Simon has co-authored more than 100 peer-reviewed journal articles and a second edition of a textbook, which has been adopted in over 20 countries, and six book chapters. Simon has been lead investigator on more than $26 Million of externally supported research and has secured nationally competitive research grants both in the United States and Australia. Prior to joining UQ, Simon served on the faculties of the Queensland University of Technology, UC Berkeley, Arizona State University, the University of Arizona and the Georgia Institute of Technology.

Dr. Lyndel Judith Bates is a Senior Lecturer in the School of Criminology and Criminal Justice and Griffith Criminology Institute at Griffith University. She is an award-winning researcher who has completed work in the areas of road policing, procedural justice, third-party policing, graduated driver licensing and novice drivers. Dr. Bates has been awarded a significant amount of competitive funding from government and industry. She has published her work widely including in the respected journals *Accident Analysis and Prevention, Policing and Society* and the *Journal of Safety Research.*

Dr. Ashleigh Filtness is a Lecturer in Transport Safety at Loughborough Design School, Loughborough University, UK. She specialises in safety implications for road user behaviour, with particular expertise in driver sleepiness, impaired driving and using driving simulators for research. Following the completion of her PhD Dr. Filtness has worked at Monash University Accident Research Centre, Australia; The Centre for Accident Research and Road Safety – Queensland, Australia; and Loughborough University, UK. During her career she has worked on a range of road safety research projects funded by both industry and competitive grant schemes. She has published her work widely in the respected journals within both the transport safety and sleep fields including *Journal of Sleep Research and Accident Analysis and Prevention.*

Professor Barry Watson is a Global Road Safety Expert and a Research Professor at the Queensland University of Technology (QUT) in Brisbane, Australia. From 2015 to 2017, he was the Chief Executive Officer of the Global Road Safety Partnership (GRSP), which is hosted by the Federation of the Red Cross and Red Crescent Societies in Geneva, Switzerland. In this role, Barry was responsible for GRSP's overall strategic direction and the management of its activities in over 35 countries worldwide. Prior to this, Barry worked in the road safety field for over 30 years in government, academic and civil society organisations. Between 2008 and 2015, he was the Director of the Centre for Accident Research & Road Safety at QUT. Barry is a member of the Executive Board of the International Council on Alcohol, Drugs & Traffic Safety and a Fellow of the Australasian College of Road Safety.

Kara Kockelman is a registered Professional Engineer and holds a PhD, MS and BS in Civil Engineering, a master's of city planning and a Minor in Economics from the University of California at Berkeley. Dr. Kockelman has been a Professor of Transportation Engineering at the University of Texas at Austin for the past 20 years. She is primary and co-author of over 140 journal articles (and one book) across a variety of subjects, nearly all of which involve transportation-related data analysis. Her primary research interests include planning for shared and autonomous vehicle systems, the statistical modelling of urban systems (including models of travel behaviour, trade and location choice), energy and climate issues (vis-à-vis transport and land use decisions), the economic impacts of transport policy and crash occurrence and consequences.

Jianming Ma holds a Ph.D. Degree in Civil Engineering from the University of Texas at Austin. Dr. Ma coordinates connected and automated vehicle activities at the Texas Department of Transportation (TxDOT), whether demonstration, pilot projects, national studies and university research. Dr. Ma has over twenty years of professional and research experience in connected and automated vehicles (CAVs), econometric modeling, human factors, intelligent transportation systems (ITS), systems engineering, traffic engineering, and traffic safety. He has over 40 technical papers and reports published in the above mentioned areas. Dr. Ma co-chairs the American Society of Civil Engineers (ASCE) CAV Impacts Committee. He is the Research Coordinator for the Transportation Research Board (TRB) Committee on Safety Data, Analysis and Evaluation (ANB20), a member of TRB's Traffic Signal Systems (AHB25).

Mitchell Cunningham is a Graduate Psychologist with a Bachelor of Science (Hons) Degree from The University of Sydney, and a Master of Research Degree in psychology from Macquarie University. Mitch currently holds the position of Behavioural Scientist (Human Factors) at the Australian Road Research Board in Sydney, Australia. Mitch has published in a number of areas including driver inattention and distraction, as well as how human drivers interact with automated vehicles. Mitch has specialist skills in the analysis, and interpretation, of behavioural and psychological data.

Dr. Michael A. Regan is the Chief Scientist-Human Factors at the Australian Road Research Board (ARRB) and holds an honorary appointment as Adjunct Professor with the School of Aviation at the University of New South Wales in Sydney, Australia. He is a past President of the Human Factors and Ergonomics Society of Australia. Mike has BSc (Hons) and PhD Degrees in Experimental Psychology and Human Factors from the Australian National University and more than 20 years' experience in transportation human factors and safety – as a Researcher, Research Manager and Policymaker. He has specialist knowledge and expertise in driver distraction and inattention and driver interaction with vehicle automation. He has authored and co-authored around 180 peer-reviewed publications, including 3 books.

Dr. Tom Brijs graduated in 1997 from the Limburg University Center (now, Hasselt University) as Engineer in Business Informatics. From 1997 until 2002, he was active at the Limburg University Center as a PhD. He obtained his PhD title in data mining in 2002. Since then, he started working for Hasselt University's Transportation Research Institute (IMOB) as an FWO Postdoctoral Researcher and Lecturer with road safety as his core research area. Currently, he is Full Professor in Road Safety, Vice Director of IMOB and coordinates scientific research in the areas of road safety, human-centered road design and driving simulation. He published over 85 scientific articles in peer-reviewed ISI-ranked academic journals or conferences of which 74 in road safety more specifically. He is a member of the Editorial Board of the Scientific Open-Access Journal *Safety* and member of the Editorial Advisory Board of *Accident Analysis and Prevention*.

Prof. Dr. Ali Pirdavani graduated in 2007 in Road and Transportation Engineering. From 2008 until 2012 he was a PhD Scholar at the Transportation Research Institute (IMOB) of Hasselt University, Belgium. He obtained his PhD in Transportation Sciences in 2012. In 2013 he obtained an FWO Postdoctoral Scholarship and continued working at IMOB with road safety as his major research domain. In 2016 he was appointed as an Assistant Professor at the Faculty of Engineering Technology of Hasselt University. Prof. Pirdavani published over 30 scientific articles in peer-reviewed ISI-ranked academic journals and international conferences over the last seven years of his academic career in the areas of road safety, infrastructure and road design, microsimulation and driving simulation. He is also the Academic Coordinator or Lecturer of several courses in the Bachelor–Master Programs in Transportation Sciences as well as Construction Engineering, such as road design and road ergonomics 1 and 2, introduction to construction engineering, research and communication: construction engineering, road safety evaluation: methods and applications, microsimulation models, internship, bachelor thesis and master thesis.

Prof. Francesca La Torre is a Full Professor of Roads Railways and Airports at the University of Florence (Italy). She obtained her PhD in 1998 at the University of Rome and she served as an Assistant Researcher at the University of Illinois at Urbana-Champaign (USA). She is a member of the EC Horizon 2020 Advisory Group for 'Smart, Green and Integrated Transport'. She is the Infrastructures Representative for Academia in ERTRAC and a Research Area Leader for Safety and Security in FEHRL. She is a member of the TG Road Safety of CEDR. She has served as member of several Committees of the US Transportation Research Board and of the US Task Force for the development of the Highway Safety Manual. She has been a member of the OECD/ITF WG on Assessment of the Effectiveness of Road Safety Measures. She has served as Coordinator in several European transnational funded projects.

Dr. Richard Tay specializes in multidisciplinary theory and evidence-based approaches to improve transportation safety, efficiency, accessibility and sustainability. He has published extensively and serves on the Editorial Boards of many journals, including *Accident Analysis and Prevention, Journal of Transportation Safety and Security, Journal of Advanced Transportation, Australian College of Road Safety Journal, Open Transportation Journal, Canadian Journal of Transportation, Asian Transport Studies, International Journal of Sustainable Transportation, International Journal of Transportation, Urban, Planning and Transport Research* and *Analytic Methods in Accident Research.* He has been invited as a Keynote Speaker and Chair of various technical sessions at many international transportation and road safety conferences. He has also been invited to many government expert panels, technical committees, advisory boards and task forces on transportation policy and road safety in USA, Canada, Australia and Asia, and invited to comment on transportation and safety issues by major news agencies around the world.

Alfonso Montella is Associate Professor (with Scientific Qualification as professor) at the University of Naples Federico II, Italy, where he Teaches Highway Design and Highway Safety. He got a PhD in Transportation Engineering at the University of Rome La Sapienza and a MSc in Civil Engineering at the University of Naples Federico II. He taught Highway Safety at USEK, Lebanon and CU, Egypt. He participated in several national and international research projects on highway design and safety and coordinated the EU-funded projects HiT4Med and HDMCuRF. He has been member of the Italian National Council of University. He is a member of the Editorial Board of Accident Analysis & Prevention, the Transportation Research Board Committee Transportation Safety Management and the Joint International Research Laboratory of Transportation Safety of Tongji University. His main areas of expertise include highway design, highway safety management, highway safety modelling, road safety audits and inspections, and drivers' behaviour investigations by driving simulator experiments.

Dr. Mohamed Abdel-Aty, PE is Trustee Chair, Pegasus Professor and Chair of the Civil, Environmental and Construction Engineering Department at the University of Central Florida. He is also the Deputy Director of the Transportation Center. His main expertise and interests are in the areas of traffic safety analysis, simulation, big data and data analytics and Intelligent Transportation Systems. In the last 20 years, Dr. Abdel-Aty has led more than 50 research projects. Dr. Abdel-Aty has published 450 papers, more than 240 in journals (Citations 9700, H-Index 51). He supervised the graduation of 62 PhD and MS students. Dr. Abdel-Aty is the Editor-in-Chief of *Accident*

Analysis and Prevention, the premier journal in safety. He and his students received multiple awards for their research from Transportation Research Board, WCTR, Intelligent Transportation Systems Florida and FL section ITE. He has been invited to deliver many keynote speeches in conferences around the world, including in Belgium, Brazil, China, Korea, Turkey, KSA, Jordan, Qatar and UAE.

Dr. Qi Shi is currently working as an Assistant Researcher at the Research Institute of Highway, Ministry of Transport (RIOH, MOT), in Beijing, China. Dr. Shi graduated from Tongji University, Shanghai, China, with a Bachelor's Degree in Transportation Engineering. He came and continued his graduate study at University of Central Florida (UCF) from 2011 to 2014 under the supervision of Dr. Abdel-Aty. He earned his PhD Degree in Civil Engineering in Fall, 2014, with a focus on traffic safety and efficiency using Big Data. After graduation, he continued his research as a postdoctoral researcher at UCF for one year. Since he came back and started working in China from 2016, he endeavoured to implement the technologies and experiences he learned in the US to solve China's transportation problems. He wishes to create a safer environment for road travellers through his research, making the world a better place for living.

Dr. Anurag Pande is Associate Professor of Civil Engineering at Cal Poly State University. Dr. Pande's research interests include traffic safety and crash data, driver behaviour, observational data applications in traffic safety, emergency evacuation, service learning and scholarship of teaching. He serves on two Transportation Research Board (TRB) Committees as a member: Safety Data, Analysis, and Evaluation (ANB 20) and Emergency Evacuation (ABR 30). He was recognized with "Young Researcher Award" by the Safety Data, Analysis, and Evaluation committee in 2007. He has co-authored more than 30 manuscripts that have been either published or are forthcoming in peer-reviewed journals. He is the Editor of the recently released 7th edition of *Traffic Engineering Handbook* (TEH) published by ITE (Institute of Transportation Engineers). Dr. Pande received his BTech in Civil Engineering from Indian Institute of Technology Bombay in Mumbai (India), and MS and PhD in Civil Engineering from University of Central Florida.

Dr. Rongjie Yu is an Associate Professor at the College of Transportation Engineering, Tongji University. His main research area is traffic safety analysis, safety improvement countermeasures and driving behaviour analysis. Focussed on the research area, he is now the PI for several projects funded by the National Natural Science Foundation of China and other organizations.

Dr. Rongjie Yu has published and co-authored 21 SCI/SSCI indexed journal papers, such as *Accident Analysis and Prevention, Transportation Research Par C*. And he is also the reviewer for a number of key academic journals in the traffic safety field.

Per Gårder is a Professor at the Department of Civil and Environmental Engineering at the University of Maine since 1992. Before that he was a Lecturer at the Royal Institute of Technology in Stockholm Sweden. He received his PhD Degree in 1982 from the University of Lund, Sweden, with a Dissertation on Pedestrian Safety at Signalized Intersections. His research interest is focussed on forecasting, designing and evaluating facilities with emphasis on traffic safety. Dr. Gårder has been conducting research since the mid-1970s in four main areas: the safety of pedestrians and bicyclists, roundabout design and safety, the safety of signalized intersections, and fatigue/distraction and the effectiveness of continuous rumble strips. Lately, he has been researching advance technology, so-called Intelligent Transportation Systems, and pedestrian safety, among other places at VTT in Finland, where he was a Fulbright Fellow for the fall of 2012. He is currently also working on traffic safety on bridges.

Professor Narelle Haworth is the Director of CARRS-Q, the Centre for Accident Research and Road Safety-Queensland in Brisbane, Australia. She has 30 years of experience in road safety research. Her special interest is in improving the safety of the most vulnerable road users – pedestrians, bicyclists and motorcyclists – and other groups such as road workers.

Jacqueline Fuller currently works for Queensland Government in strategic policy. She has a range of experience in research and in public policy, with a background in criminology and criminal justice. She is currently undertaking her PhD at Griffith University in Brisbane, Australia.

Fred Mannering is currently a Professor of Civil and Environmental Engineering (with a courtesy appointment in economics) at the University of South Florida. He previously held academic appointments as the Charles Pankow Professor of Civil Engineering at Purdue University, Professor and chair of Civil Engineering at the University of Washington and Assistant Professor at the Pennsylvania State University. He received his Undergraduate Degree from the University of Saskatchewan, Masters from Purdue University and Doctorate from the Massachusetts Institute of Technology. Dr. Mannering's expertise is in the application of statistical and econometric methods to study a variety of subject areas including highway safety, transportation economics, automobile demand and travel behaviour. He has published over 130 refereed journal articles, 2 text books, over 60 other publications (conference proceedings, project reports, book reviews and commentaries) and has given over 120 invited lectures and presentations at professional conferences.

Mohammed Quddus is a Professor of Intelligent Transport Systems (ITS) and Head of Transport Studies Group within the School of Civil and Building Engineering at Loughborough. Professor Quddus received his PhD in ITS from Imperial College London in 2006. Over the last 15 years, Professor Quddus has conducted cutting-edge research leading to innovative, influential and transformative outcomes in the area of transport and safety modelling as well as in ITS. His seminal papers on map-matching algorithms have been very highly cited by researchers worldwide. He has directed more than 20 national and international research projects funded by the UK Engineering and Physical Sciences Research Council, Highways England, Department for Transport (UK) and the European Union. So far, he has authored/co-authored over 80 journal articles, 70 conference papers, 4 book chapters and 12 technical reports.

Dr. Srinivas Reddy Geedipally received his Doctorate from Texas A&M University and has been with Texas A&M Transportation Institute since 2005. He is currently the Associate Research Engineer in the Center for Transportation Safety and has more than 11 years of experience in traffic safety research. He has more than 45 papers published in high-standard international journals and conferences. Dr. Geedipally has participated in numerous traffic safety research projects with state and federal governments and international sponsors. He has been a key contributor in the development of the *Highway Safety Manual*. He is a registered Professional Engineer in the state of Texas. Dr. Geedipally is a two-time recipient of the Young Researcher Award and a Fred Burggraf Award winner from Transportation Research Board.

John N. Ivan is a Professor of Civil and Environmental Engineering at the University of Connecticut and served as Associate Head of the Department for 10 years. He served in 2016 as a Research Civil Engineer at the Turner Fairbank Highway Research Center of Federal Highway Administration in the Office of Safety Research and Development. In 2009 he was a visiting researcher at Lund University, Sweden, and in 2002–2003 he was a Fulbright Senior Scholar at the Institute for Transport Studies at the University of Karlsruhe in Germany, and a Research Engineer at the Texas Transportation Institute at Texas A&M University. He has earned BS, MS and PhD Degrees in Civil Engineering at Carnegie Mellon University, Massachusetts Institute of Technology and Northwestern University, respectively. He is an Associate Editor of *Accident Analysis and Prevention*, and in 2011 was elected for the Connecticut Academy of Science and Engineering.

Dr. Karthik C. Konduri is an Assistant Professor in the Department of Civil and Environmental Engineering at the University of Connecticut. He received his BTech in Civil Engineering from the Institute of Technology at Banaras Hindu University, India, in 2004 followed by his MS in Civil Engineering from the University of Kentucky, in 2006 and finally his PhD in Civil Engineering from Arizona State University in 2012. His primary research interest is in the field of transportation systems with an emphasis on econometric and statistical modelling methodologies. Papers he co-authored have received best paper awards including the 2015 Transportation Research Board's Committee ANB20 (Safety Data Analysis and Evaluation) Young Researcher Paper Award and the 2012 Transportation Research Board's Pyke Johnson Award for the best paper in the field of transportation systems planning and administration. He also serves as the associate editor of the *Transportation Letters* journal.

Amir Pooyan Afghari is a PhD Candidate in Transportation Engineering at the University of Queensland in Australia. He received his Bachelors in Civil Engineering from Isfahan University of Technology (Iran) and his Masters in Transportation Engineering from Concordia University (Canada). Amir has worked at several research centres around the world and has been involved in globally developed research projects. To name a few, he participated in a stated preference study at Ferdowsi University of Mashhad in Iran, a structural design study at the University of Wuppertal in Germany, an automated safety study at Ecole Polytechnique de Montreal and McGill University in Canada. He is currently conducting fundamental research in transportation safety and crash data analysis at the University of Queensland in Australia. Amir's research interests cover a variety of topics in transportation engineering including transportation data analysis, transportation safety, econometrics, discrete choice models, intelligent transportation systems and transportation asset management.

Dr. Md. Mazharul Haque is a Senior Lecturer in Transportation Engineering at the Queensland University of Technology in Australia. He is a specialist in statistical and econometrics applications in transportation engineering and traffic safety. His research interests include traffic safety analysis, blackspot identification techniques, microscopic traffic flow models, travel behaviour and human factors and driving behaviour. Dr. Haque has published more than

50 peer-reviewed research articles in top-class journals and international conferences. His journal papers are published in journals like *Accident Analysis & Prevention*, *Transportation Research Part B: Methodological*, *Transportation Research Part C: Emerging Technologies*, *Safety Science* and *Transportation Research Record*. Dr. Haque is an Associate Editor of the *ASCE – Journal of Transportation Engineering*, a prestigious journal within civil engineering published by the American Society of Civil Engineers. He is also a member of the Committee on Statistical Methods (ABJ80), Transportation Research Board of the National Academics, USA.

Dr. Andrew P. Tarko is Professor of Civil Engineering, Lyles School of Civil Engineering, Purdue University, and Director of Purdue Center for Road Safety at Purdue University and Research Director of the International Laboratory on Transportation Safety at Tongji University, Shanghai, China. His research interest includes transportation safety with focus on fundamentals of safety measurement and on safety management methods. Dr. Tarko teaches transportation courses at Purdue University and short courses and seminars in USA and overseas. He is a member of Editorial Boards of several international journals. Dr. Tarko was a Principal Investigator of research grants and awards for the total amount of nearly eight million dollars. His team has developed two safety management computer tools implemented to transportation practice. Dr. Tarko has published many journal and conference papers, chapters in six books and numerous research reports. His team received the AASHTO recognition for high-value research in 2012, 2013 and 2017.

Bhagwant Persaud, a Professor of Civil Engineering at Ryerson University, conducts research in statistical methods in highway safety analysis. Dr. Persaud has been, and is currently involved, in several safety-related research projects for the Federal Highway Administration, the Transportation Research Board (TRB) and many North American road agencies. He has authored or co-authored numerous peer-reviewed papers, three of which won awards for TRB's Annual Meeting outstanding paper in the field of operation, safety and maintenance of transportation facilities. He has held appointments on two TRB committees: as a member of 'Statistical Methods' and as Co-Chair and now Emeritus Member of 'Safety Data, Analysis and Evaluation'. He has been a member of the TRB Task Force for the *Development of a Highway Safety Manual* and has been involved in research projects for several chapters of this manual.

Rune Elvik is a Political Scientist from the University of Oslo. He attained the Degree of Doctor of Political Science (Dr. Polit) in 1993 and the Degree of doctor of Philosophy (Dr. Philos) in 1999. In 2007, he was awarded a Doctoral Degree by Aalborg University in Denmark. His main areas of research include evaluation of the effects of road safety measures, research synthesis by means of meta-analysis and cost-benefit analysis. During the years 1997–2004 he was an associate editor of *Accident Analysis and Prevention*. From 2005 to 2013 he was one of the Editors-in-Chief of the journal. From 1999 to 2008 Elvik was a member the Committee for Safety Data, Analysis and Evaluation (ANB20) of the Transportation Research Board.

ACKNOWLEDGEMENTS

This book would never have come to fruition without the contributions from numerous influential and inspirational individuals. This book project was an arduous endeavour fraught with setbacks and various delays beyond our control, but perseverance prevailed.

First and foremost, we would like to thank all of the authors who participated in the writing of this book. Without their time, consider efforts and perseverance, this book would not be possible. The authors are also very grateful to the following individuals at Emerald Publishing Group: Jennifer McCall, Rachel Ward and Cristina Irving Turner. Particular thanks goes to Dr. Stephen Ison, from Loughborough University, who first contacted us about writing this book 4 years ago and provided important feedback and encouragement throughout the preparation of the manuscript, along with a dose of good humour which was always welcome and appreciated. Dr. Jon Shaw from Plymouth University provided invaluable feedback at the beginning of the project. The production of this book would not have been possible without the meticulous input and feedback of Mrs. Hannah Murphy, who spent several hours editing the preliminary drafts of each and every chapter. Finally, in the spirit of peer review and to achieving a quality manuscript, the book has greatly benefitted from the following reviewers:

Tom Brijs
Mitchell Cunningham
Rune Elvik
Per Gårder
Mazharul Haque
Narelle Haworth
Paul Jovanis
Francesca La Torre
Jianming Ma
Fred Mannering
Bhagwant Persaud

Ali Pirdavani
Mohammed Quddus
Michael Regan
Nicholas Saunier
Venky Shankar

The constructive comments provided by these individuals surpassed our expectations and reminded us of what a wonderfully supportive and collaborative discipline to which we have chosen to associate ourselves.

CHAPTER 1

INTRODUCTION

Dominique Lord and Simon Washington

ABSTRACT

Purpose – *This chapter first provides the motivation for writing this book. It then describes the challenges involved with assessing societal safety through the analysis of transport system crashes. It concludes with a summary of the contents of the remainder of the book, identifying how various dimensions of the transport system challenges are addressed.*

Methodology/Approach – *This chapter discusses important real-world and methodological challenges that practitioners, academics and researchers face in making a more sustainable highway system through a reduction in the number and severity of transport network crashes resulting in fatalities, injuries and property damage.*

Findings – *The chapter first describes important challenges, such as complexity of the driving task, the challenges of engineering transport systems for humans, unanticipated effects that arise from differences between driver safety and security, the co-mingling of mobility modes of travel, and challenges in evaluating road safety. The chapters are separated into five general themes: driver behaviour, the transportation network, vulnerable road users, methods for understanding and predicting safety performance, and methods for evaluating safety impacts of countermeasures.*

Safe Mobility: Challenges, Methodology and Solutions
Transport and Sustainability, Volume 11, 1–10

ISSN: 2044-9941/doi:10.1108/S2044-994120180000011001

Practical Implications – *Comprehending the challenges associated with road crashes is a first step in making the roadway system more sustainable. This book provides a broad and understandable description of these challenges and how they can be overcome by academics and practitioners working in transport network safety management.*

Originality – *This book presents a clear understanding and offers insights about the challenges and potential solutions that can be brought to bear to make a more sustainable and safe transport system, whether it is located in an urban or rural area, and for a wide variety of functional classifications and designs. The topics covered in this book are intended to be useful and applied to tackle transport system management anywhere in the world.*

Keywords: Road crashes; road safety; societal risk; risk management; transport engineering; driver behaviour; pedestrians; transport network management; bicyclists

MOTIVATION FOR THE BOOK

Injuries to people and property arising from crashes on road transport networks around the world are staggering by any measure. Road crashes result in 1.33 million deaths annually and 78.2 million non-fatal injuries warranting medical care (Global Road Safety Facility at the World Bank & the Institute for Health Metrics and Evaluation, 2014). Road crashes are among the top 10 leading causes of death between the age of 1 and 59 – and exceed those arising from HIV, tuberculosis and malaria. Pedestrians using the transport network represent 35% of road injury deaths globally. The economic burden of crashes is astonishing. In the United States alone, for example, road crashes are estimated to have caused more than US$871 billion in economic loss and societal harm in 2010 (Blincoe, Miller, Zaloshnja, & Lawrence, 2015). Globally, it is estimated that 3% of gross domestic product (GDP) is lost to road crashes (fatal and non-fatal) and can be as high as 5% for low- and middle-income countries (WHO, 2015). Finally, road crashes have increased by 46% over the most recent two decades, highlighting the growing importance of the global road safety problem, reflecting in part increased mobility in developing countries and emphasising the importance of doubling down to tackle the problem (Global Road Safety Facility at the World Bank and the Institute for Health Metrics and Evaluation, 2014).

As a result of the health burden on global societies, significant questions arise about the sustainability of transport networks as they are currently built, operated and managed. This book is devoted to presenting a thorough review of what is currently known about global road safety and is authored by invited road safety experts from around the world with expertise in their targeted chapters. The intent of the book is to provide a broad, yet detailed reference that presents state-of-the-practice guidance regarding both academic and practical knowledge in transport network safety. The book is intended to be used by practitioners, academics, researchers and students interested in transport system safety. The hopeful and intended outcome of the book is to increase the level of knowledge on road safety contexts, issues and challenges; share what can currently be done to address the variety of issues; and understand what needs to be done to make further gains in road safety.

CHALLENGES IN REDUCING ROAD CRASHES

Improving road safety, or reducing the number of injuries and fatalities that arise during the pursuit of mobility on the transport network, is an enormous challenge. There is a large number of reasons why practitioners and academics struggle to reduce the negative societal impacts of road crashes. These challenges include the complexity of the driving task, the challenges of engineering transport systems for humans, the unanticipated effects that arise from differences between driver safety and security, the co-mingling of mobility modes of travel, and challenges in evaluating road safety. While all of these challenges are briefly discussed below with the intent to provide an introduction to these topics, additional detail is provided in the relevant chapters within the book.

Complexity of Driving Task

Driving is a phenomenally complex task. Driving requires that we are not impaired in any way, that we employ the use of all of our senses, that attention is paid at precisely all of the key moments, that we accurately predict what those around us are intending to do, and we possess sufficient coordination and manual dexterity leading to appropriate driving inputs. It requires us to continually monitor an extremely large number of inputs and not be distracted by things that may be unhelpful to the driving task. It requires us to appropriately manage the effect of our moods while driving, and to responsibly manage the use of prescription drugs, alcohol and illegal drugs. Because driving demands are proportionate to the complexity of the driving environment, drivers are not required to provide constant vigilance – leading to a requirement for drivers to correctly anticipate when cognitive loads and subsequent driving inputs are relatively high or low. Different segments of the network also require different levels of skills and vigilance. High-speed highways are very different than parking lots, while urban city driving is different than driving near to schools. Driving also requires constant monitoring and prediction of the behaviour of other drivers, pedestrians and cyclists. Considering the routine activities of life, driving is very likely to be the most complex activity undertaken by the majority of people on a regular basis.

Engineering Transport Systems for Humans

A high level of safety is not the only attribute of a transport system that engineers strive to deliver through design. Transport systems are also designed to deliver efficient throughput, minimise congestion, support multiple users, link multiple modes of travel, provide equity of mobility and link important transport destinations. Designing all elements of these systems to be as safe as possible while simultaneously delivering other performance indicators often places road designers into difficult trade-off decision contexts regarding safety. As an example, providing efficient access for pedestrians and cyclists on transport networks is likely to put more cyclists and pedestrians at risk of conflict with motor vehicles. Similarly, providing high-speed travel on high-capacity roads increases the risk of injury and death when crashes occur. These trade-offs in transport network design cannot be ignored, as they govern the context of decision making with regard to major transport investments.

Differences between Driver Safety and Security

How drivers interpret and assess crash risk is complex, with large variability in driver responses to the designed road network. One driver may perceive crash risk as low while driving at night, maintaining posted speeds, while another driver may assess crash risk as high and select much lower driving speeds. Some drivers feel secure (with respect to risk assessment) following other drivers according to a 1-second time headway, while others feel secure driving with 2-second headways. When encountering reduced visibility, some drivers may slow their speeds significantly, while others may only slightly adapt their driving to the changed conditions. These differences in driver responses to risk have been noted in a large number of observational and simulator studies and create challenges in the smooth operation of transport networks.

Co-Mingling of Mobility Modes

Transport systems are meant to serve a variety of users, including pedestrians, cyclists, motorists, motorcyclists and public transport users. The crash risk of these user groups varies drastically across the network and network design features. Fundamentally, pedestrians and cyclists lack the safety features that adorn modern vehicles. Safety features like airbags, dynamic stability control, anti-lock brakes and safety restraints work collectively and aggressively to protect motor vehicle occupants in crashes, yet do little to protect pedestrians and cyclists. The newest vehicles are designed to minimise injury risk with pedestrians, whilst some vehicles have pedestrian and cyclist warning detectors built into their systems. By and large, however, most vehicles on the road do little to assist in the protection of the transport system's vulnerable road users and are formidable obstacles when collisions do occur.

Challenges in Evaluating Road Safety

While operating transport systems safely is difficult enough, conducting rigorous scientific studies to quantify and understand crash risk in the transport network is fraught with challenges. The following partial list highlights some of the many challenges confronting the evaluation of road safety:

1. Road crashes are relatively rare events with respect to how much time is spent driving. In essence, this rarity translates to a difficulty in establishing crash trends that are reliable. Moreover, crash causes are quite varied and thus further compound the difficulty in establishing recognisable trends.
2. It is difficult to determine the actual causes of crashes. While most crashes will have a 'critical event' that triggers a sequence of events, it is often difficult to identify the 'trigger' and all of the essential events that followed. Without a reliable explanation of crash causes available at the 'scene of the crash', most of the conclusions drawn are extracted from statistical analysis of readily available site and driver features, which may or may not be directly related to a crash cause.

3. Road crashes are predominately caused by driver errors, which may inappropriately be attributed to roadway features upon subsequent analysis. It is estimated that about 93% of road crashes are caused by human error in part or in total (NHTSA 2008). Conversely, a crash where the primary trigger was inattention by inappropriately be attributed to a roadway design feature, as the *inattention* status of a driver is likely to be unavailable.
4. Because crashes are rare and analysts rely on detecting reliable trends to make safety recommendations, data are often aggregated and as such lose the resolution necessary to capture crash-specific conditions. Average annual daily traffic, for example, is often used to capture exposure to risk but rarely is indicative of the conditions at the actual times of crashes.
5. Evaluation studies aimed at assessing the impact on road safety before and after treatment are threatened by the lack of experimental control on the myriad of factors that can affect road crashes.
6. Simulator studies, while extremely useful for establishing relative risk (i.e., task A is riskier than task B), are notoriously difficult for establishing absolute risk. Moreover, there is not a straightforward method or theoretical underpinning for translating simulator risk to actual driving crash risk.
7. Crash surrogate measures that measure crash 'precursor' events, such as near-misses, are extremely promising tools for evaluating road safety, but are in their infancy in terms of relating crash surrogates to actual crash risk, are sometimes difficult to measure, and are likely to not be generalizable across countries, states and regions – and thus require local calibration.
8. Real-time crash prediction, with the aim of taking a proactive and immediate response to crash mitigation, struggles with the establishment of theoretical relationships between crash risk and microscopic traffic conditions. Moreover, while real-time prediction may become reliable in predicting conditions 'ideal' for crashes to occur, the methods suffer from high false-positive rates. The consequence of having many false positives is losing the confidence of the motoring public in assisting to mitigate traffic risk.

STRUCTURE OF THE BOOK

Chapters are grouped into both human behaviour and engineering design aspects of transport network operation, and road crash evaluation methodologies. All chapters aim to provide current state-of-the-practice knowledge on several questions: what do we know, what do we not know, and how we are going to find out? The chapters are grouped under five general themes: driver behaviour, the transportation networks, vulnerable road users, methods for understanding and predicting safety performance, and methods for evaluating safety impacts of countermeasures.

Because the transport network is an engineered facility that relies on the diligent, consistent behaviour of pedestrians, cyclists and people operating motor vehicles (e.g., cars, motorcycles, buses, commercial vehicles, etc.), there are predictable and often undesirable behaviours that contribute to the road toll. Some of the behaviours are intentional, while others are unintended but are often equally risky. The transport systems we build also influence behaviour and safety performance. Some portions of the network are easier to navigate safely compared to others, and much is known about the traffic and design influences of the network on safety. The chapters that constitute the remainder of this book are organised as follows.

Opening the theme on driver behaviour, Chapter 2 discusses the role of driver licensing on motor vehicle crashes, covering a variety of topics, including graduated driver licensing programs, the role of parents in licensing, compliance and enforcement of licensing, driver testing and unlicensed driving. Graduated driver licensing programs have been shown to be effective, unlike many other driver training programs. Best practice in graduated driver licensing can assist in mitigating the risk of the young and inexperienced driving cohort.

Chapter 3 examines the role of aggressive driving and speeding in motor vehicle crashes, and the role that speeding-related crashes play in the global crash picture. How public attitudes, personal behaviours, vehicle performance, roadway design and laws and policies influence speeding and speed-related crashes are also discussed. Solutions to aggressive driving may include practical enforcement, intentional speed-related roadway design elements, education campaigns and automated driving.

Chapter 4 examines the role of driver distraction in motor vehicle crashes. Driver distraction may lead to reduced driving performance, detracting from the essential driving task and leading to increased crash risk. Empirical research supports the risk associated with some types of distraction – such as mobile phone use whilst driving that is rampant in many parts of the world – requiring an improved understanding of distraction. The links between distraction research and crashes are sparse and require further research. Solutions to distraction offering promise may include the use of technology, driver licensing, and education and training.

On the transport network theme, Chapter 5 presents the evidence around the safety performance of suburban and urban arterial roads – where a large proportion of motor vehicle crashes occur. Traffic characteristics and design features affect the performance of these facilities. While some of the features have known effects, others are inconsistent and are counter-intuitive.

Chapter 6 presents the collective knowledge around controlled access, typically high-speed facilities. While these facilities typically enjoy the highest design-standards of all roads, and also are typically prohibited from use by pedestrians and cyclists, crashes on these facilities tend to be severe.

Rural and urban intersection crashes are described in Chapter 7. The factors contributing to crashes at these locations are diverse and complex. Despite a large number of quantitative studies looking at these network locations, some factors reveal inconsistent effects on safety. Engineering design treatments need to be examined holistically to understand potential unintended effects and should be designed to satisfy Pareto improvement social equity criterion.

Chapter 8 focuses on the safety performance of roundabouts, which offer operational and environmental advantages, but may pose challenges to different road users. Specific features of roundabouts are very important in determining the way they operate, particularly with respect to pedestrian and cyclist safety.

Chapter 9 focuses on progress and issues remaining with real-time traffic operations and safety. The chapter focuses on the methods used to make sense of real-time traffic operations as they relate to safety, the limitations on data sources for implementing such programs and issues of inter-jurisdictional transferability.

Pedestrians, accounting for 35% of persons killed on transport networks globally, are the topic of Chapter 10, which opens the theme on vulnerable road users. Sidewalks are needed on all arterials and collector roads with pedestrian traffic greater than 50 per day to achieve acceptable levels of road safety. Elderly pedestrians require lows speeds and narrow streets to cross safely. Posted speeds yielding 90% speeds of no more than 30 km/h are a target for pedestrian-friendly and safe environments. While pedestrian travel is highly desirable from health liveability perspectives, we need to design more conducive and safe environments for pedestrians.

Another vulnerable road-user group is cyclists, the topic of Chapter 11. While cycling offers a sustainable and affordable solution for fulfilling mobility needs it poses some challenges with respect to safety – particularly when mixed with motor vehicle traffic. A variety of cycling infrastructure treatments are shown to be effective in improving cycling safety, including both on-road and off-road treatments. Getting the infrastructure right is critical for increasing cycling participation rates.

Chapter 12 is the first chapter to focus on crash evaluation methodologies. Cross-sectional data based on information gleaned from crash reports form the basis of many safety studies and require careful guidance on how these data should be analysed. Commonly applied methods are presented in this chapter, along with important data features that represent more pesky yet important features of cross-sectional crash data analysis.

Time-series methods for assessing crash data are presented in Chapter 13. Because crashes are often observed over time, and safety investments often take time to accrue benefits, time-series methods can be extremely useful. While much of the book is focused on road safety, air travel as a transport mode is also of significant importance with respect to risk management. Examples in air safety are provided in this chapter to illustrate the state-of-the-practice techniques in time-series methods applied to assess crashes.

Because crashes are rare events, crash data often are characterised by many zeroes (lots of sites observed in a sample record zero crashes during the observation period). Characterising these data well is important to draw correct inferences from the data. Chapter 14 describes the most current thinking around dealing with these challenging crash datasets, and recommends methods for making sense of crashes from datasets with many zero responses.

Crash severity is an important characteristic of crashes. Because crash severity is the most accurate indicator of the societal harm caused by road crashes, methods designed to model and understand crash severity are discussed in Chapter 15. Data limitations hinder the ability to undertake large and accurate severity analyses. Several different methods explained in this chapter are useful for modelling and understanding what transport network features are related to road crash severity.

Chapter 16 presents the underpinnings of transport network screening, or safety management practice. This practice involves a large number of assumptions and makes significant use of crash modelling discussed in Chapters 12, 14 and 17. A variety of methodologies have been used over the years, with the Empirical Bayes method for identifying hot spots the most reliable.

Surrogate measures of safety are discussed in Chapter 17. Surrogate measures of safety are a rapidly growing subfield in transport safety. The fundamental dilemma of needing to wait for crashes to occur to gain information about how to prevent them is a primary motivator for developing this subfield of research. Identifying the pre-crash movements of vehicles (and other vehicles, pedestrians, cyclists, etc.) that relate to crashes that are expected to eventuate is the goal of this approach. Much of the current work in this area is described, along with where this subfield is headed.

In the last theme related to the safety impacts of countermeasures, Chapter 18 focuses on before–after safety evaluations. These types of investigations are thought to be the most reliable of all safety investigations, yet remain fraught with challenges. The chapter outlines the methods used to conduct these studies, and identifies the salient features of these studies and solutions to overcome their technical difficulties.

Chapter 19 concludes the book, appropriately, with an overview of meta-analytic methods. These methods are used to make sense of a body of research – capitalising on the enormous resources that have already been invested to study a particular safety feature or problem. It presents the methodologies used to compile, assess and make sense of a sample of studies with varied findings across the studies. Because the overall sample size grows with multiple studies, there can be significant benefit in pooling the results to draw meaningful safety conclusions. This chapter provides a roadmap for conducting such studies in a rigorous way.

Overall, this book offers students, practitioners, academics and researchers both an in-depth and broad understanding of the issues and challenges related to road crashes and provides solutions that can be used to overcome these limitations in order to promote a more sustainable transport system by reducing the number and severity of road crashes.

REFERENCES

Blincoe, L., Miller, T. D., Zaloshnja, E., & Lawrence, B. A. (2015). *The economic and societal impact of motor vehicle crashes, 2010* (Revised). Technical Report DOT HS 812 013. U.S. Department of Transportation, National Highway Traffic Safety Administration, Washington, DC.

Global Road Safety Facility at the World Bank & the Institute for Health Metrics and Evaluation. (2014). *Transport for health: The global burden of disease from motorized transport.* Seattle, WA; Washington, D.C.: IHME. Retrieved from http=://documents.worldbank.org/curated/en/984261468327002120/pdf/863040IHME0T4H0ORLD0BANK0compressed.pdf

NHTSA (2008). *National motor vehicle crash causation survey: Report to Congress.* Report DOT HS 811 059. U.S. Department of Transportation, National Highway Traffic Safety Administration, Washington, DC.

WHO (2015). *Global status report on road safety 2015.* Geneva, Switzerland: World Health Organization, WHO Press. Retrieved from http://www.who.int/violence_injury_prevention/road_safety_status/2015/en/

DRIVER BEHAVIOUR: CHALLENGES AND SOLUTIONS

CHAPTER 2

DRIVER EDUCATION AND LICENSING PROGRAMS

Lyndel Judith Bates, Ashleigh Filtness and Barry Watson

ABSTRACT

Purpose – *Driver education and licensing are two mechanisms used to reduce crash rates. The purpose of this chapter is to provide an overview of these countermeasures and consider how simulators can be used to augment more traditional approaches.*

Approach – *A literature review was undertaken evaluating key concepts in driver licensing including graduated driver licensing (GDL), the role of parents in licensing, compliance and enforcement, driver testing and how the driver licensing system impacts on levels of unlicensed driving. Literature regarding driver education for individuals who have and not yet obtained a licence was also reviewed.*

Findings – *GDL is a successful countermeasure for reducing the crash rates of young novice drivers as it limits their exposure to higher risk situations. The support for driver education initiatives is mixed. As there are big differences between education programs, there is a need to consider each*

Safe Mobility: Challenges, Methodology and Solutions
Transport and Sustainability, Volume 11, 13–36

ISSN: 2044-9941/doi:10.1108/S2044-994120180000011002

program on its own merits. Driving simulators provide a safe environment for novices to gain experience. In particular, they may be bifacial for development of hazard perception and visual scanning skills.

Practical Implications – *GDL systems should be introduced where appropriate. Existing systems should be strengthened where possible by including additional, best-practice and restrictions. When considering driver education as a countermeasure, the type of program is very important. Education programs that have been shown to increase crashes should not be introduced. Further research and development are necessary to ensure that driver education and licensing adequately equip novice drivers with the skills necessary to drive in the 21st century.*

Keywords: Driver licensing; driver education; driver testing; simulators; graduated driver licensing; hazard perception training

INTRODUCTION

This chapter is the first of three chapters focused on driver behaviour impacts on transport network safety and crash risk. While this chapter focuses on driver licensing and education, it complements Chapter 3, which focuses on aggressive driving and speed, and Chapter 4, which delves into driver distraction.

Driver licensing and education are countermeasures used to facilitate greater safety of road users. There are extensive variations in the way in which these countermeasures are implemented. This chapter focuses on car drivers and discusses the evidence base available for various forms of driver licensing and education. As well as the more traditional forms of licensing and education, this chapter considers the emergence of simulators and their use in this area.

DRIVER LICENSING

Almost all motorised societies have implemented driver licensing systems. Generally speaking, the four major objectives of driver licensing are to: (1) ensure that drivers are competent enough to commence driving, maintain their driving privileges and proceed to a higher class of licence; (2) acknowledge the obligations associated with driving and discourage inappropriate behaviour that may result in that person losing their licence;

(3) administer a more graduated entry to driving that reduces the potential impact of high-risk road use and (4) provide a method that can be used to identify drivers in order to monitor and manage their behaviour on the road (Watson et al., 1996).

Novice Driver Licensing

Young, newly licensed drivers are over-represented in car crashes (Bates, Davey, Watson, King, & Armstrong, 2014; Shope, 2006). As a result, jurisdictions throughout the world have focused considerable attention on designing and applying licensing to reduce the crash risk of inexperienced drivers.

Minimum Driver Licensing Age

Younger drivers are more likely to crash than older drivers (McKnight & McKnight, 2003). This is related to their lack of experience and tendency to drive in high-risk conditions. Additionally, younger drivers are immature, lack risk-perception abilities and overestimate their driving skills (Bates, Davey et al., 2014). Although it is difficult to know if age or inexperience is a more important contributing factor to crash risk, evidence is emerging, which suggests that experience is a more significant factor in crash risk when compared with age (McCartt, Mayhew, Braitman, Ferguson, & Simpson, 2009).

The age at which individuals can obtain a licence differs across jurisdictions. Delaying when individuals are able to obtain their driver licence limits the learner's exposure to risky driving situations as well as allowing them time to mature (Simons-Morton, 2007). This development can occur both by increasing the amount of time that must be spent as a learner driver and by increasing the minimum age at which individuals can obtain a learner licence (McCartt, Farmer, & Jenness, 2010; Williams, 2009).

Graduated Driver Licensing

Graduated driver licensing (GDL) systems aim to reduce the number of novice driver crashes by limiting the risks for the group, rather than for individual drivers, while still enabling them to gain experience in lower-risk situations (Shults, 2010). Generally speaking, there are three stages within a GDL system: learner, intermediate or provisional and full (Bates, Allen et al., 2014).

Learner Licence

The learner phase enables new drivers to learn on the road while being supervised by an experienced driver. This provides an opportunity to learn about vehicle handling, perceiving hazards and other drivers' behaviour (Foss, 2007). Supervised driving is inherently different from unsupervised driving as it provides a learning opportunity and allows accumulation of experience (Scott-Parker, Bates, Watson, King, & Hyde, 2011).

While it is possible to obtain the basic motor skills required for driving relatively quickly, higher-order skills such as perception, attention and judgement develop over several years (Sagberg & Bjørnskau, 2006). Research suggests that more supervised practice undertaken by a learner driver reduces their risk of crashing once they commence driving by themselves (Gulliver, Begg, Brookland, Ameratunga, & Langley, 2013).

Most learner drivers use a combination of both professional driving instruction and practice with parents and friends to learn to drive (Bates, Watson, & King, 2009; Nyberg, Gregersen, & Wiklund, 2007). Research with young Norwegian drivers identified that professional driving instruction and supervised practice with family and friends complement each other. Tronsmoen (2011) found that professional driving instructors tend to teach the more demanding aspects of driving, while private supervisors tend to allow the learner to develop their driving skills. The combination of both types of supervisors may be beneficial for learner drivers as each tends to focus on different aspects of the process.

In some jurisdictions, including various states within the United States (US) and Australia, learners must practice for a certain number of hours or record that practice in a log book. The requirements vary, with different states in the US requiring between 0 and 65 hours (Insurance Institute for Highway Safety, 2015). In Australia the requirements are greater, with learner drivers in Victoria and New South Wales required to complete 120 hours, while those in Queensland must record 100 hours (Senserrick, 2009). It appears that mandating a number of supervised hours of driving practice does increase the amount of practice undertaken (Bates, Watson, & King, 2010). While there is little research basis for the selection of particular time requirements, there is some research support for lower crash rates for novice drivers that complete close to 120 hours of supervised driving practice (Gregersen et al., 2000). It appears that there is higher parental involvement with the supervision of learner drivers if licensing systems require them to accumulate 50 or more hours of practice (Jacobsohn, Garcia-Espana, Durbin, Erkoboni, & Winston, 2012). One potential drawback of mandating the number of hours of practice is that it may indicate to these learners and their parents that this driving practice is sufficient to acquire all the necessary driving skills (Foss, 2007).

Provisional or Intermediate Licence

A key element of the most effective GDL systems is restricting driving in high-risk circumstances for the first few months or years after the new driver receives their licence (Preusser & Tison, 2007). Thus, the provisional phase limits novice drivers' exposure to high-risk situations such as at night, with passengers or after drinking alcohol (Preusser & Leaf, 2003).

When compared with older drivers, younger drivers are more likely to crash at night (Preusser & Tison, 2007). As a result, night-driving restrictions have been introduced in several jurisdictions (Williams & Mayhew, 2003) and have been shown to be effective in reducing fatalities for this group (Carpenter & Pressley, 2013). The effectiveness of this restriction is affected by when it starts, the support of parents and the availability of exemptions (Williams, 2007). Night-driving restrictions that start before midnight are more effective even when the compliance rate is about half (Jones, Begg, & Palmer, 2012).

When young drivers are accompanied by young passengers, they are more likely to crash (Bates, Davey et al., 2014), with the risk increasing with more passengers (Williams, 2007). When a passenger restriction is in place drivers are unable to take passengers under a particular age in the car with them. However, family members are generally exempt from this restriction (Williams, 2007). Research suggests that this restriction is effective in reducing crashes by provisionally licensed drivers (Fell, Todd, & Voas, 2011) even when compliance rates are approximately 50% (Jones et al., 2012). However, young drivers are less likely to comply with a passenger restriction than with a night-driving restriction (Williams, 2007).

Despite young people drinking alcohol and driving less frequently than adults they do have an increased risk when they drive under the influence (Bates, Davey et al., 2014; Shope & Bingham, 2008). Licensing systems often restrict the percentage of alcohol allowed in the bloodstream for all drivers. However, within GDL systems the blood alcohol content is frequently lower for new drivers (Bates, Allen et al., 2014). There is a more stringent blood alcohol content restriction for young drivers in Australia, New Zealand, the US and Canada (Carpenter, 2006), with research evidence suggesting that this reduces crash risk for new drivers (Carpenter & Pressley, 2013).

When compared to older drivers, young drivers are more likely to use their mobile phone while driving (Gauld, Lewis, & White, 2014), which, as research suggests, increases the risk of crashing (Cazzulino, Burke, Muller, Arbogast, & Upperman, 2014), including when drivers use hands-free mobile phone devices (McEvoy et al., 2005). The introduction of a mobile phone ban is designed to reduce distraction for new drivers (Williams, 2007) – see Chapter 4 for a comprehensive discussion of distracted driving. Within Australia and the US there are restrictions on mobile phone use for either all drivers or newly licensed drivers (Senserrick, 2009; Williams, 2007). These restrictions appear to have little impact on the use of mobile phones by young drivers (Foss, Goodwin, McCartt, & Hellinga, 2009), thus the method used to enforce this restriction is important in maximising its effectiveness (Foss et al., 2009).

Vehicle power restrictions refer to limitations placed on the type of car that a newly licensed driver can use (Keall & Newstead, 2013). For instance, new drivers may be restricted to a certain number of cylinders or a power-to-weight ratio, as young drivers in vehicles with above-average engine power tend to have more risky attitudes toward driving than other young drivers (Clarke, Ward, & Truman, 2002). However, research suggests that the effectiveness of this restriction is quite limited, as very few newly licensed drivers drive high-powered vehicles. Even with 100% compliance with this restriction, researchers estimate that the reduction in injury rates would range from 0.4% in New Zealand to 2.5% in the Australian states of Queensland and Victoria (Keall & Newstead, 2013).

P-plates or decals indicate to others that the vehicle's driver does not yet have a full licence (Curry, Pfeiffer, Localio, & Durbin, 2013). The rationale behind the use of P-plates is that they will increase compliance with (Triggs & Smith, 1996), and enforcement of, other restrictions (Williams & Shults, 2010). It may also encourage the driver to limit their risk-taking behaviours (Senserrick & Whelan, 2002). One study evaluated the decal law within New Jersey and found that it positively affected provisional drivers' safety and reduced their crash rate by 9% (Curry et al., 2013). However, this requirement is not popular with young people (Williams & McCartt, 2014).

Role of Parents

The level of support that parents provide to new drivers is a key factor in the effectiveness of GDL systems (Williams & Shults, 2010). Parents involved in the learning process tend to be strong supporters of GDL (Brookland & Begg, 2011; Campbell, Chaudhary, Saleheen, Borrup, & Lapidus, 2009), although the licensing system tends not to systematically encourage their involvement (Simons-Morton, 2007). Parental involvement in the learner phase of GDL is essential for most learner drivers to obtain sufficient driving practice (Harrison, 2004), given that many learner drivers believe that professional driving instructors are too expensive (Scott-Parker, 2015). Mothers take greater responsibility for supervising learner drivers while they accrue driving practice when compared with fathers (Bates, Watson, & King, 2013).

Once a new driver starts to drive unaccompanied, parents continue to influence their behaviour (Allen, Murphy, & Bates, 2015; Prato, Toledo, Lotan, & Taubman-Ben-Ari, 2010), with some authors suggesting that parental involvement is more important at this stage (Simons-Morton, Ouimet, & Catalano, 2008). Parents can play an active role in supporting the aims of GDL systems by enforcing the GDL restrictions for provisionally licensed drivers (Brookland, Begg, Langley, & Ameratunga, 2014). Research suggests that when parents impose stricter limits younger drivers report that they engage in less-risky driving behaviour and less frequently commit traffic offences or are involved in crashes (Ginsburg, Durbin, Garcia-Espana, Kalicka, & Winston, 2009).

Education programs to develop parents' skills and abilities to manage their child's progress through the GDL system are being developed. Indeed, there are many educational programs available for parents (Williams, Tefft, & Grabowski, 2012). One example is the Checkpoints Program in the US, which aims to encourage parents to limit their child's driving under high-risk conditions when they are first licensed (Williams & Shults, 2010). Research suggests that it influences parental limit setting, teenage risky-driving behaviour and traffic offences during the first 12 months of driving (Simons-Morton et al., 2008).

Compliance and Enforcement

As novice drivers progress through their provisional driver licence they become less compliant with the road laws (Scott-Parker et al., 2012). As examples, 17% of learner drivers drove without a supervisor in North Carolina; while in Nova Scotia and California approximately 40% of drivers holding an intermediate licence occasionally violated the night-driving restriction, although only 12–15% of drivers reported doing this often (Mayhew, Simpson, Ferguson, & Williams, 1998; Williams, Nelson, & Leaf, 2002). Despite the lack of full compliance in some parts of the system the GDL laws are still successful in reducing crash risk for novice drivers (Senserrick & Whelan, 2003).

Generally speaking, the most common method used to modify driver behaviour in order to reduce the incidence of crashes is traffic law enforcement

(Bates, Soole, & Watson, 2011). However, in the case of GDL this may be difficult, particularly if police officers are unable to recognise which driving restrictions apply to specific licences (Hedlund, 2007). If police officers do not fully understand GDL laws they may be placed in situations where they attempt to enforce driving laws where the novice driver has a greater understanding of the law (Goodwin, Wells, Foss, & Williams, 2006). Two small studies have been conducted in Queensland, Australia, which suggest that there is a limited deterrent effect of police enforcing provisional driving laws and that provisional drivers are more affected by parental enforcement (Allen et al., 2015; Bates, Darvell, & Watson, 2017). GDL systems support parents to set limits for their children, and they clearly identify the high-risk situations and signal to parents what is appropriate driving behaviour (Hartos, Simons-Morton, Beck, & Leaf, 2005).

Driver Testing

Although current driving tests appear to have limited validity in terms of identifying drivers who are more likely to crash in the future, they serve an important function by encouraging new drivers to obtain education and training and to establish a base level of competency for entering into the driver licensing system. There are four main types of driver testing for novice drivers: theory or knowledge, on-road driving, hazard perception and exit (Senserrick & Williams, 2015).

In many places around the world individuals are required to pass a written test in order to obtain a learner permit and a practical driving test to drive independently (Mayhew, Williams, & Pashley, 2014). Written tests generally focus on the road rules. There appears to be little association between passing a written test and driving performance or attitudes (Senserrick & Williams, 2015).

On-road driving tests generally focus on vehicle skills and the driver's ability to control the vehicle in traffic. The research evidence regarding the effect of these tests on crashes is mixed, with some research from the United Kingdom (UK) indicating that fewer errors on the tests are associated with lower crash rates. However, the research from Europe is mixed (Senserrick & Williams, 2015). One Australian study suggests that failing a practical driving test four times is linked to higher risk of being involved in a crash (Boufous, Ivers, Senserrick, & Stevenson, 2011).

Passing a hazard perception test is a condition for advancing through GDL systems in some jurisdictions (Engstrom, Gregersen, Hernetkoski, Keskinen, & Nyberg, 2003). Hazard perception tests are not required as part of novice driver testing in the US or New Zealand. However, they are present at various stages of the licensing process in the UK and Australia (Senserrick & Williams, 2015). It appears that failing a hazard perception test twice is linked to a higher risk of crash involvement (Boufous et al., 2011).

An exit test aims to ensure that provisional drivers are able to hold a full driving licence (Ferguson, 2003) and may also indicate to the driver that they are still developing their driving skills (Williams & Mayhew, 2003). Exit tests are available in a range of formats, including knowledge tests, hazard perception tests or on-road driving tests (Senserrick & Whelan, 2003). There is limited research evidence regarding the effects of exit tests on crash risk (Haire, Williams, Preusser, & Solomon, 2011).

Monitoring the Impact of Driver Licensing Changes on Unlicensed Driving

Unlicensed driving remains a serious problem in many countries, despite ongoing improvements in traffic law enforcement practices and technology. In the US, for example, over 10% of drivers involved in fatal crashes were found to not hold a valid licence, while approximately 20% of all fatal crashes involved at least one unlicensed driver (Scopatz et al., 2003). Unlicensed driving represents a major problem for road safety in two respects. Firstly, it undermines the effectiveness of GDL systems by preventing the allocation of demerit points and reducing the impact of licence loss, which otherwise has been demonstrated to be an effective deterrent to illegal driving behaviour (Nichols & Ross, 1990; Siskind, 1996). Secondly, there is a growing body of evidence linking unlicensed driving to a cluster of high-risk behaviours, including drink driving, speeding (see Chapter 3), failure to wear seat belts and motorcycle use (Watson, 2004a). Consistent with this, studies have indicated that unlicensed drivers are at a higher risk of crashing than licensed drivers, and their crashes also tend to be more severe (DeYoung, Peck & Helander, 1997; Watson, 2004b).

While many unlicensed drivers may reduce their overall driving exposure in order to avoid detection, this avoidance does not appear to result in safer driving. While it remains possible that unlicensed drivers tend to act more cautiously than they would otherwise, as a cohort they are not as safe as licensed drivers. Consequently, the potential impacts of new driver licensing and testing requirements on participation in the licensing system need to be closely considered during the policy development phase. Similarly, trends in unlicensed driving should be monitored following the introduction of new driver licensing and testing requirements through the use of crash data and periodic on-road surveys (see Watson et al., 2011).

ROLE OF DRIVER EDUCATION

While many consider the terms 'driver education' and 'driver training' interchangeable, there are theoretical distinctions between the two (Watson et al., 1996). Driver training tends to focus on developing a person's vehicle control skills. In contrast, driver education may include driver training but is predominately focused on skills and attitudes required to become a safe driver (Langford, 2002). Driver education exists in many different forms (Beanland, Goode, Salmon, & Lenne, 2013). Therefore, the profession needs to develop a better understanding of the processes used in various types of education and training and determine whether these programs are effective or not (Watson, 2003).

There is research to suggest that some forms of education or training can lead to new drivers becoming overconfident, taking risks and, consequently, higher crash rates (Lonero & Mayhew, 2010). In fact, a common criticism of driver education and training is that it has limited capacity to reduce crashes (Hatakka, Keskinen, Gregersen, Glad, & Hernetkoski, 2002). While early evaluations of training and education indicate that they reduce crash risk for the trainees, these studies had methodological flaws, limiting confidence in the results (Zhao et al., 2006).

One common theoretical framework applied to driver education and training is the Goals for Driver Education framework (also known as the GADGET matrix). The framework can be used to both evaluate existing driver education initiatives and plan new programs (Hatakka et al., 2002). In this framework there are four hierarchical levels which are, from lowest to highest: vehicle manoeuvring, mastery of traffic situations, goals and context of driving (trip-related), and goals for life and skills for living (general). Within each level are three categories: knowledge and skills, risk increasing factors and self-evaluation (Berg, 2006; Hatakka et al., 2002).

The licensing system appears to influence the type of driver education and training undertaken by novice drivers (Bates, Watson, & King, 2006). However, research suggests that reducing the amount of time a novice driver spends in the GDL system because they attend a driver education course may be counterproductive as it does not reduce crashes or offences (Hirsch, Maag, & Laberge-Nadeau, 2006). For example, a time reduction is available in New Zealand; however, an analysis suggested that when novice drivers completed an approved education course after 12–18 months' driving experience on a provisional licence, there was greater benefit in the novice driver completing the full 18 months on a restricted licence (Lewis-Evans, 2010). There is also some evidence to suggest that when parents select a driver education program for their child they are less likely to provide supervision (Hirsch et al., 2006).

Pre-Licence Training

Before obtaining a driver licence a person needs to learn basic driving skills, which may include both procedural skills and higher-order cognitive skills (Beanland et al., 2013). This pre-licence training can occur in several ways, including school-based driver training, simulator training and resilience training. Additionally, learner drivers often receive professional driving instruction.

School-Based Driver Training

The DeKalb County study is the largest experimental study of driver education and training provided within high schools (Young, 2002). The study aimed to identify the effect of driver education and training on crashes. In the study, over 16,000 drivers were randomly assigned to one of three groups. The first group received the Safe Performance Curriculum, which was the most advanced education and training available in the US at the time. The second group received the Pre-Driver Licensing curriculum, which taught the basic skills required to pass the driving test. The final group was a control group and did not receive any driver education through the school (Lund, Williams, & Zador, 1986). The original analysis suggested that there were no differences between the groups in terms of crash rates or traffic offences and that there was a short-term benefit for the driver education groups, with those students demonstrating better driving skills and lower crash rates when compared with the control group during the first 6 months of driving (Lund et al., 1986). However, this analysis only used data from individuals who had obtained their licence. Subsequent analysis suggested that those students who were in the Safe Performance Curriculum group obtained their driver licence earlier and consequently had higher crash rates and traffic offences than those in the control group. Those in the Pre-Driver Licensing group were also more likely to receive their driver licence earlier than the control group, although there was no difference between the two groups in terms of crash rates (Lund et al., 1986). This finding suggests that providing driver education and training in schools may lead to individuals obtaining their driver licence earlier than they would otherwise, which increases their exposure to crash risk (Lonero & Mayhew, 2010).

Resilience Training

Training that aims to reduce risky-driving behaviours in novice drivers is known as resilience training (Beanland et al., 2013). This type of training does not focus specifically on driving skills or abilities but rather on interpersonal skills that will help the young person to engage in safe behaviours. This may include the young person not getting into vehicles where the driver is impaired or supporting them to resist peer pressure that may encourage them to drive unsafely. One example of this type of program is *You Hold the Key*, which aims to increase safe driving and responsible passenger behaviour amongst teenagers (15–19 years) within the US. The program, which was implemented in schools for 10 weeks, involved interactive discussions and lessons, safety promotion education, cooperative learning, videos and presentations (King, Vidourek, Love, Wegley, & Alles-White, 2008). Students learnt about potential driving hazards, resistance skills and strategies that may help to reduce crashes. The program was evaluated using a pre- and post-test design, and it was found that there was a positive difference in young people's intended behaviours, including in the areas of seatbelt use and drink driving situations immediately after the program and that these differences still existed 6 months later. There also appears to be a positive effect of resilience programs on crashes. Researchers who compared a resilience program with a general education program for high school students in the Australian state of New South Wales identified a 44% reduction in crashes for those who undertook the resilience training (Senserrick et al., 2009).

Post-Licence Education

The aim of post-licence education is to expand the current set of driving skills for individuals who are already licensed. This type of education and training often emphasises managing difficult situations, such as skid handling or advanced braking, which are not normally included in pre-licence training (Beanland et al., 2013). It is difficult to make an overall assessment of post-licence education and training given that it is delivered in so many different ways.

Procedural Skills Training

The focus of procedural skills training is to develop and enhance a driver's ability to operate a vehicle (Beanland et al., 2013). There is limited research evidence to suggest that procedural skills training is effective in reducing crashes. For example, the introduction of skid training for novice drivers led to an increase in the number of crashes in program participants (Katila, Keskinen, Hatakka, & Laapotti, 2004). Some studies within Australia suggest that driver training does affect vehicle handling immediately following the training. However, these effects may be highly context-specific and there is no research regarding the longer-term impacts (Petersen & Barrett, 2009; Petersen, Barrett, & Morrison, 2006, 2008). It is possible that the safety messages provided in short courses may be undermined by other influences that shape new drivers' behaviour, such as parental, peer and other social influences (Williams, 2006).

Hazard Perception Skills Training and Education

Hazard perception skills are considered some of the most important for driving. Training and education that focuses on these skills is more likely to be beneficial (Christie, 2001; Senserrick et al., 2009). Despite this, novice drivers have a lesser ability to detect hazards and react more slowly to hazardous situations when compared with more experienced drivers (Isler, Starkey, & Williamson, 2009). There are two main types of hazard perception training: part-task or commentary. Part-task training generally involves completing activities in response to video clips that show hazards (Fisher, Pollatsek, & Pradhan, 2006; Weiss, Petzoldt, Bannert, & Krems, 2013). A novice driver who verbalises or receives commentary of the hazard perception task as it happens is undergoing commentary training (McKenna, Horswill, & Alexander, 2006).

Insight Training

Insight training is another type of cognitive training that is designed to prevent overconfidence in newly licensed drivers by raising awareness of their driving skills and teaching them to adjust their behaviour according to the intricacies of various situations (Senserrick & Swinburne, 2001). Insight training does appear to be similar, conceptually, to other forms of cognitive driver education such as resilience training. While limited research has focused on insight training for novice drivers it appears that insight training may affect sub-groups of drivers differently – meaning that it is important that appropriate groups of drivers undertake this type of training (White, Cunningham, & Titchener, 2011).

USE OF SIMULATORS IN LICENSING AND EDUCATION

Driving simulators have recently become a potentially useful tool for novice driver education. They allow novices to experience a range of situations and practice their skills without risk of crashing (Flach, Dekker, & Stappers, 2008).

Simulator managers can control of the environment in order that specific teachable moments are encountered and repeated. Driving instructors claim that controlling presentation of teachable moments increases the efficiency of teaching, suggesting that specific learning content can be achieved in 1 hour of simulator training that would otherwise take 3 hours of on-road training (Kappe, van Emmerik, van Winsum, & Rozendom, 2003).

When considering the potential for simulators, two important aspects are fidelity and validity. Fidelity relates to the realism of the simulator including physical fidelity (the way the simulator looks and behaves) and psychological fidelity (the way a driver relates to the simulator). Validity relates to how well driving behaviours in a simulator transfer to real-world driving. The interaction between these two concepts means that it is possible to have a low-fidelity simulator, which has high validity. High-fidelity simulators are expensive; if deciding to purchase a simulator, an education centre should consider how to get the highest validity within the fidelity they can afford. For example, risk awareness and perception training administered using a low-fidelity, PC-based, basic simulator has been shown to improve visual scanning behaviour both in subsequent simulator testing and in real-world driving (Fisher, 2008; Fisher, Narayanaan, Pradhan, & Pollatsek, 2003; Fisher et al., 2006; Pradhan, Fisher, & Pollatsek, 2006).

Education

Hazard perception and visual scanning are the skills most widely trained using simulators. Overall, research evidence suggests that hazard perception simulator training results in skill improvement, at least in the short term (e.g., Carpentier, Wang, Jongen, Hermans, & Brijs, 2012; Chapman, Underwood, & Roberts, 2002; Pradhan et al., 2006; Regan, Triggs, & Godley, 2000a, 2000b; Thomas et al., 2011). However, few studies have undertaken follow-up investigations and even fewer have investigated the transfer of skills from the simulator to on-road driving. Those completing follow-ups have reported greater hazard perception performance in simulator-trained, compared to untrained, individuals 4 days (Pradhan et al., 2006) and 4 weeks (Carpentier et al., 2012; Regan et al., 2000b) after training. Simulator training of hazard perception skills appears to be effective when follow-up occurs in a driving simulator; however, it is less clear whether there are benefits for on-road driving. For example, 3–6 months after completing simulator hazard perception training, there was no difference in on-road driving performance despite persistent differences in the simulator (Chapman et al., 2002). In contrast, during an on-road follow-up participants who had completed visual scanning-focused simulator training, with emphasis on driver distraction (see Chapter 4), demonstrated safer driving behaviour (shorter glances away from the road) than non-simulator-trained drivers (Thomas et al., 2011). Part of the difficulty of follow-up evaluation is that both trained and untrained novices will continue to learn to drive in the intervening period; therefore, skills are likely to improve for both groups.

It is less common to use simulators for teaching procedural skills. This is partly associated with the high costs of faithfully replicating the driving experience. In many cases it is considered more cost-effective to teach these skills in a real car (Pollatsek, Vlakveld, Kappe, Pradhan, & Fisher, 2011). A notable exception is in the Netherlands, where simulators are used for procedural skills training. Driving instructors report simulators as beneficial for procedural skills training because they are then able to focus on higher-order cognitive skills and spend less time on basic skills during driving lessons (Fisher, Caird, Rizzo, & Lee, 2011).

Despite the emerging evidence of the benefits of driving simulators for driver education, there is concern over the lack of behavioural consequence (Caird & Horrey, 2011). As crashing in a simulator provides no physical danger, it is possible that users do not engage with the same level of caution as they would in real-world driving, which may potentially lead to overconfidence. It is important that simulators do not induce overconfidence in novice drivers as this is correlated to increased crash risk (e.g., Weiss et al., 2013).

One proposed benefit of simulator education is the exposure of novices to situations which may be dangerous on real roads thereby exposing the novice's limitations and discouraging dangerous behaviour. An example of this is the Forward Concentration and Attention Learning (FOCAL) program (Thomas et al., 2011), where novices are instructed to complete a simulator drive while simultaneously completing a map-based navigation task. Following the task the novice is replayed footage of the simulated road which is blacked out every time they took their eyes off the road. In an on-road follow-up the novices who had undertaken FOCAL training spent less time with their eyes away from the road than novices who had not completed the training.

It is likely that some skills are more suited to simulator education than others. Simulators can offer the opportunity to practice and augment already acquired skills. This could be a valuable asset for novice driver training, particularly where supervised practice is not always possible. However, the simulator itself is a tool, as such the outcomes of training will be as much (if not more) influenced by the processes and goals of the education program as a whole. Currently, there is a lack of simulator research controlling for educational aims and curriculum, only altering the point of implementation (simulator vs. on-road). Without such a controlled investigation it is not possible to separate the influence of the education material from that of the delivery method.

Driver Testing

Simulators are not routinely used to assess novice drivers' procedural skills but have the potential to address some of the weaknesses of on-road practical driving tests. For example, an on-road examiner cannot control traffic situations encountered during testing; additionally, there may be inter-examiner differences in interpreting safe driving (Kappe, de Penning, & Marsman, 2010). The controllable nature of simulators would allow all examinees to be tested under the same situations and evaluated against the same criteria. If such testing were to be considered, it is vital that the simulator test be able to distinguish between sufficiently competent and incompetent drivers. A vital first step has already been demonstrated, whereby simulator assessment can accurately distinguish between experienced and novice drivers (Damm, Nachtergaele, Meskali, & Berthelon, 2011), suggesting future simulator testing of novice drivers may be an option. Indeed, there is currently at least one example of experienced driver testing using a simulator. Within the context of heavy vehicle driver training in Ecuador, experienced drivers may complete their assessment for a truck driver licence on a simulator (instead of a closed training circuit) (Allen, Woon, Park, & Grant, 2010).

While it appears unlikely that simulator assessment will replace novice drivers' practical driving assessments in the near future, driving simulators may have potential applications for assessing isolated procedural skills which, although integral to driving, are rarely encountered during practical driving assessments.

FUTURE DIRECTIONS

A future challenge for driver education and training will be to keep up with advances in technology. Increasingly, new car models include autonomous technology features (e.g., parking assist). Assuming technological development continues, it is predicted that autonomous vehicles will become a common feature in the future (Fagnant & Kockelman, 2015). There is an increasing volume of autonomous vehicle research; however, to date there is a dearth of literature surrounding implications for driver education and training. It is vital that this gap is addressed in order that driver education and licensing continue to be relevant and appropriate. Presently, driver education has been slow to adapt and respond to the increasing popularity of in-vehicle technology. The most noticeable formal change in this area came in 2015, when the UK Government began trialling the inclusion of 'independent driving using satellite navigation' (Sat Nav) as a component of driver licensing (Driver & Vehicle Standards Agency, 2015). As part of the licence testing, drivers are required to demonstrate independent driving following Sat Nav directions given by a TomTom Go. Arguably, this change could be seen as a step forward in preparing novices for the modern driving environment; however, it should be noted that this development was implemented 13 years after the launch of the first TomTom Sat Nav in Europe (TomTom International, 2017), demonstrating the slow pace of licencing change compared to the introduction of technology. In the context of autonomous vehicles, increasing research is needed to ensure that driver education and licensing are able to adequately equip novice drivers with the skills necessary to drive in the 21st century.

CONCLUSIONS

Driver licensing is a key mechanism used by authorities to ensure driver competency, encourage appropriate driving behaviour while monitoring inappropriate behaviour and introduce new drivers to the system. GDL systems involving learner, provisional and open licences have been shown to be an effective way of reducing novice driver crashes. As there are many jurisdictional differences authorities should carefully consider what components they need to include to maximise crash reductions.

The licensing system influences the type of driver education undertaken by drivers. There is a significant range of driver education programs that can be delivered at different points within the licensing system. Research suggests that some types of driver education, such as resilience and situation awareness, may be more effective at reducing crashes and altering attitudes compared with other programs.

There is increasing emerging evidence as to the value of driving simulators in novice driver education and testing. However, further research is needed regarding the long-term, on-road safety benefits. If simulators are considered for novice driver education, it should be remembered that the simulator itself is a tool; the outcomes of education will only be as good as the content and purpose of the education program itself, regardless of the fidelity and validity of the simulator used.

There is a significant range of driver licensing systems, testing regimes and forms of driver education and training in use throughout the world. Given this, it is critical that any changes to these programs are based on research and best practice to maximise outcomes.

REFERENCES

Allen, R. W., Woon, S. J., Park, G. D., & Grant, J. K. (2010). The development of a low cost driver licensing simulator. Paper presented at the Europe Driving Simulation Conference, Actes INRETS.

Allen, S., Murphy, K., & Bates, L. (2015). What drives compliance? The effect of deterrence and shame emotions on young drivers' compliance with road laws. *Policing and Society, 27*(8), 884–898.

Bates, L., Allen, S., Armstrong, K., Watson, B., King, M., & Davey, J. (2014). Graduated driver licensing: An international review. *Sultan Qaboos University Medical Journal, 14*(4), 403–412.

Bates, L., Darvell, M., & Watson, B. (2017). Young and unaffected by road policing strategies: Using deterrence theory to explain provisional drivers' (non)compliance. *Australian and New Zealand Journal of Criminology, 50*(1), 23–38. doi:10.1177/0004865815589824

Bates, L., Davey, J., Watson, B., King, M., & Armstrong, K. (2014). Factors contributing to young driver crashes: A review. *Sultan Qaboos University Medical Journal, 14*(3), 297–305.

Bates, L., Soole, D., & Watson, B. (2011). The effectiveness of traffic policing in reducing traffic crashes. In T. Prenzler (Ed.), *Policing and security in practice: Challenges and achievements*. Basingstoke: Palgrave Macmillan.

Bates, L., Watson, B., & King, M. (2006). Competing or complementing: Driver education and graduated driver licensing. Paper presented at the Road Safety Research, Policing and Education Conference, Gold Coast.

Bates, L., Watson, B., & King, M. (2009). Driving and licensing experiences of learner drivers in two Australian states prior to changes in the licensing laws. *Journal of the Australasian College of Road Safety, 20*(3), 51–55.

Bates, L., Watson, B., & King, M. (2010). Required hours of practice for learner drivers: A comparison between two Australian jurisdictions. *Journal of Safety Research, 41*, 93–97.

Bates, L., Watson, B., & King, M. (2013). Mothers vs fathers as learner driver supervisors: Time commitment, driving activities and perceptions of risk. Paper presented at the 2013 Australasian Road Safety Research, Policing & Education Conference, Brisbane, Australia.

Beanland, V., Goode, N., Salmon, P., & Lenne, M. (2013). Is there a case for driver training? A review of the efficacy of pre- and post-licence driver training. *Safety Science, 51*, 127–137.

Berg, H.-Y. (2006). Reducing crashes and injuries among young drivers: What kind of prevention should we be focusing on? *Injury Prevention, 12*(Suppl. 1), i15–i18.

Boufous, S., Ivers, R., Senserrick, T., & Stevenson, M. (2011). Attempts at the practical on-road driving test and the hazard perception test and the risk of traffic crashes in young drivers. *Traffic Injury Prevention, 12*(5), 475–482. doi:10.1080/15389588.2011.591856

Brookland, R., & Begg, D. J. (2011). Adolescent, and their parents, attitudes towards graduated driver licensing and subsequent risky driving and crashes in young adulthood. *Journal of Safety Research, 42*(2), 109–115.

Brookland, R., Begg, D., Langley, J., & Ameratunga, S. (2014). Parental influence on adolescent compliance with graduated driver licensing conditions and crashes as a restricted licensed driver: New Zealand drivers study. *Accident Analysis & Prevention*, *69*, 30–39. doi:dx.doi.org/10.1016/j.aap.2013.06.034

Caird, J. K., & Horrey, W. J. (2011). Twelve practical and useful questions about driving simulation. In D. L. Fisher, M. Rizzo, J. K. Caird, & J. Lee (Eds.), *Handbook of driving simulation for engineering, medicine and psychology* (pp. 5-1–5-16). Boca Raton, FL: CRC Press.

Campbell, B., Chaudhary, N., Saleheen, H., Borrup, K., & Lapidus, G. (2009). Does knowledge of teen driving risks and awareness of current law translate into support for stronger GDL provisions? Lessons learned from one state. *Traffic Injury Prevention*, *10*(4), 320–324.

Carpenter, C. (2006). Did Ontario's zero tolerance and graduated licensing law reduce youth drunk driving? *Journal of Policy Analysis and Management*, *25*(1), 183–195.

Carpenter, D., & Pressley, J. C. (2013). Graduated driver license nighttime compliance in U.S. teen drivers involved in fatal motor vehicle crashes. *Accident Analysis & Prevention*, *56*, 110–117.

Carpentier, A., Wang, W., Jongen, E. M. M., Hermans, E., & Brijs, T. (2012). Training hazard perception of young novice drivers – A driving simulator study. Paper presented at the TRB 2013 Annual Meeting.

Cazzulino, F., Burke, R., Muller, V., Arbogast, H., & Upperman, J. (2014). Cell phones and young drivers: A systematic review regarding the association between psychological factors and prevention. *Traffic Injury Prevention*, *15*(3), 234–242. doi:10.1080/15389588.2013.822075

Chapman, P., Underwood, G., & Roberts, K. (2002). Visual search patterns in trained and untrained novice drivers. *Transportation Research Part F: Traffic Psychology and Behaviour*, *5*(2), 157–167.

Christie, R. (2001). The effectiveness of driver training as a road safety measure: An international review of the literature. Paper presented at the 2001 Road Safety Research, Policing and Education Conference, Adelaide.

Clarke, D., Ward, P., & Truman, W. (2002). *In-depth accident causation study for young drivers* (TRL542). Prepared for Road Safety Division, Department for Transport, Local Government and the Regions. Crowthorne, United Kingdom: TRL Limited.

Curry, A., Pfeiffer, M., Localio, R., & Durbin, D. (2013). Graduated driver licensing decal law: Effect on young probationary drivers. *American Journal of Preventive Medicine*, *44*(1), 1–7.

Damm, L., Nachtergaele, C., Meskali, M., & Berthelon, C. (2011). The evaluation of traditional and early driver training with simulated accident scenarios. *Human Factors: The Journal of the Human Factors and Ergonomics Society*, *53*(4), 323–337.

DeYoung, D. J., Peck, R. C., & Helander, C. J. (1997). Estimating the exposure and fatal crash rates of suspended/revoked and unlicensed drivers in California. *Accident Analysis and Prevention*, *29*(1), 17–23.

Driver & Vehicle Standards Agency. (2015). *Improving the driving test*. Retrieved from https://www.gov.uk/government/publications/improving-the-driving-test/improving-the-driving-test

Engstrom, I., Gregersen, N. P., Hernetkoski, K., Keskinen, E., & Nyberg, A. (2003). *Young novice drivers, driver education and training: Literature review*. VTI rapport 491A. Linköping, Sweden: Swedish National Road Administration.

Fagnant, D. J., & Kockelman, K. (2015). Preparing a nation for autonomous vehicles: Opportunities, barriers and policy recommendations. *Transportation Research Part A: Policy and Practice*, *77*, 167–181.

Fell, J., Todd, M., & Voas, R. B. (2011). A national evaluation of the nighttime and passenger restriction components of graduated driver licensing. *Journal of Safety Research*, *42*, 283–290.

Ferguson, S. A. (2003). Other high-risk factors for young drivers – how graduated licensing does, doesn't, or could address them. *Journal of Safety Research, 34*(1), 71–77.

Fisher, D. L. (2008). *Evaluation of PC-based novice drivr risk awareness*. Washington, DC, USA: Department of Transportation, National Highway Traffic Safety Administration.

Fisher, D. L., Caird, J. K., Rizzo, M., & Lee, J. (2011). Handbook of driving simulation for engineering medicine and psychology: An overview. In D. L. Fisher, M. Rizzo, J. K. Caird, & J. Lee (Eds.), *Handbook of driving simulation for engineering, medicine and psychology* (pp. 1-1 – 1-16). Boca Raton, FL: CRC Press.

Fisher, D. L., Narayanaan, V., Pradhan, A., & Pollatsek, A. P. (2003). Using eye movements in driving simulators to evaluate effects of PC-based risk awareness training. Paper presented at the Human Factors and Ergonomics Society Annual Meeting.

Fisher, D. L., Pollatsek, A. P., & Pradhan, A. (2006). Can novice drivers be trained to scan for information that will reduce their likelihood of a crash? *Injury Prevention, 12*(Suppl. 1), i25–i29.

Flach, J. M., Dekker, S., & Stappers, P. J. (2008). Playing twenty questions with nature (the surprise version): Reflections on the dynamics of experience. *Theoretical Issues in Ergonomics Science, 9*, 125–154.

Foss, R. (2007). Improving graduated driver licensing systems: A conceptual approach and its implications. *Journal of Safety Research, 38*(2), 185–192.

Foss, R., Goodwin, A. H., McCartt, A., & Hellinga, L. A. (2009). Short-term effects of a teenage driver cell phone restriction. *Accident Analysis & Prevention, 41*, 419–424.

Gauld, C., Lewis, I., & White, K. (2014). Concealing their communication: Exploring psychosocial predictors of young drivers' intentions and engagement in concealed texting. *Accident Analysis & Prevention, 62*, 285–293. doi:http://dx.doi.org/doi:10.1016/j.aap.2013.10.016

Ginsburg, K., Durbin, D., Garcia-Espana, J., Kalicka, E., & Winston, F. K. (2009). Associations between parenting styles and teen driving, safety-related behaviors and attitudes. *Pediatrics, 124*, 1040–1051.

Goodwin, A. H., Wells, J. K., Foss, R., & Williams, A. (2006). Encouraging compliance with graduated driver licensing restrictions. *Journal of Safety Research, 37*, 343–351.

Gregersen, N. P., Berg, H.-Y., Engstrom, I., Nolen, S., Nyberg, A., & Rimmo, P.-A. (2000). Sixteen years age limit for learner drivers in Sweden – An evaluation of safety effects. *Accident Analysis and Prevention, 32*, 25–35.

Gulliver, P., Begg, D., Brookland, R., Ameratunga, S., & Langley, J. (2013). Learner driver experiences and crash risk as an unsupervised driver. *Journal of Safety Research, 46*, 41–46.

Haire, E. R., Williams, A., Preusser, D., & Solomon, M. (2011). *Driver license testing of young novice drivers*. Washington, DC, USA: Department of Transportation, National Highway Traffic Safety Administration.

Harrison, W. A. (2004). Investigation of the driving experience of a sample of Victorian learner drivers. *Accident Analysis and Prevention, 36*, 885–891.

Hartos, J. L., Simons-Morton, B., Beck, K. H., & Leaf, W. A. (2005). Parent-imposed limits on high-risk adolescent driving: Are they stricter with graduated driver licensing? *Accident Analysis & Prevention, 37*(3), 557–562.

Hatakka, M., Keskinen, E., Gregersen, N. P., Glad, A., & Hernetkoski, K. (2002). From control of the vehicle to personal self-control: Broadening the perspectives to driver education. *Transportation Research Part F: Traffic Psychology and Behaviour, 5*(3), 201–215.

Hedlund, J. (2007). Novice teen driving: GDL and beyond. *Journal of Safety Research, 38*(2), 259–266.

Hirsch, P., Maag, U., & Laberge-Nadeau, C. (2006). The role of driver education in the licensing process in Quebec. *Traffic Injury Prevention, 7*, 130–142.

Insurance Institute for Highway Safety. (2015). Teenagers: Driving carries extra risk for them. Retrieved from http://www.iihs.org/iihs/topics/laws/graduatedlicenseintro

Isler, R. B., Starkey, N. J., & Williamson, A. (2009). Video-based road commentary training improves hazard perception of young drivers in a dual task. *Accident Analysis & Prevention, 41*, 445–452. doi:10.1016/j.aap.2008.12.016

Jacobsohn, L., Garcia-Espana, J., Durbin, D. R., Erkoboni, D., & Winston, F. K. (2012). Adult-supervised practice driving for adolescent learners: The current state and directions for interventions. *Journal of Safety Research, 43*(1), 21–28.

Jones, S. J., Begg, D. J., & Palmer, S. R. (2012). Reducing young driver crash casualties in Great Britain – Use of routine police crash data to estimate the potential benefits of graduated driver licensing. *International Journal of Injury Control and Safety Promotion, 20*(4), 321–330.

Kappe, B., de Penning, L., & Marsman, M. (2010). Driver performance assessment in driving simulators. Paper presented at the Driving Simulation Conference 2010 Europe, Paris, France.

Kappe, B., van Emmerik, M., van Winsum, W., & Rozendom, A. (2003). Virtual instruction in driving simulators. Paper presented at the Driving Simulator Conference, Dearborn, MI.

Katila, A., Keskinen, E., Hatakka, M., & Laapotti, S. (2004). Does increased confidence among novice drivers imply a decrease in safety? The effects of skid training on slippery road accidents. *Accident Analysis & Prevention, 36*(4), 543–550.

Keall, M. D., & Newstead, S. (2013). The potential effectiveness of young driver high-performance vehicle restrictions as used in Australia. *Accident Analysis & Prevention, 52*, 154–161.

King, K., Vidourek, R., Love, J., Wegley, S., & Alles-White, M. (2008). Teaching adolescents safe driving and passenger behaviours: Effectiveness of the *You Hold the Key* teen driving countermeasure. *Journal of Safety Research, 39*, 19–24.

Langford, J. (2002). Using the research to reduce novice driver crashes. *Proceedings of the Developing Safer Drivers and Riders Conference*. Brisbane: ACRS and Travelsafe Committee.

Lewis-Evans, B. (2010). Crash involvement during the different phases of the New Zealand Graduated Driver Licensing System (GDLS). *Journal of Safety Research, 41*, 359–365.

Lonero, L. P., & Mayhew, D. (2010). *Large-scale evaluation of driver education: Review of the literature on driver education evaluation*. Washington, DC, USA: AAA Foundation for Traffic Safety.

Lund, A., Williams, A., & Zador, P. (1986). High school driver education: Further evaluation of the DeKalb County study. *Accident Analysis & Prevention, 18*, 349–357.

Mayhew, D., Simpson, H., Ferguson, S. A., & Williams, A. (1998). Graduated driver licensing in Nova Scotia: A survey of teenagers and parents. *Journal of Traffic Medicine, 26*(1–2), 37–44.

Mayhew, D., Williams, A., & Pashley, C. (2014). *A new GDL framework: Evidence base to integrate novice driver strategies*. Ottawa, Canada: Traffic Injury Research Foundation (TIRF).

McCartt, A., Farmer, C., & Jenness, J. (2010). Perceptions and experiences of participants in a study of in-vehicle monitoring of teenage drivers. *Traffic Injury Prevention, 11*(4), 361–370.

McCartt, A., Mayhew, D., Braitman, K., Ferguson, S. A., & Simpson, H. (2009). Effects of age and experience on young driver crashes: Review of recent literature. *Traffic Injury Prevention, 10*(3), 209–219.

McEvoy, S., Stevenson, M., McCartt, A., Woodward, M., Haworth, C., Palamara, P., & Cercarelli, R. (2005). Role of mobile phones in motor vehicle crashes resulting in hospital attendance: A case-crossover study. *British Medical Journal, 331*, 428–430.

McKenna, F., Horswill, M., & Alexander, J. (2006). Does anticipation training affect drivers' risk taking? *Journal of Experimental Psychology: Applied, 12*(1), 1–10. doi:10.1037/1076-898X.12.1.1

McKnight, A. J., & McKnight, A. S. (2003). Young novice drivers: careless or clueless? *Accident Analysis & Prevention*, *35*(6), 921–925.

Nichols, J. L. & Ross, H. L. (1990). The effectiveness of legal sanctions in dealing with drinking Drivers. *Alcohol, Drugs and Driving*, *6*(2), 33–60.

Nyberg, A., Gregersen, N. P., & Wiklund, M. (2007). Practicing in relation to the outcome of the driving test. *Accident Analysis & Prevention*, *39*, 159–168.

Petersen, A., & Barrett, R. (2009). Postural stability and vehicle kinematics during an evasive lane change manoeuvre: A driver training study. *Ergonomics*, *52*(5), 560–568. doi:10.1080/00140130802419495

Petersen, A., Barrett, R., & Morrison, S. (2006). Driver-training and emergency brake performance in cars with antilock braking systems. *Safety Science*, *44*(10), 905–917. doi:10.1016/j.ssci.2006.05.006

Petersen, A., Barrett, R., & Morrison, S. (2008). Enhanced postural stability following driver training is associated with positive effects in vehicle kinematics during cornering. *Human Factors: The Journal of the Human Factors and Ergonomics Society*, *50*(1), 159–172. doi:10.1518/001872008X250593

Pollatsek, A. P., Vlakveld, W., Kappe, B., Pradhan, A., & Fisher, D. L. (2011). Driving simulators as training and evaluation tools: Novice Drivers. In D. L. Fisher, M. Rizzo, J. K. Caird, & J. Lee (Eds.), *Handbook of driving simulation for engineering, medicine and psychology* (pp. 30-31 – 30-18). Boca Raton, FL: CRC Press.

Pradhan, A., Fisher, D. L., & Pollatsek, A. P. (2006). Risk perception training for novice drivers: Evaluating duration of effects on training on a driving simulator. *Transportation Research Record, 1969*, 58–64.

Prato, C., Toledo, T., Lotan, T., & Taubman-Ben-Ari, O. (2010). Modeling the behavior of novice young drivers during the first year after licensure. *Accident Analysis & Prevention*, *42*, 480–486.

Preusser, D., & Leaf, W. A. (2003). Provisional license. *Journal of Safety Research*, *34*, 45–49.

Preusser, D., & Tison, J. (2007). GDL then and now. *Journal of Safety Research*, *38*(2), 159–163.

Regan, M. A., Triggs, T. J., & Godley, S. T. (2000a). Evaluation of a novice driver CD-ROM based training program: A simulator study. Paper presented at the Human Factors and Ergonomics Society Annual Meeting.

Regan, M. A., Triggs, T. J., & Godley, S. T. (2000b). Simulator-based evaluation of the DriveSmart novice driver CD-ROM training product. Paper presented at the Road Safety Research Policing and Education Conference.

Sagberg, F., & Bjørnskau, T. (2006). Hazard perception and driving experience among novice drivers. *Accident Analysis & Prevention*, *38*(2), 407–414.

Scopatz, R. A., Hatch, C. E., DeLucia, B. H., & Tays, K. A. (2003). *Unlicensed to kill: The sequel*. Washington, DC: AAA Foundation for Traffic Safety.

Scott-Parker, B. (2015). Experiences of teen drivers and their advice for the learner licence phase. *Traffic Injury Prevention*, *16*(2), 109–115. doi:10.1080/15389588.2014.909594

Scott-Parker, B., Bates, L., Watson, B., King, M., & Hyde, M. (2011). The impact of changes to the graduated driver licensing program in Queensland, Australia on the experiences of learner drivers. *Accident Analysis & Prevention*, *43*, 1301–1308.

Scott-Parker, B., Watson, B., King, M. J., & Hyde, M. K. (2012). Young, inexperienced and on the road – Do novice drivers comply with road rules? *Transportation Research Record: Journal of the Transportation Research Board of the National Academies,* No. 2318, Transportation Research Board of the National Academies, Washington, D.C., 98–106.

Senserrick, T. (2009). Australian graduated driver licensing systems. *Journal of the Australasian College of Road Safety*, *20*(1), 20–26.

Senserrick, T., Ivers, R., Boufous, S., Chen, H.-Y., Norton, R., Stevenson, M., ... Zask, A. (2009). Young driver education programs that build resilience have potential to reduce road crashes. *Pediatrics*, *124*, 1287–1292.

Senserrick, T., & Swinburne, G. C. (2001). *Evaluation of an insight driver-training program for young drivers*. Melbourne: Monash University Accident Research Centre.

Senserrick, T., & Whelan, M. (2002). *Graduated driver licensing systems*. Melbourne: Monash University Accident Research Centre.

Senserrick, T., & Whelan, M. (2003). *Graduated driver licensing: Effectiveness of systems and individual components*. Report no. 209. Melbourne: Monash University Accident Research Centre.

Senserrick, T., & Williams, A. (2015). *Summary of literature of the effective components of graduated driver licensing systems* (AP-R476-15). Sydney: Austroads.

Shope, J. T. (2006). Influences on youthful driving behavior and their potential for guiding interventions to reduce crashes. *Injury Prevention*, *12*(Suppl. 1), i9–i14.

Shope, J. T., & Bingham, C. R. (2008). Teen driving: Motor-vehicle crashes and factors that contribute. *American Journal of Preventive Medicine*, *35*(3, Suppl. 1), s261–s271.

Shults, R. A. (2010). Foreword to 'Graduated driver licensing research, 2007-present: A review and commentary'. *Journal of Safety Research*, *41*, 75. doi:10.1016/j.jsr.2010.04.001

Simons-Morton, B. (2007). Parent involvement in novice teen driving: Rationale, evidence of effects, and potential for enhancing graduated driver licensing effectiveness. *Journal of Safety Research*, *38*(2), 193–202.

Simons-Morton, B., Ouimet, M. C., & Catalano, R. F. (2008). Parenting and the young driver problem. *American Journal of Preventive Medicine*, *35*(3, Suppl. 1), s294–s303.

Siskind, V. (1996). Does licence disqualification reduce reoffence rates? *Accident Analysis and Prevention*, *28*(4), 519–524.

Thomas, F. D., Pollatsek, S., Pradhan, A., Divekar, G., Blomberg, R. D., Reagan, I., & Fisher, D. L. (2011). *Field and simulator evaluations of a PC-based attention maintenance training program*. Washington, DC, USA: Department of Transportation, National Highway Traffic Safety Administration.

TomTom International. (2017). *Our company*. Retrieved from http://corporate.tomtom.com/history.cfm

Triggs, T. J., & Smith, K. B. (1996). *Young driver research program: Digest of reports and principal findings of the research* (CR 164). Canberra: Federal Office of Road Safety.

Tronsmoen, T. (2011). Differences between formal and informal practical driver training as experienced by the learners themselves. *Transportation Research Part F: Traffic Psychology and Behaviour*, *14*, 176–188.

Watson, B. (2003). Research priorities in driver training: bridging the gap between research and practice. Paper presented at the 2003 Road Safety Research, Policing and Education Conference, Sydney.

Watson, B. (2004a). *The psychosocial characteristics and on-road behaviour of unlicensed drivers. PhD thesis*. Brisbane, Australia: Queensland University of Technology.

Watson, B. (2004b). The crash risk of disqualified/suspended and other unlicensed drivers. In Oliver, Williams & Clayton (Eds.), *Proceedings of the 17th International Conference on Alcohol, Drugs & Traffic Safety (T2004)*, Glasgow: International Council on Alcohol, Drugs & Traffic Safety.

Watson, B., Armstrong, K., Watson, A., Wilson, A., Livingstone, K., Sultana, K. & O'Donnell, L. (2011). A roadside survey of unlicensed driving in Queensland. In: *Proceedings of the 2011 Australasian Road Safety Research, Policing & Education Conference*. Perth: Western Australian Government.

Watson, B., Fresta, J., Whan, H., McDonald, J., Dray, R., Bauermann, C., & Churchward, R. (1996). *Enhancing driver management in Queensland*. Brisbane: Queensland Transport.

Weiss, T., Petzoldt, T., Bannert, M., & Krems, J. (2013). Calibration as side effect? Computer-based learning in driver education and the adequacy of driving-task-related self-assessments. *Transportation Research Part F: Traffic Psychology and Behaviour*, *17*, 63–74. doi:dx.doi.org/10.1016/j.trf.2012.10.001

White, M., Cunningham, L., & Titchener, K. (2011). Young drivers' optimism bias for accident risk and driving skill: Accountability and insight experience manipulations. *Accident Analysis & Prevention*, *43*, 1309–1315. doi:10.1016/j.aap.2011.01.013

Williams, A. (2006). Young driver risk factors: Successful and unsuccessful approaches for dealing with them and an agenda for the future. *Injury Prevention*, *12*(Suppl. 1), i4–8.

Williams, A. (2007). Contribution of the components of graduated licensing to crash reductions. *Journal of Safety Research*, *38*(2), 177–184.

Williams, A. (2009). Licensing age and teenage driver crashes: A review of the evidence. *Traffic Injury Prevention*, *10*(1), 9–15.

Williams, A., & Mayhew, D. (2003). *Graduated licensing: A blueprint for North America*. Arlington: Insurance Institute for Highway Safety.

Williams, A., & McCartt, A. (2014). Views of New Jersey teenagers about their state's policies for beginning drivers. *Journal of Safety Research*, *48*, 1–6. doi:dx.doi.org/10.1016/j.jsr.2013.10.001

Williams, A., Nelson, L. A., & Leaf, W. A. (2002). Responses of teenagers and their parents to California's graduated licensing system. *Accident Analysis and Prevention*, *34*, 835–842.

Williams, A., & Shults, R. A. (2010). Graduated driver licensing research, 2007-present: A review and commentary. *Journal of Safety Research*, *41*, 77–84.

Williams, A., Tefft, B., & Grabowski, D. C. (2012). Graduated driver licensing research, 2010-present. *Journal of Safety Research*, *43*, 195–203.

Young, K. (2002). *School-based road safety education: A review*. Melbourne: Monash University Accident Research Centre.

Zhao, J., Mann, R. E., Chipman, M., Adlaf, E., Stoduto, G., & Smart, R. G. (2006). The impact of driver education on self-reported collisions among young drivers with a graduated license. *Accident Analysis & Prevention*, *38*(1), 35–42.

CHAPTER 3

AGGRESSIVE DRIVING AND SPEEDING

Kara Kockelman and Jianming Ma

ABSTRACT

Purpose: *This chapter synthesises a variety of findings on the topic of aggressive driving and delivers a suite of strategies for moderating such behaviours. Examples and formal definitions of aggressive driving acts are given, along with specific techniques for reducing excessive speed and other aggressive behaviours.*

Methodology: *Key references from the literature are summarised and discussed, and two examples detailing how multi-parameter distributions and models compare with the negative binomial distribution and model are presented.*

Findings: *Speeding is the most common type of aggressive driving, and speeding-related crashes represent a high share of traffic deaths. Speeding relates to many factors, including public attitudes, personal behaviours, vehicle performance capabilities, roadway design attributes, laws and policies. Anonymity, while encased in a vehicle, and driver frustration, due to roadway congestion or other issues, contribute to aggressive driving.*

Safe Mobility: Challenges, Methodology and Solutions
Transport and Sustainability, Volume 11, 37–55

ISSN: 2044-9941/doi:10.1108/S2044-994120180000011003

Research implications: *More observational data are needed to quantify the effects of the contributing factors on aggressive driving.*

Practical implications: *Driver frustration, intoxication and stress can lead to serious crashes and other traffic problems. They can be addressed, to some extent, through practical enforcement, design decisions and education campaigns.*

Keywords: Aggressive driving; speeding; traffic law enforcement; roadway safety

INTRODUCTION

This chapter is the second of three chapters focussed on driver behaviour impacts on transport network safety and crash risk. While this chapter focusses on speeding and aggressive driving, it complements Chapter 2, which focusses on driver licensing and education, and Chapter 4, which delves into driver distraction.

Among the many driver errors that are typically at play in pre-crash contexts, aggressive driving is a serious issue, accounting for nearly 7 million crashes every year in the United States (US) alone (National Highway Traffic Safety Administration [NHTSA], 2014) and over half of all US fatal crashes (AAA Foundation for Traffic Safety [AAAFTS], 2009). Aggressive driving is believed to have been rising in response to rising traffic congestion, which frustrates many drivers (Neuman et al., 2003). Rising congestion is largely due to more motorists driving more kilometres or miles every year, with increases in vehicle-miles travelled outstripping new roadway lane-mile provisions by one to two orders of magnitude (NHTSA, 2000). At the same time, the level of law enforcement present on major roadways and elsewhere in the system may be falling, due to shifting police priorities. Consequentially, many locations find positive correlations in crash counts, congestion and aggressive driving, so better enforcement of aggressive driving may ameliorate many roadway issues simultaneously (NHTSA, 2000). Among the many aggressive behaviours drivers can demonstrate, excessive speed, a topic emphasised in this chapter, is the most common (AAAFTS, 2009). A suite of strategies for combatting aggressive behaviours are discussed towards the end of this chapter.

What Is Aggressive Driving?

Aggressive driving is not easy to define. Most national policies and definitions generally exclude violent, 'road rage' situations, where intent is to harm, and criminal statutes can be invoked (Tasca, 2000). Aggressive driving behaviours (sometimes classified as risky or dangerous driving behaviours; Richer & Bergeron, 2012; Richard et al., 2012) are labelled as traffic offences, rather than criminal offences, and include: excessive speed; failure to signal a lane change; running a red light, stop sign or railroad signal; failure to yield at a yield sign; passing another on the outer-lane side (depending on whether cars drive on the right or left in particular countries); tailgating; weaving in and out of traffic and other unsafe lane changes; passing on the shoulder; preventing others from passing; yelling or gesturing; and sustained horn honking, headlight flashing and other '(d)isplays of annoyance or hostility' (Tasca, 2000, p. 2). Such behaviours are often provoked by travel delays, due to unexpected traffic obstructions and/or poorly coordinated signals. Mina, Verma, Balhara, and Ul-Hasan's (2014) study of Indian drivers found that police presence is a strong moderating factor, curbing anger and driver aggression.

As Tasca (2000, p. 2) notes that many drivers admit to exhibiting such behaviours from time to time. Great Britain's Automobile Association's 1995 survey of 526 drivers (Joint, 1997) found that 62% reported having engaged in aggressive tailgating over the prior year, 59% had engaged in 'high beaming' those they wished to pass, 48% had received rude or obscene gestures from other motorists, 21% had been 'deliberately blocked by another vehicle' and 21% had received some verbal abuse while motoring (Wiesenthal, Roseborough, & Wickens, 2015). A more recent AAAFTS (2008) survey found that 78% of Americans considered aggressive driving to be a serious problem, yet nearly half reported driving at least 15 mph [24 kilometres per hour (km/hr)] over the speed limit in the past month. Such responses underscore the notion that many drivers do not have the same high standards for their own driving as they do for others with whom they share the roadways, and/or they overestimate their ability to manage risks.

Tasca's (2000, p. 2) extensive research led him to recommend the following definition: 'A driving behaviour is aggressive if it is deliberate, likely to increase the risk of collision and is motivated by impatience, annoyance, hostility and/or an attempt to save time.' In contrast, NHTSA (2015) defines aggressive driving as when 'an individual commits a combination of moving traffic offenses so as to endanger other persons or property'. Wiesenthal et al. (2015)

propose a definition based on a driver's actions, rather than intent, since motivation is so difficult to measure and establish in practice. Richer and Bergeron (2012) note that intentional aggressive driving versus risky and dangerous driving practices can emanate from very different personality profiles. Some communities have defined aggressive driving as 'three or more moving violations as part of a single continuous sequence of driving acts, which is likely to endanger any person or property' (NHTSA, 2000, p. 6).

Who Drives Aggressively Most Often?

NHTSA (2000) notes that aggressive driving deserves the kind of attention that impaired or driving under the influence (DUI) driving received decades ago, to ensure that aggressive driving is not 'socially acceptable'. Tasca's (2000, p. 3) review of self-reported, survey-based literature suggests a greater likelihood of aggressive driving behaviours in those who are young, male, 'in traffic situations which confer anonymity and/or where escape is highly likely' (e.g., at night or along a freeway), already angry (often due to other, non-traffic events), 'believe they possess superior driving skills', 'generally disposed to sensation-seeking or aggressiveness in other situations' and/or facing unexpected traffic congestion. Similarly, Ellison-Potter, Bell, and Deffenbacher (2001) found that aggressive driving is more often associated with contextual or situational variables (like sensing one's relative anonymity on a busy freeway and events that stimulate driver aggression) than with a driver's general disposition. Events that trigger aggression often occur on the roadway itself, like feeling another driver has cut one off; such situations can result in retaliatory actions (Wiesenthal et al., 2015).

Shinar (1998) notes that more extroverted, 'Type A' personalities tend to have lower thresholds for tolerating frustration and, therefore, are more likely to demonstrate aggressive driving behaviours. Those driving in settings where negative consequences are less likely (e.g., contexts offering greater driver anonymity and escape) and those sensing that others are driving improperly (e.g., travelling relatively slowly in the left-most lane of a multilane, divided highway or waiting too long to accept a gap in traffic and leave the stop bar) are more likely to demonstrate aggression. Shinar (1998) distinguishes aggressive drivers (which is a long-lived and relatively infrequent, but intrinsic disposition) from aggressive driving (which is more frequent, but abates when the annoyance or frustration is removed). Exceeding the speed limit may or may not be considered aggressive; if done in the presence of other motorists so the safety of others is at risk by not driving faster it is generally considered aggressive, even if the speeder is, for example, simply trying to avoid being late for an important meeting (Tasca, 2000).

THE ROLE OF SPEEDING

As noted above, many behaviours constitute aggressive driving. Speeding is a common, costly and relatively aggressive behaviour for many drivers. In fact, speeding was a causal factor in over 30% of US motor vehicle fatalities in 2010 and resulted in more than 10,000 deaths, nearly US$60 billion in economic costs and US$210 billion overall societal harm (Blincoe, Miller, Zaloshnja, & Lawrence, 2015; FARS, 2012). Speeding is aggressive when a driver significantly and dangerously exceeds the speed of those around them.

Speeding contributes to both the frequency and severity of traffic crashes (Kockelman & CRA International Inc., 2006). It impairs a driver's ability to react to emergencies, like sudden tire blowouts, unsafe manoeuvres by other travellers, roadway hazards and dangerous weather conditions. Speeders need additional time and distance to slow and stop their vehicles, and can create fear and confusion for those around them – including pedestrians and cyclists (see Chapters 10 and 11).

Perhaps most importantly, the laws of physics (kinetic energy $= 0.5 \times$ mass $\times$ velocity^2) suggest that speed is one of the most critical factors affecting crash severity (Transport Research Board, 1998). This is a key basis for the idea that 'Speed Kills', and the notion that a speed limit increase will naturally lead to higher numbers of traffic fatalities and serious injuries. In some contrast, Lave (1985) has argued for a 'Variance Kills' theory, as first proposed by Solomon (1964) and supported by Cerillo (1968). The rural road data analysed by these researchers appear to indicate that crash likelihood increases with an individual's deviation from that roadway's average speed. Several researchers (Fowles & Loeb, 1989; Levy & Asch, 1989; Snyder, 1989) have attempted to refute Lave's argument by enhancing his data and model specification. Interestingly, as Lave (1989) notes in his reply/rebuttal, their findings provide evidence for both theories (both speed and its variance are dangerous). Somewhat later, Davis (2002) noted how various correlations in aggregate data of speed variations and crash rates could be simply an ecological fallacy.

Evans (2004) reports how data from Australia and Sweden show sizable crash rate increases with speed, via the following (approximate) relationship:

$$\frac{\mathrm{CR}'}{\mathrm{CR}^0} \approx \left(\frac{V'_{50\%}}{V^0_{50\%}}\right)^p$$

where the exponent p is roughly 4 for fatal crash rates and 2 for all crash types, **CR** denotes crash rates (after and before the change in speeds) and

$V_{50\%}$ denotes the median (50th percentile) speeds of the traffic (after and before speed changes).

This relationship suggests that just a 1% increase in speed (so $V'/V^0 = \mathbf{1.01}$) will result in a 4% higher fatal crash rate. In terms of an occupant's own risk, Evans (2004) reports that research estimating the probability of not surviving the crash goes up by 4–12%. If speeds fall just 3%, Evans (2004) expects a 13% reduction in one's probability of dying, which is similar to the benefits provided by airbag use. Such results may change quite a bit with the advent of smarter, more automated, connected and (eventually) self-driving vehicles.

Kockelman and CRA International Inc.'s (2006) National Cooperative Highway Research Program report traced the many separate and interrelated effects of speed limit increases in the US on: (1) driver speed choice, (2) roadway crash rates (per mile-travelled on that road) and (3) injury severity outcomes for those crashes. Their data suggest that crash rates increase only slightly with a speed limit increase, but added that injury severities are notable, as indicated in Table 1. While higher speeds are more dangerous, everything else constant – due to the kinetic energy at play, a 2005 NHTSA study found that around half of all fatal speeding crashes in the US occur on roads with a posted speed limit of 50 mph (80 km/hr) or lower (Liu, Chen, Subramanian, & Utter, 2005). Malyshkina and Mannering's (2008) look at the state of Indiana's crash rates and severities after speed limits went up in 2005, found more severe crashes along non-interstates, but no statistically significant changes along interstate highways in that state at that time.

Speeding's Ties to Demographic Attributes and Driving Contexts

As alluded to above, observable driver characteristics often correlate with aggressive behaviours. Context is also important. For example, Evans (2004)

Table 1. Safety Effects Associated with a 10 mph (16 km/hr) Speed Limit Increase on High Speed Roads.

Increase in Speed Limit (mph)	Change in Average Driving Speed (mph)	Average Change in Total Crash Count (%)	Change in Probability of Fatal Injury (%)	Total Change in Predicted Fatal Injury Count (%)
55–65	+3	+3.3	+24	+28
65–75	+3	+0.64	+12	+13

Source: Kockelman & CRA International Inc., 2006
Note: Results come from typical high-speed roadway settings.

suggests that February is the month with the least severe crashes (measured as fatalities per injury) because of it's more difficult driving conditions (due to the presence of snow, ice and storms in many parts of the US), resulting in lower speeds. While young males tend to have the best reaction times, visual acuity and a relatively high knowledge about vehicles and driving, they also exhibit a higher crash rate because they tend to take more risks (Evans, 2004). As evident in many studies, Schroeder, Kostyniuk, and Mack (2013) also found that young drivers tend to speed more frequently, and speeding tends to be less likely as one ages. Inexperience is a serious issue and graduated licensing strategies (now the norm across the US) certainly help reduce such risk-taking and presumably more aggressive behaviours among younger drivers (see Chapter 2). Road rage (defined by NHTSA as a 'physical assault of a person or vehicle as a result of a traffic incident') involves using a vehicle as one's weapon and appears to be most common among relatively young males (American Safety Council, 2015).

Richard et al. (2012) analysed driver speeding behaviour using naturalistic driving data (with cameras watching many drivers and their traffic conditions over a number of months in the US), as well as data from focus groups. They noticed that speeding is relatively common among drivers and that young male drivers are more likely to speed than other demographic groups studied. They also concluded that speeding is not just associated with a small group defined by age (i.e., young), gender (i.e., male) and/or attitude (i.e., recklessness or risk-taking propensity). Most people admitted that they sped at least some of the time, but they normally did not think they were driving unsafely when they sped. Speeding behaviour is strongly related to a number of factors, like situational factors (e.g., type and length of trip, presence of others in the vehicle, likelihood of limit enforcement, opportunities to speed, road conditions and geographic location), demographic factors and a driver's 'personality' (including a person's motivations, beliefs and attitudes towards speeding). They found that personality factors were generally stronger predictors of speeding tendencies than demographic attributes, and speeding was more common during the morning peak period (or 'rush hour'), as well as on weekends.

Rural driving also comes with higher speeds, thanks to less congestion and less complex navigation, resulting in significantly higher fatality rates (Evans, 2004; Kockelman & CRA International Inc., 2006; Levine, Kim, & Nitz, 1995a, 1995b). Evans (2004) believes that road improvements tend to result in higher speeds and less congestion, and thus often noticeably higher fatality rates, but higher-design roadways tend to be safer, per mile travelled (Kockelman & CRA International Inc., 2006), and actual speed effects are tricky to ascertain exactly, because actual vehicle speed is generally unknown at the time of crash (Levine, Kim, & Nitz, 1995c).

Evans (2004) argues that speedometers are very important in helping cue drivers to their actual speed, helping them avoid failures in speed judgment. Of course, speed limiters and global positioning system (GPS)-based warnings of one's location-based, speed-limit violations will be useful. Many people wonder why vehicles are designed and sold to go well over 100 mph (160 km/hr), if 100 mph (160 km/hr) is already well above legal speed limits in nearly every nation. Speed governors (common on fleet-managed trucks and lorries) and automated speed enforcement (ASE) strategies (along roadways) seem wise but are often unpopular with the motoring public when applied to their own personal vehicles.

Chen and Kockelman's (2013) survey of over 1,000 Americans found that the least popular crash-reduction measures lie in addressing speeding violations: only 40.5% of respondents admitted to regularly driving above the speed limit on freeways, but over half (54.3%) responded that they support ASE technologies (while 27.3% opposed such enforcement). Among ASE supporters, 10 mph (16 km/hr) was the average speed over the speed limit for which automatic ticketing would be acceptable, and this jumped to 15 mph (24 km/hr) for those who opposed ASE. Use of speed governors or limiters on all vehicles was the least popular policy, with 37% approving and 41% opposing. Those who support governors supported an average maximum speed of 83 mph (133 km/hr), and opponents favoured 106 mph (170 km/hr).

As Chen and Kockelman (2013) note, despite the overall negative view of such speed reduction measures, ASE cameras and on-vehicle speed governors have been shown to reduce injury crashes by 11–58% and 8–34%, respectively. Saving lives and property can make terrific economic and social sense even when individuals are reluctant to espouse such changes in current laws and vehicle designs. However, the debate may be moot, as highly automated and self-driving vehicles begin production: such vehicles may be designed, by law, to obey speed limits and other traffic rules.

The Role of Gender

Evans (2004, p. 212) believes that the fastest drivers on roads are to blame for most fatal crashes and make the 'main contribution to risk'. He provides tales of speed violation rates across US drivers' ages, genders and ethnicities using 1998 data, suggesting that men were receiving tickets for speeding twice as often as women were. However, as Kweon and Kockelman (2003) note, men tend to drive almost 50% longer (total) distances than women do. And they may drive more often in settings that lead to higher police observation (e.g., during the night and along high-speed facilities). Evans (2004) wrote that younger drivers (ages 16–22) received the most violations (per year of driving) among the three age groups (young, middle-age and older), even though they drive significantly less distance, per year, than older persons do (Kweon & Kockelman, 2003). Kweon and Kockelman (2006) analysed the US Motor Vehicle Occupant Safety Surveys to estimate that males, college-educated and higher-income individuals tend to travel at higher speeds on highways, and are more likely to drink and drive (or at least admit to it); those of higher income and education are also more likely to wear seat belts, everything else constant. Levine (2011) examined the Houston metro area's 25,240 crashes (between 1991 and 2001) and found that more males exhibit riskier driving behaviour than females. He also noticed that males are more likely to have crashes in the central city, which tends to have higher crash rates per vehicle-miles travelled, as compared to other parts of the region. Moreover, NHTSA's safety statistics and survey results consistently show male drivers are more likely to speed and more involved in fatal crashes while intoxicated than female drivers (see, e.g., NHTSA, 2008; Schroeder et al., 2013). Fig. 1 illustrates such findings.

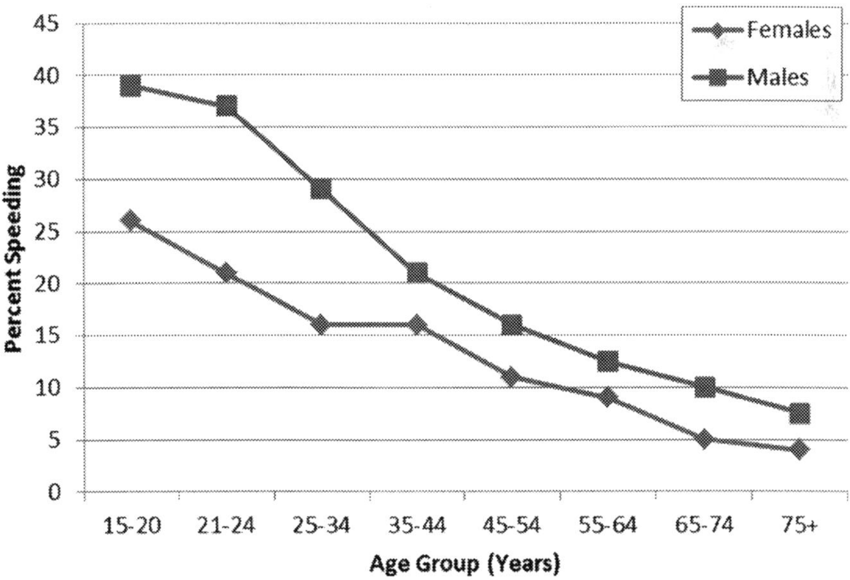

Fig. 1. Speeding Drivers in Fatal Crashes by Age and Gender, 2006 US Data (*Source*: NHTSA, 2008).

Evans (2004) notes that a Scottish investigation reported males being more likely to slow down in light rain, versus female drivers, and that risk-taking or aggressive-driving behaviours (such as higher speed choices, tailgating vehicles in front, driving unbelted and running red lights) tend to fall with a driver's age. According to Evans, earlier research had suggested that those dying in crashes or receiving more driving violations are much more likely to be in legal troubles and/or have personality disorders, be less academically successful and more antisocial. They also tend to more often have poor credit histories, smoke cigarettes and so forth. Evans (2004, p. 224) wrote that 'Cars are used as an outlet for the independence, rebelliousness, and peer acceptance needs of newly licensed adolescents.' He also notes that some past studies find higher crash rates and ticketing rates among children of parents with such issues.

Evans (2004) observed that males' testosterone levels peak quite abruptly around age 20 and then decline, much like severe crash rates and arrests (per capita) do when plotted versus a person's age. If one plots the male-to-female ratios, these follow a very similar profile, rising to roughly 4 to 1 on all three indicators (average testosterone levels, crash rates and arrests per capita) at around age 20.[1] Such similarities may suggest that testosterone is a 'common origin' of such problematic and dangerous behaviours, but Kweon and Kockelman's (2006) work found the same crash rates for males and females, across different age groups. Moreover, females are more likely to die in the same crash (see, e.g., Kahane, 2013; Kockelman & CRA International Inc., 2006; Wang & Kockelman, 2005), everything else constant (including seating position in the vehicle), thus women may be at more risk per mile driving.

Driving Under the Influence

DUI is also considered aggressive driving. NHTSA (2013) notes that about 31% of all US motor vehicle traffic fatalities involve alcohol-impaired driving. Evans (2004) notes that people with higher blood-alcohol-content (BAC) levels are more likely to die when involved in a crash. For example, those with a BAC of 0.08% (i.e., 8 g of alcohol per 100 mL of blood) are estimated to be 73% more likely to die than those with a 0% BAC in the same crash, largely because the crashes that result under DUI conditions are less controlled (e.g., at higher speed and later at night, with less thoughtful driver response). In 2012, 7% of drivers with a BAC of 0.08% or higher involved in fatal crashes had a prior conviction for DUI; only 1% of drivers in fatal crashes with no alcohol had a prior conviction for DUI (NHTSA, 2013). Drunk drivers often experience impaired perceptions that may result in risky behaviour such as speeding (Syrcle & White, 2006). According to NHTSA (2010b), drivers with a 0.08% or higher BAC in fatal crashes tend to have more prior driving convictions (e.g., speeding) than drivers with lesser or no alcohol content. DUI behaviour often correlates with aggressive driving behaviours.

OPPORTUNITIES FOR MODERATING AGGRESSIVE DRIVING AND MANAGING SPEED

Unfortunately, aggressive driving is associated with higher rates of crash involvement and is a factor in a high share of recorded crashes (AAAFTS, 2009; Blows, Ameratunga, Ivers, Lo, & Norton, 2005; Mann et al., 2007; Wells-Parker et al., 2002). Moderating aggressive driving requires a multi-prong approach, including driver education, engineering via roadway design details, laws and enforcement (Neuman et al., 2003).

Speed limits are standard practice around the globe, and they can be made variable (e.g., to reflect changing weather and traffic conditions, in real time) in coming years, thanks to greater use of connected/communicating vehicles and relatively precise location information provided via GPS. Variable speed limit strategies have been widely used in some European countries (Mirshahi et al., 2007) to improve traffic flow and safety, and possibly reduce emissions and conserve fuel. Proper use of the variable speed limit strategy may also improve driver compliance with speed limits – a common form of aggressive driving.

Other countermeasures include higher traffic fines, with mandated reporting of violations to insurance companies and loss of driving licences, along with roadside units to display vehicle speeds or other poor behaviours, making violations more obvious to drivers. Speed governors for those caught speeding may also prove a useful policy tool to pursue (much like ignition locks on vehicles of those who have been caught driving while intoxicated), but there is no evidence of such strategies in place yet. Roadway design strategies, automated enforcement and driver education options are covered in more detail below.

Driver Education

'Driver education' encompasses the idea that drivers know the rules of the road, including what constitutes aggressive driving versus appropriate driving behaviour, what the speed limits are from location to location, and the risks and consequences of poor driving behaviours. Unfortunately, Sivak et al.'s (2007) review of the literature suggests that standard driver education has not had significant success in taming speeding violations. Perhaps improvements in vehicle safety and performance (e.g., in the form of side air bags and electronic stability controls) is making drivers feel safer at higher speeds, thus simply maintaining current speed-violation levels is a measure of success.

More useful training, in the domain of avoiding aggressive driving, may come from finding ways to help drivers recognise that many traffic situations are not intentional, and that any retaliatory actions are offensive. For example, being abruptly cut off may be due to the offending driver being new to driving or having to get to the hospital to see a very ill family member; similarly, being slowed in traffic by a lead driver may be due to engine issues, an elderly driver lacking confidence or good night-time eyesight. In other words, poor driving behaviours do not always constitute bad intent by the drivers of those lead vehicles.

Wiesenthal et al. (2015) also recommend training drivers to reduce stress, via breathing exercises en route (or while roadside), carpooling or public transit use, and other helpful practices, both psychological and physical. They also recommend using the media to communicate the downsides of aggressive driving (such as fatal crash details) and possibly using lotteries with incentives to reward those who obey speed limits regularly and such. Finally, simply labelling all cars and trucks, or just those of repeat offenders, with 'How's my driving?' stickers, including phone numbers to call for reporting purposes, can pay dividends (by reducing drivers' perceptions of anonymity on motorways). Smart phone applications can be designed and developed to provide feedback on people's driving behaviour, which potentially helps promote safe driving.

Engineering: Roadway and Vehicle Design

Wiesenthal et al. (2015) recommend traffic calming techniques for diminution in aggressive driving. Standard treatments are narrower lanes, mid-block chicanes and traffic humps, rumble strips, raised pedestrian platforms and other obstacles to high-speed driving. All of these can help, at least in lower-speed locations (like residential streets). Smart, Cannon, Howard, Frise, and Mann (2005) suggest designing vehicles to (temporarily) lower horn volume after repeated horn use, and send audible messages to drivers who are following too closely, for example.

Technologies can also be used to monitor the speed of vehicles and help law enforcement agencies to better deploy their officers to enforce speed limits. For example, connected vehicle technologies allow vehicles to talk to each other and to roadside devices, to relay their positions, to avoid collisions and to share other valuable information. Connected vehicles can warn drivers of speeders and other problematic driving nearby and help law enforcement agencies pinpoint areas where vehicle speeds are frequently exceeding speed limits or other laws and decide where they should step-up patrols to catch violators.

Radar (or Lidar) and cameras are commonly used to detect and deter speeding and other traffic offences, such as red-light running, illegal rail crossings and toll violations. Radar or Lidar detectors measure the speed of vehicles, cameras capture images of vehicles speeding and committing other violations and citations (along with images) are mailed to the vehicle owner. Such automated enforcement technologies have proven effective in ensuring compliance with speed limits and other traffic laws and thus enhancing traffic safety. For example, Friedman, Hedeker, and Richter (2009) found that speed cameras and reduced speed limits could reduce operating speeds and thus reduce fatal injuries. Ellison-Potter et al. (2001) suggest that the use of 'camera radar' warning signs would reduce anonymity and thus reduce aggressive driving behaviours such as speeding. Automated enforcement does not intend to replace traditional law enforcement officers; instead, it augments enforcement levels at locations where traffic enforcement officers cannot be deployed safely.

Currently, rather few jurisdictions use automated enforcement technologies, like speed cameras, which are limited or prohibited in several US states. To help obtain public support for such policies, automated enforcement can easily be limited to particular neighbourhoods, fines can be lower than standard speeding tickets issued directly by police officers, and citations may not show on driving records. In addition, radio and laser detection used by motorists to avoid such detection should be prohibited.

Further Suggestions

In order to address aggressive driving, relevant laws must be on the books and be ready to apply. NHTSA (2000, p. 7) stresses the importance of reviewing what statues already exist in a city or state to see if they appropriately address 'the problem of careless or negligent driving'. Such laws regularly already exist, and rigorous enforcement of such laws then becomes critical.

NHTSA (2000) urges that convictions should result in significant 'point loss' on one's driving record (which affects insurance rates and may result in compulsory education and training) and/or suspension of one's driving licence, with more severe penalties for repeat offenders and those whose actions result in injury to others. NHTSA also recommends using the news media (for regular public service announcements and other education campaigns), real-time radar and video (on enforcement vehicles) to record key evidence and thus facilitate legal prosecution, red-light cameras with automated photos and red-light indicators (on the back sides of signals) to alert police of conditions just downstream of signal lights (which facilitates safe stopping of red-light violators).

NHTSA notes that special laws may be needed to allow use of unmarked vehicles (including those with cherry-picking lift units, to provide better video of traffic violations in hot-spot locations). NHTSA (2000) also recommends a single, shared message across multiple jurisdictions, agencies and other stakeholders to relay the message far and wide, since drivers cross such boundaries regularly and may not perceive the risks of being arrested and/or being injured. They recommend including business leaders, insurers, law enforcement officers (including judges and prosecutors), victims of aggressive driving incidents, medical personnel and many others in crafting the messages, laws and enforcement strategies and deploying them.

NHTSA (2000) also suggests including strategies like a 'tip line' (for citizens to report licence plates and other details of violations), providing partners with a 'stock speech' and other presentation materials, offering police-car ride-alongs to journalists and judges and conducting enforcement in high-visibility locations during rush hour. Collecting and comparing performance metrics over time is valuable in gauging and reporting how effective actions are. Metrics include the share of fatal crashes in a region where speeding and reckless driving are cited as factors, share of vehicles exceeding the posted speed limit by 15 mph (24 km/hr) or 20 mph (32 km/hr) or more on major roadways, annual number of speeding citations given and share of drivers who report speeding.

Example Applications of Anti-Aggressive Driving Strategies

Noteworthy examples of aggressive driving enforcement programs within the US include Massachusetts' use of unmarked and non-traditional vehicles, with in-card video cameras and radar units, laws to request immediate licence suspension and compulsory training of aggressive drivers (NHTSA, 2000).

Officers in Albuquerque, New Mexico, have used trucks with extendable buckets to hold officers aloft with radar guns in hand, who then radio about violators to downstream patrol vehicles. A Colorado *Two Seconds for Safety* campaign urged motorists to use a two-finger peace or victory sign to convey a quick message of thank you or please excuse me and/or remind fellow motorists of the value of a two-second-headway (inter-car interval timing). The Colorado State Patrol's two-page annual report about deaths on the roadway, DUI arrests and safety messages can be folded and handed to motorists at traffic stops (NHTSA, 2000). Newly elected judges in Ohio were offered 'ride-alongs' inside planes to witness traffic issues from the air. Neuman et al. (2003) discuss the role of habitual offenders and policy champions, and offer additional details and strategy examples in their NCHRP 500 report.

CONCLUSIONS

Aggressive driving behaviours may include multiple driving offences, such as driving faster than conditions allow, following too closely, making unsafe lane changes and running red lights (NHTSA, 2010a). Aggressive driving not only tends to increase the likelihood and severity of crashes but also interrupts traffic flow and adds to congestion. Aggressive driving can be caused by a driver's personal physical and psychological condition (e.g., immaturity and/or temporary stress) and/or by their environment (e.g., a congested commute and/or poorly timed traffic signals). According to NHTSA's 2002 survey, aggressive driving behaviours are perceived to be common, but they are difficult to measure (NHTSA, 2004). Aggressive driving is unsafe driving that disregards safety and courtesy. The AAAFTS (2009) found that most fatal crashes involve aggressive driving behaviours, such as running red lights or improperly changing lanes. Speeding is one of the most common forms of aggressive driving and is a contributing factor in about one-third of fatal crashes.

As discussed in this chapter, speeding is the most common type of aggressive driving and speeding-related crashes represent a high share of traffic deaths. Speeding relates to many factors, including public attitudes, personal behaviours, vehicle performance capabilities, roadway design attributes, laws and policies. Driver education, speed zoning and speed enforcement are top recommendations for combating excessive speed choices and other forms of driving aggression. Emerging technologies also offer much hope.

By avoiding conflicts and crashes, and reducing demands on drivers (Fagnant & Kockelman, 2013), new and emerging technologies (like blind-spot monitoring, lane centring, fully self-driving vehicles and smart-cycle intersections) will moderate driver frustrations and presumably reduce aggressive behaviours. Self-driving (autonomous) vehicles will be designed to safely drive themselves, and may allow for shorter headways, higher capacities and less roadway congestion. Such vehicles may well be programmed to never break the speed limit, change lanes without signalling, run red lights or demonstrate other unsafe or hostile behaviours. Moreover, without steering wheels and foot pedals before them, drivers become passengers and can focus on more meaningful and pleasant pursuits en route. Radio-connected vehicles with cameras can report on and document other drivers' aggressive driving behaviours, potentially conventionally driven. Smart intersection control units may eventually double intersection capacities and keep traffic moving (Dresner & Stone, 2008). If self-driving vehicles are one day mandated, aggressive driving may remain possible only on special tracks, where competitors race against the clock or one another. Until then, millions of crashes annually and upwards of 50% of all roadway fatalities (AAAFTS, 2009) remain a globally costly result of aggressive driving choices.

NOTE

1. After age 50, testosterone ratios in males versus females rise to roughly 8 to 1, but absolute testosterone levels are down, and crash rate and arrest ratios are much less than 8 to 1 for males versus females in that age group (Severson & Barclay, 2015).

ACKNOWLEDGEMENTS

The authors appreciate the careful review and valuable feedback given by the reviewer, along with editorial support and review by Dr. Dominique Lord and Dr. Simon Washington.

REFERENCES

AAA Foundation for Traffic Safety (AAAFTS). (2008). *Traffic Safety Culture Index*. Washington, DC: AAA Foundation for Traffic Safety. Retrieved from http://www.aaafoundation.org/pdf/2008TSCIndexFinalReport.pdf.

AAA Foundation for Traffic Safety (AAAFTS). (2009). *Aggressive driving: Research update.* Washington, DC: AAA Foundation for Traffic Safety. Retrieved from https://www.aaafoundation.org/sites/default/files/AggressiveDrivingResearchUpdate2009.pdf.

American Safety Council. (2015). *Aggressive driving and road rage.* Retrieved from http://www.safemotorist.com/articles/road_rage.aspx.

Blincoe, L. J., Miller, T. R., Zaloshnja, E., & Lawrence, B. A. (2015). *The economic and societal impact of motor vehicle crashes, 2010 (Revised).* Report no. DOT HS 812 013. Washington, DC: National Highway Traffic Safety Administration. Retrieved from http://www-nrd.nhtsa.dot.gov/pubs/812013.pdf.

Blows, S., Ameratunga, S., Ivers, R. Q., Lo, S. K., & Norton, R. (2005). Risky driving habits and motor vehicle driver injury. *Accident Analysis and Prevention, 37*(4), 619–624.

Cerillo, J. A. (1968). Interstate system accident research study II (Interim Report II). *Public Road, 35*(5), 71–75.

Chen, D., & Kockelman, K. (2013). Americans' crash histories and opinions on safety policy. *Transportation Research Record, 2364*, 12–22.

Davis, G. (2002). Is the claim that 'variance kills' an ecological fallacy? *Accident Analysis and Prevention, 34*(3), 343–346.

Dresner, K., & Stone, P. (2008). A multiagent approach to autonomous intersection management. *Journal of Artificial Intelligence Research, 31,* 591–656.

Ellison-Potter, P., Bell, P., & Deffenbacher J. (2001). The effects of trait driving anger, anonymity, and aggressive stimuli on aggressive driving behavior. *Journal of Applied Social Psychology, 31*(2), 431–443.

Evans, L. (2004). *Traffic safety.* Bloomfield Hills, MI: Science Serving Society.

Fagnant, D., & Kockelman, K. (2013). *Preparing a nation for autonomous vehicles: Opportunities, barriers and policy recommendations.* Eno Center for Transportation Report. Retrieved from http://www.caee.utexas.edu/prof/kockelman/public_html/ENOReport_BCAofAVs.pdf.

FARS. (2012). *Fatal accident reporting system data.* Washington, DC: National Highway Transportation Safety Administration, U.S. Department of Transportation. Retrieved from http://www.nhtsa.gov/FARS.

Fowles, R., & Loeb, P. D. (1989). Speeding, coordination and the 55 mph limit: Comment. *The American Economic Review, 79*(4), 916–921.

Friedman, L. S., Hedeker, D., & Richter, E. D. (2009). Long-term effects of repealing the national maximum speed limit in the United States. *American Journal of Public Health, 99*(9), 1626–1631.

Joint, M. (1997). Road rage. In AAA Foundation for Traffic Safety (Ed.), *Aggressive driving: Three studies.* Retrieved from https://www.aaafoundation.org/sites/default/files/agdr3study.pdf.

Kahane, C. J. (2013). *Injury vulnerability and effectiveness of occupant protection technologies for older occupants and women.* Report no. DOT HS-811-766. Washington, DC: National Highway Traffic Safety Administration.

Kockelman, K. M., & CRA International, Inc. (2006). *Safety impacts and other implications of raised speed limits on high-speed roads.* NCHRP 17-23 Report. Transportation Research Board.

Kweon, Y.-J., & Kockelman, K. (2003). Overall injury risk to different drivers: Combining exposure, frequency, and severity models. *Accident Analysis and Prevention, 35*(4), 313–321.

Kweon, Y.-J., & Kockelman, K. (2006). Driver attitudes and choices: Speed limits, seatbelt use, and drinking-and-driving. *Journal of the Transportation Research Forum, 45*(3), 39–56.

Lave, C. (1985). Speeding, coordination and the 55 mph limit. *The American Economic Review, 75*(5), 1159–1164.

Lave, C. (1989). Speeding, coordination and the 55 mph limit: Reply. *The American Economic Review*, *79*(4), 926–931.

Levine, N. (2011). Spatial variation in motor vehicle crashes by gender in the Houston, Texas, metropolitan area. *Transportation Research Board Conference Proceedings*, *4*(46), 12–25. Retrieved from http://onlinepubs.trb.org/onlinepubs/conf/cp46v2.pdf.

Levine, N., Kim, K. E., & Nitz, L. H. (1995a). Spatial analysis of Honolulu motor vehicle crashes: I. Spatial patterns. *Accident Analysis and Prevention*, *27*(5), 663–674.

Levine, N., Kim, K. E., & Nitz, L. H. (1995b). Spatial analysis of Honolulu motor vehicle crashes: II. Zonal generators. *Accident Analysis and Prevention*, *27*(5), 675–685.

Levine, N., Kim, K. E., & Nitz, L. H. (1995c). Daily fluctuations in Honolulu motor vehicle accidents. *Accident Analysis and Prevention*, *27*(6), 785–796.

Levy, D. T., & Asch, P. (1989). Speeding, coordination and the 55 MPH limit: Comment. *The American Economic Review*, *79*(4), 913–915.

Liu, C., Chen, C. C. L., Subramanian, R., & Utter, D. (2005). *Analysis of speeding-related fatal motor vehicle crashes*. Technical Report no. DOT HS-809-839. Washington, DC: National Highway Traffic Safety Administration. Retrieved from http://www.nhtsa.gov/staticfiles/nti/enforcement/pdf/809839.pdf.

Malyshkina, N. V., & Mannering, F. (2008). Effect of increases in speed limits on severities of injuries in accidents. *Transportation Research Record, 2083*, 122–127.

Mann, R. E., Zhao, J., Stoduto, G., Adlaf, E. M., Smart, R. G., & Donovan, J. E. (2007). Road rage and collisions involvement. *American Journal of Health Behavior, 31*(4), 384–391.

Mina, S., Verma, R., Balhara, Y., & Ul-Hasan, S. (2014). Road rage: Prevalence pattern and web based survey feasibility. *Psychiatry Journal, 897493*, http://dx.doi.org/10.1155/2014/897493. Retrieved from http://www.hindawi.com/journals/psychiatry/2014/897493/#B33.

Mirshahi M., Obenberger, J., Fuhs, C. A., Howard, C. E., Krammes, R. A., Kuhn, B. T., … Yung, J. L. (2007). *Active traffic management: The next step in congestion management*. Report #FHWA-PL-07-012. Washington, DC: Federal Highway Administration, US Department of Transportation.

Neuman, T. R., Pfefer, R., Slack, K. L., Hardy, K. K., Raub, R., Lucke, R., & Wark, R. (2003) *Guidance for implementation of the AASHTO strategic highway safety plan, Volume 1: A guide for addressing aggressive-driving collisions*. NCHRP Report 500. Washington DC: Transportation Research Board. Retrieved from http://safety.transportation.org/htmlguides/AggDrvr/assets/ADguide.pdf.

NHTSA. (2000) *Aggressive driving enforcement: Strategies for implementing best practices*. Report no. US DOT-HS-809-031. Washington, DC: National Highway Transportation Safety Administration.

NHTSA. (2004). *Safe routes to school: Practice and promise*. Washington, DC: National Highway Traffic Safety Administration. Retrieved from www.nhtsa.dot.gov/people/injury/pedbimot/bike/Safe-Routes-2004/.

NHTSA. (2008). *Traffic safety facts 2006*. Report no. DOT HS 810 809. Washington, DC: National Highway Traffic Safety Administration. Retrieved from http://www-nrd.nhtsa.dot.gov/Pubs/810809.pdf.

NHTSA. (2010a). *Countermeasures that work: A highway safety countermeasure guide for state highway safety offices* (5th ed.). Report no. DOT HS811 258. Washington, DC: National Highway Traffic Safety Administration.

NHTSA. (2010b). *Traffic safety facts 2009; Alcohol-impaired driving*. Report no. DOT HS 811 385. Washington, DC: National Highway Traffic Safety Administration. Retrieved from http://www-nrd.nhtsa.dot.gov/Pubs/811385.PDF.

NHTSA. (2013). *Traffic safety facts 2012: Alcohol-impaired driving*. Report no. DOT HS 811 870. Washington, DC: National Highway Traffic Safety Administration.

NHTSA. (2014). *Traffic safety facts 2012 data*. Report no. DOT HS 812 032. Washington, DC: National Highway Traffic Safety Administration Report. Retrieved from http://www-nrd.nhtsa.dot.gov/Pubs/812032.pdf.

NHTSA. (2015). *Stop aggressive driving*. Washington, DC: National Highway Transportation Safety Administration. Retrieved from http://www.nhtsa.gov/Aggressive.

Richard, C. M., Campbell, J. L., Lichty, M. G., Brown, J. L. Chrysler, S., Lee, J. D., … Regale, G. (2012). *Motivations for speeding, Volume 1: Summary report*. Report no. DOT HS 811 658. Washington, DC: National Highway Traffic Safety Administration. Retrieved from http://www.nhtsa.gov/staticfiles/nti/pdf/811658.pdf.

Richer, I., & Bergeron, J. (2012) Differentiating risky and aggressive driving: Further support of the internal validity of the Dula Dangerous Driving Index. *Accident Analysis & Prevention, 45,* 620–627.

Schroeder, P., Kostyniuk, L., & Mack, M. (2013). *2011 National survey of speeding attitudes and behaviors*. Report no. DOT HS 811 865. Washington, DC: National Highway Traffic Safety Administration.

Severson, A., & Barclay, R. S. (2015). Testosterone levels by age. *Healthline*. Retrieved from http://www.healthline.com/health/low-testosterone/testosterone-levels-by-age#overview1.

Shinar, D. (1998). Aggressive driving: The contribution of the drivers and the situation. *Transportation Research Part F, 1*(2), 137–160.

Sivak, M., Luoma, J., Flannagan, M. J., Bingham, C. R., Eby, D. W., & Shope, J. T. (2007). Traffic safety in the U.S.: Re-examining major opportunities. *Journal of Safety Research, 38,* 337–355.

Smart, R., Cannon, E., Howard, A., Frise, P., & Mann, R. (2005). Can we design cars to present road rage? *International Journal of Vehicle Information & Communication Systems, 1,* 44–55.

Snyder, D. (1989). Speeding, coordination and the 55 mph limit: Comment. *The American Economic Review, 79*(4), 922–925.

Solomon, D. (1964). *Accidents on main rural highways related to speed, driver, and vehicle*. Washington, DC: Federal Highway Administration.

Syrcle, J., & White, W. (2006). *The Illinois DUI risk reduction project: ASUDS-RI pilot phase I* (Statistical summary). Springfield, IL: Center for Legal Studies, University of Illinois-Springfield.

Tasca, L. (2000). *A review of the literature on aggressive driving research*. Retrieved from http://www.stopandgo.org/research/aggressive/tasca.pdf.

Transportation Research Board. (1998). *Managing speed – Review of current practice for setting and enforcing speed limits*. Special Report 254. Washington, DC: National Academy Press.

Wang, X., & Kockelman, K. (2005). Occupant injury severity using a heteroscedastic ordered logit model: Distinguishing the effects of vehicle weight and type. *Transportation Research Record, 1908*, 195–204.

Wells-Parker, E., Ceminsky, J., Hallberg, V., Snow, R. W., Dunaway, G., Guiling, S., … Anderson, B. (2002). An exploratory study of the relationship between road rage and crash experience in a representative sample of US drivers. *Accident Analysis and Prevention, 34*(3), 271–278.

Wiesenthal, D., Roseborough, J., & Wickens, C. (2015). Behavioral approaches for combating aggressive driving (Chap. 8). In D. Teodorovic (Ed.), *Routledge handbook of transportation.* Taylor & Francis. New York.

CHAPTER 4

DRIVER DISTRACTION AND INATTENTION

Mitchell L. Cunningham and Michael A. Regan

ABSTRACT

Purpose – *Driver distraction and other forms of driver inattention remain significant road safety problems. The purpose of this chapter is to explore recent developments in theoretical and empirical research on driver distraction and inattention and provide the reader with a sense for, and understanding of, the key issues.*

Methodology – *Key references from the literature are reviewed and discussed.*

Findings – *First, we discuss one way of conceptualising the distinction between driver distraction and other forms of inattention, as well as the mechanisms which may underlie these forms of inattention. Second, we underscores how driver distraction may derive from a plethora of sources, and how the potential for performance degradation deriving from driver interaction with these sources may be moderated by a range of factors. Third, we review recent literature on the types of impairments in driving performance and safety associated with driver distraction. Fourth, we outline recent literature on driver distraction and inattention in the realm of*

Safe Mobility: Challenges, Methodology and Solutions
Transport and Sustainability, Volume 11, 57–82

ISSN: 2044-9941/doi:10.1108/S2044-994120180000011004

highly automated vehicles that will drive the transport future. Finally, we discuss some promising strategies aimed at preventing and mitigating the impact of driver distraction.

Research implications – *There are many gaps in the driver distraction literature that need to be addressed. In addition, further research needs to be undertaken to examine the role of driver distraction in the realm of highly automated vehicles.*

Practical implications – *The findings point towards of a range of injury prevention countermeasures that have potential to prevent and mitigate driver distraction.*

Keywords: Driver distraction; distraction; inattention; human factors; road safety; traffic safety

INTRODUCTION

This chapter is the third of three chapters focussed on driver behaviour impacts on transport network safety and crash risk. Whilst this chapter focusses on driver distraction and inattention, it complements Chapter 2, which examines driver education and licensing, and Chapter 3, which focusses on aggressive driving and speed.

Safe driving requires a complex amalgamation of behaviours. It requires the driver to successfully perform simultaneously several complex activities: route finding, route following, controlling speed and lateral position, avoiding collisions, obeying road rules and monitoring the status of vehicle systems (Brown, 1986; Michon, 1985).

Despite the often-significant attentional resources that such activities demand, drivers are often observed to engage in other activities that have potential to divert attention away from activities critical for safe driving. It is perhaps not surprising that empirical research studies reveal that lapses of attention, including driver distraction, play a significant role in road crashes, near-crashes and other safety-critical events (e.g., Hickman, Hanowski, & Bocanegra, 2010; Klauer, Dingus, Neale, Sudweeks, & Ramsey, 2006; Olson, Hanowski, Hickman, & Bocanegra, 2009). Data from so-called 'naturalistic driving studies', for example, suggest that inattention and distraction are contributing factors in around 65% and 23% of crashes and near-crashes, respectively (Klauer et al., 2006). In response, there has been a significant increase in recent years on research aimed at shedding light on the mechanisms that underlie distraction, at understanding its impact on driving performance and safety, and in informing the development of countermeasures to prevent and mitigate its effects (e.g., Regan, Lee, & Young, 2009). Despite advancements in the literature, however, much remains unknown about the link between distraction, crash risk and crashes.

This chapter explores developments in theoretical and empirical research on driver inattention and distraction. Firstly, driver distraction will be defined in light of its relationship with other forms of driver inattention and theories of attention. Secondly, the chapter will highlight potential sources of driver distraction and their corresponding modal properties. Thirdly, the chapter will illustrate how a number of factors may moderate the impact of driver distraction on driver performance and safety. Fourthly, the effect of driver distraction on both driving performance and safety will be discussed. Finally, the chapter will examine new interfaces between distraction and driving that are emerging as modern motor vehicles become increasingly autonomous. We start by considering what is meant by the term 'driver distraction'.

DEFINITIONS AND THEORIES OF DRIVER INATTENTION AND DISTRACTION

Definitions

The terms 'driver inattention' and 'driver distraction' are sometimes used interchangeably (Caird & Dewar, 2007). Within the transport safety field both driver inattention and driver distraction are inconsistently defined. For example, Victor, Engstrom, and Harbluk (2008) define driver inattention as the 'inappropriate selection of the information that is relevant for safe driving, either by the selection of irrelevant information (driving and non-driving related) or by the lack of selection of driving information (e.g., eyes closed due to fatigue and low arousal or effort)' (p. 137). One of the most widely cited definitions of driver inattention is that offered by Regan, Hallett, and Gordon (2011): 'insufficient, or no attention, to activities critical for safe driving' (p. 1775). The authors note that driver inattention can derive from different mechanisms (including distraction), not all of which are necessarily controlled by the driver. Some sources of driver distraction (e.g., another driver driving erratically through the traffic stream), for example, may capture involuntarily the attention of a driver.

A recent taxonomy of driver inattention proposed by Regan et al. (2011) is helpful in understanding how driver distraction relates conceptually to driver inattention and the various mechanisms through which driver inattention may arise. Regan et al. propose the following mechanisms of inattention:

- *Driver-restricted attention*: 'Insufficient or no attention to activities critical for safe driving brought about by something that physically prevents (due to biological factors) the driver from detecting (and hence from attending to) information critical for safe driving' (p. 1775). An example may be having a micro-sleep and, in doing so, failing to attend to critical information on the road.
- *Driver-misprioritised attention*: 'Insufficient or no attention to activities critical for safe driving brought about by the driver focussing attention on one aspect of driving to the exclusion of another, which is more critical for safe driving' (p. 1775). An example, here, would be a driver who looks over their shoulder for too long while merging with freeway traffic and fails to attend to a vehicle ahead suddenly braking.
- *Driver-neglected attention*: 'Insufficient or no attention to activities critical for safe driving brought about by the driver neglecting to attend to activities critical for safe driving' (p. 1775). An example may be a driver neglecting to scan for trains at a railway level crossing (because they are rarely or never seen).
- *Driver-cursory attention*: 'Insufficient or no attention to activities critical for safe driving brought about by the driver giving cursory or hurried attention to activities critical for safe driving' (p. 1776). An example may be a driver who performs only a quick and incomplete head check when merging and, in doing so, ends up colliding with a merging car.
- *Driver-diverted attention*: 'The diversion of attention away from activities critical for safe driving towards a competing activity, which may result in insufficient or no attention to activities critical for safe driving' (p. 1776). This is synonymous with driver distraction. There are two sub-categories of driver-diverted attention:
    - Non-driving-related, which 'is the diversion of attention away from activities critical for safe driving towards a competing non-driving-related activity' (p. 1776). An example of this may be when a driver attempts to compose a text message on their mobile phone while driving.
    - Driving-related, which 'is the diversion of attention away from activities critical for safe driving towards a competing driving-related activity' (p. 1776). An example of this may be a driver fixated on a low-fuel-warning light who fails to detect a braking vehicle ahead.

Although conceptually similar, driving-related distraction can be distinguished from *driver-misprioritised attention*. The former denotes the improper allocation of attention to non-safety critical activities (e.g., monitoring fuel gauge and reading a text message), while the latter involves the improper allocation of attention between multiple safety-critical events (e.g., monitoring forward traffic but neglecting to monitor adjacent traffic when changing lanes; Regan et al., 2011).

The conceptual difference between driver distraction and inattention is clear in the Regan et al. (2011) taxonomy. The former involves some competing activity (e.g., looking at a mobile phone while driving) that diverts attention away from activities critical for safe driving, whereas the latter occurs in the absence of any competing activity. Noteworthy is that not all definitions of driver distraction acknowledge that not all events or objects that divert attention away from activities critical for safe driving necessarily degrade performance (see Regan et al., 2011 for review).

Theories of Driver Distraction and Inattention

A number of psychological theories attempt to account for why humans are limited in their ability to perform two tasks concurrently or temporally in close succession (Pashler, 1994). Driver distraction may be conceptualised as a form of dual-task interference, in which the performance of the primary task (driving) can be interfered with by the concurrent performance of some secondary task (e.g., reading a text message on a phone). Pashler (1994) and Gladstones, Regan and Lee (1989) argue that this interference may be due to a central processing bottleneck within the attentional system, which allows only one task at a time to be attended to, thus delaying and impairing the performance of one or both tasks. On the other hand, Kahneman (1973) purports that the attentional system is not limited by a bottleneck mechanism per se but by the availability of only a finite amount of attentional resources. Proponents of this limited attentional 'pool' theory claim that the performance of a secondary task will only impair primary task performance if the total amount of attentional resources they require does not exceed the total supply of the pool (Kahneman, 1973).

In contrast to this 'single' pool view, Multiple Resource Theory (MRT) (Horrey & Wickens, 2004) proposes that multiple processing resources (or channels) exist in the brain for processing different task properties. Here, each resource channel is capacity-limited, and dual-task interference occurs if two tasks demand and exceed the resources of a single channel. These channels are defined by four dimensions: processing stage (i.e., does the task require perceptual, cognitive or motor resources?), processing code (i.e., does the task require spatial or verbal resources?), perceptual modalities (i.e., does the task require visual or auditory resources?) and visual channel (i.e., does the task require focal or ambient vision?). In driving, MRT would predict that, since driving primarily demands visual perception as opposed to auditory perception, a secondary task that is visual (e.g., looking at a phone) will cause greater dual-task interference with driving than one that is auditory (e.g., listening to a phone conversation).

SOURCES OF DISTRACTION

Studies have shown that drivers are often engaged in secondary activities at some point during their drive (e.g., Fitch et al., 2013), and these secondary activities can be both within and outside the vehicle. Sources of distractions include, but are not limited to, objects (e.g., roadside advertising), events (e.g., a crash scene), passengers, other road users and stimuli internal to the driver's mind (e.g., thinking and daydreaming; Regan et al., 2009).

Despite the existence of an almost infinite number of potential sources of distraction, all may be classified according to a finite set of 'triggering modal properties', that is, by the stimulus properties of the source of distraction that attracts attention (Hallett, Regan & Bruyas, 2011). According to Hallett et al. (2011), a source of distraction can be: (a) visual (e.g., looking at a billboard), (b) auditory (e.g., hearing a screaming child), (c) olfactory (e.g., smelling smoke coming from the car engine), (d) gustatory (e.g., tasting an overly sour candy), (e) tactile (e.g., feeling a spider crawl on one's leg) and (f) cognitive (e.g., daydreaming or emotional experiences; see for review Cunningham & Regan 2016). The source of distraction may possess one or more of these modal properties. In turn, the modal properties of a competing activity created by driver interaction with a source of distraction may elicit a finite set of behavioural, or 'triggered', responses while driving (Hallett et al., 2011), that is, the competing activity may induce behavioural responses that are visual ('eyes off road'), cognitive ('mind off road') and manual ('hands off vehicle controls'). These behavioural, or triggered, responses are commonly referred to in the literature as visual distraction, cognitive distraction and manual distraction, respectively. However, defining types of distraction based on the impact of a competing activity on driving is problematic, as it leads to an artificially truncated repertoire of distraction types, and fails to acknowledge that when one's eyes are off the road, their minds are also invariably off the road.

These behavioural responses to competing activities may interfere with activities critical for safe driving (Hallet et al., 2011). For example, the sound of a crash in the vicinity of a car may trigger in the driver a cognitive response (mind off road; "what was that?"), followed by a visual response (eyes off road) as visual attention is diverted from the roadway ahead toward the accident scene. As a consequence, a driver may fail to process important traffic information ahead, for example a braking lead vehicle, and may even crash into it. This crash outside the car may also have auxiliary effects. For example, it might induce further cognitive distraction if the driver continues to think and dwell over it for some time after passing the crash scene. In this case, a driver allowing their mind to wander may impair their ability to detect hazardous information in their visual field ('look but fail to see' e.g., a crossing pedestrian). The ramifications of such types of distractions are further highlighted in the following sections.

Case Study: Distraction from Billboards and Roadside Advertising

Roadside advertisements and billboards, for obvious reasons, are intended to be as conspicuous and attention-grabbing to drivers as possible. Given that billboards are one of the easiest objects in the road environment to control, a wealth of empirical research has been devoted to exploring their influence on driver distraction and road safety. As such, this section provides a brief review of the empirical literature pertaining to the potential for billboards to produce driver distraction.

It is important to define first what a billboard is and the mechanisms by which billboards may distract drivers in their vicinity. Decker et al. (2014) define a billboard as 'any off-premises external-to-vehicle advertising sign that is permanently constructed along a roadway and conveys visual information' (p. 235). Billboards can be further categorised based on how their content is conveyed, and include passive billboards, which convey static information, and active billboards, which convey dynamic information (e.g., flashing and digitally moving messages, displays that change mechanically).

According to the attention framework proposed by Trick and Enns (2009), a billboard may capture driver attention via two different processes. Firstly, a billboard may distract a driver if they choose to look at it (i.e., top-down processing), such as when a driver is distracted when searching for their favourite advertising billboard in a familiar driving environment. Conversely, a billboard may involuntarily capture the driver's visual attention (i.e., bottom-up processing), often due to characteristics such as bright colouring and flashing lights, which are physical properties that are particularly effective at grabbing attention (Trick & Enns, 2009). Trick and Enns further contend that this latter mechanism of attentional processing may be of special concern in a driving context as it is often automatic and the distraction source is very difficult to ignore.

A theme in recent empirical road-safety research is that, where passive billboards are distracting, dynamic billboards appear to be even more so. In one of the more methodologically rigorous studies in this area, Dukic, Ahlstrom, Patten, Kettwich and Kircher (2013) used an instrumented vehicle study to examine the effects of dynamic billboards on the visual behaviour of drivers in comparison to passive billboards and other normal road safety signs. It was found that dynamic billboards generally attracted longer visual dwell times and a greater number of eye fixations than all other passive signs and advertising. These results are consistent with the findings of a recent meta-analysis (Decker et al., 2014) purporting that dynamic billboards tend to attract a greater number of glances that are considered dangerously long (i.e., ≥ 2 seconds; Klauer et al., 2006) compared to passive billboards. Decker et al. (2014) conclude that dynamic billboards are especially potent at attracting visual attention as more resources and time are required to decode the dynamic messages they convey; and dynamic billboard characteristics, such as 'flashing' and other transient features, are particularly good at reorienting visual attention to that location.

Advertising billboards may pose a risk to drivers when they attract eye glances lasting longer than 0.75 seconds, as this is regarded as the minimum perception–reaction time required respond to an unexpected braking lead vehicle (Smiley, Smahel, & Eizenman, 2004). In addition, glances longer than 2.0 seconds have been suggested as being particularly dangerous and have been shown to be associated with a two-fold increase in crash risk (Klauer et al., 2006). Judging by the visual metrics noted by both Dukic et al. (2013) and Decker et al. (2014), both passive and active billboards have the potential to compete for visual attention, which may leave insufficient resources for activities critical for safe driving. Dynamic billboards appear to be relatively more capable of doing so.

When reviewing the empirical evidence from before-after studies, however, a clear relationship between the presence of billboards and crashes has not been established. A recent study by Hawkins, Kuo and Lord (2014) examined crashes before and after the installation of 135 135) electronic billboards across four states in the United States (US), including California (6 sites), North Carolina (33 sites), Ohio (63 sites) and Washington (33 sites). In each state, and in aggregate, a safety effect of the digital billboards was not detected. After conducting a Bayesian analysis with regression to the mean corrections, the authors concluded: 'there is no statistically significant change in crash frequency associated with installing on-premise digital signs…'

Another before–after study by Izadpanah, Reza, Koo, and Hadayeghi (2014) examined 10 digital signs in Canada and found a statistically nonsignificant reduction of crashes of 4%, concluding 'there is not enough evidence to suggest that these signs have any impact on safety'.

A before–after study conducted by Smiley et al. (2005) examined eight sites in Toronto, Canada, and concluded 'Overall there was no effect on total collisions'. Finally, a study by Yannis, Papadimitriou, Papantoniou, and Voulgari (2013) examined nine electronic billboards in a before–after study, including sites where five billboards were removed – thus looking to detect an improvement in safety upon removal. Their conclusion was 'there is no statistical correlation between road accidents and advertising signs at the microscopic level'.

So, while the studies that focus on gaze behaviour and attention diversion suggest a negative safety effect of billboards, before–after studies have yet to substantiate a potential negative effect on crashes. There may be two possible explanations for this contradiction: (1) there is no safety effect of billboards or (2) the safety effect has been too small to detect with before–after studies conducted to date. These differences highlight the need for continued research on strengthening the ties between surrogate safety measures (i.e., braking, accelerations and gaze behaviour) and road crashes. Further research is likely to shed additional light on the effects of billboards on driver safety.

MODERATORS OF DRIVER DISTRACTION

Young, Regan, and Lee (2008) have postulated that the extent to which a competing activity distracts a driver and impairs driving performance depends on four moderating factors: (1) driver characteristics; (2) driving task demand; (3) competing task demand and (4) the driver's ability to self-regulate in order to prepare for, or respond to, distraction caused by a competing activity (Young et al., 2008).

Driver Characteristics

The extent to which a competing activity distracts a driver and impairs driving performance may be influenced by a number of driver characteristics, including age, gender, driving experience, driver state, degree of practice/familiarity with the competing activity and personality (Regan et al., 2011).

Age and driving experience are examples of individual differences that have received considerable empirical attention in the literature. For example, Regan et al. (2011) note that less experienced drivers are less able to perform competing activities without compromising activities critical for safe driving as they have only partially automated driving skills. This claim is supported by studies demonstrating that younger drivers are not as efficient as more experienced drivers in processing critical driving information (Wikman, Nieminen, & Summala, 1998), as well as by instrumented vehicle studies showing that a range of secondary activities interfere with novice drivers' driving performance but not with that of more experienced drivers (e.g., Klauer et al., 2014).

Another driver characteristic that may moderate driver distraction is driver state, for example, a driver's level of alcohol intoxication. Simulated driving studies have shown, for example, that impairments in driving performance due to engagement with a secondary activity are exacerbated when under the influence of alcohol (Rakauskas et al., 2008; Verster et al., 2009). In a more recent simulated driving study, Harrison and Fillmore (2011) showed that, while driver engagement with a simple two-choice reaction time task did not impair driving performance of drivers when sober, the introduction of alcohol significantly degraded driving performance. These findings suggest that seemingly 'easy' secondary tasks within a vehicle (e.g., changing the radio channel) may cause significant interference with the driving task when a driver is under the influence of alcohol, as the capacity to process information in this state is severely curtailed (Fillmore, 2003). This increased propensity for a driver to be distracted by other activities when intoxicated may provide deeper insight into the extremely high rates of alcohol-related crashes (National Highway Traffic Safety Administration [NHTSA], 2000).

Driving Task Demand

Driving task demand refers to the load that the driving task itself places on the driver. Factors such as traffic conditions, weather conditions, road conditions/design, the number and type of vehicle occupants, the vehicle cockpit interface design and vehicle speed, together or separately, may influence driving task demand (Regan et al., 2011). Generally speaking, the performance of a secondary task will be more likely to produce interference, and hence distraction, when the driving environment itself is resourcefully taxing (Young et al., 2008). For example, research has shown that drivers tend to be more hesitant to engage in secondary tasks in poor weather, on winding roads, in heavy traffic, at night or close to schools (Young & Lenné, 2010), indicating that drivers may perceive such driving contexts as heavily demanding. More difficult driving contexts demand more attention in order to safely navigate through them, meaning that engaging in a secondary task is less likely to be able to be timeshared without drawing attention away from and compromising the performance of activities critical for safe driving.

Secondary Task Demand

The demands of the secondary, competing, task will be influential in determining the degree of interference it will create with the primary driving task (Regan et al., 2009). In general, the more demanding a secondary activity, the less likely it is that it can be timeshared without degrading activities critical for safe driving (Regan et al., 2011). Factors that may influence secondary task demand include its resource compatibility with the driving task (e.g., whether it is visual–manual (VM) versus auditory–vocal), its complexity, whether it can be ignored and how interruptible it is (Regan et al., 2009).

Due to technological advancements, engagement with secondary tasks, such as infotainment devices (e.g., smartphones) with highly demanding VM interfaces, inadvertently imposes high cognitive demands that can undermine driving performance (Horrey & Wickens, 2006; Strayer, Drews, & Johnston, 2003; Tsimhoni, Smith, & Green, 2004). For example, a recent simulated driving study found that engaging with MP3 players with longer song playlists (i.e., more 'demanding' as there were more songs to search through) was associated with greater variability in vehicular speed, lane position and longer eye glances compared to simpler MP3 players (i.e., less 'demanding' due to having fewer songs to search through; Lee, Roberts, Hoffman, & Angell, 2012). This finding is especially relevant today as newer technologies have the capacity to hold much more information than conventional devices, and interactions with them may require relatively more attentional resources (e.g., searching through an MP3 playlist of 1,000 songs compared to 20 songs). The increased level of attention that such secondary activities may require will likely augment the interference with activities critical for safe driving.

Self-regulation

Self-regulation refers to how well a driver can change their behaviour to maintain adequate driving performance in the face of competing tasks (Young & Regan, 2007). For example, self-regulation can be *preparatory* in nature, which involves a driver preparing themselves a priori for the anticipated effects of potential distraction (e.g., turning off a cell phone before driving in case it rings during the drive).

Self-regulation can also be *reactive*, in which the driver behaves strategically in a way that compensates for the effects of an existing source of distraction (e.g., asking a passenger to stop talking when navigating a turn at a busy intersection). The extent to which both these processes occur is likely governed by the same moderating factors above (e.g., younger and more inexperienced drivers are more likely to overestimate driving skill and thus feel it is not necessary to prepare for and reduce their exposure to sources of distraction).

There are a number of specific ways in which a driver may commonly attempt to self-regulate in response to distraction. For example, drivers may partake in behaviours perceived as being 'safer', such as reducing speed (Engström, Johansson, & Östlund, 2005; Törnros & Bolling, 2005) or increasing headway to a lead vehicle to compensate for engaging in secondary tasks (Fitch et al., 2013; Hosking, Young, & Regan, 2009). Another simple and common self-regulation strategy drivers tend to use includes taking shorter glances away from the road when driving demand increases (e.g., keeping their eyes on the road when road is foggy; Tivesten & Dozza, 2014; Tsimhoni et al., 2004).

Tivesten and Dozza (2014) used naturalistic driving data to determine how driving task demand moderated drivers' self-regulation in terms of their engagements with VM phone tasks. They found that drivers increased engagement with the VM task when driving task demand was low, for example, when not navigating sharp turns, roundabouts (see Chapter 8), etc. Through this self-regulation strategy of modulating their use of phones in response to changing roadway demands, drivers may reduce the level of interference between driving performance and interacting with the VM phone task.

DRIVER DISTRACTION, DRIVING PERFORMANCE AND SAFETY

Distraction can affect driver behaviour and undermine driving performance. Reported deficits of driving performance widely vary due to methodological variability between studies, and can include degraded lane keeping, degraded speed control, increased reaction time, missed traffic signals, shorter or longer inter-vehicle distances, unsafe gap acceptances, reduced situational awareness, poorer visual scanning, reduced horizontal field of view and missed checks (e.g., mirror checks; Bayly, Young, & Regan, 2008; Horberry & Edquist, 2008). The type of impairment depends upon the type of distraction that brings about the impairment (see for review Cunningham, Regan, & Imberger, 2017).

Because driving is primarily a visual task, visual distraction is often cited as being especially dangerous for safe driving performance (Young & Lenné, 2010).

Visual distraction can cause drivers to look away from the roadway and has been found to lead to large and frequent lane deviations, abrupt steering movements and slow responses to lead vehicles braking (Dingus, Antin, Hulse, & Wierwille, 1989; Zhang, Smith, & Witt, 2006). Moreover, the diversion of visual attention away from the roadway may lead to failures in detecting hazardous events and increase the reaction time to safety-critical events (Regan et al., 2011). Perhaps not surprisingly, such indices of impaired driving performance are strongly correlated with off-road glance patterns (Dingus et al., 1989; Dingus, Hulse, McGehee, & Manakkal, 1994; Zhang et al., 2006).

Compared to visual distraction, the effects of cognitive distraction on driving performance may be more covert. Like visual distraction, cognitive distraction can undermine a driver's sensitivity to critical cues and traffic signals in the roadway environment (e.g., 'looked at but not seen'; Strayer & Johnston, 2001; Strayer et al., 2003). This idea is further supported by a meta-analysis of 23 studies which found that cognitive distraction caused by using auditory e-mail systems, performing math calculations or having hands-free cell phone conversations delayed driver response to hazards by an average of 130 ms (Horrey & Wickens, 2006). In addition, drivers that are cognitively distracted tend to have longer fixations and a denser gaze concentration in the centre of the road context, which may result in failures to perceive important information in peripheral vision (Recarte & Nunes, 2003; Regan et al., 2011).

Although driver distraction has been shown to be related to degraded driving performance using a range of methodologies, exactly how these impairments translate into real-world safety critical events (e.g., crashes) is not clearly understood. However, there is evidence that they do, and recent estimates are alarming. For example, in the US, which has relatively sophisticated data collection systems, distraction was a contributing factor in around 10% of fatal crashes and 18% of injury crashes in 2013 (NHTSA, 2015). In addition, The Australian National Crash In-depth Study (Beanland, Fitzharris, Young, & Lennac, 2013) found that 57.6% of crashes showed evidence of driver inattention as a contributing factor between 2000 and 2011 in New South Wales and Victoria. Distraction accounted for around 18% of these inattention crashes. The most frequent sources of distraction recorded were in-vehicle distractions, such as interactions with passengers and mobile phones; these accounted for 20% of distraction crashes. Although it is beyond the scope of this chapter to review in detail all of the literature pertaining to the impact of driver distraction and driver inattention on driver safety (see, e.g., Regan et al., 2009, for a review), the findings from a number of seminal naturalistic driving studies are highlighted here.

Naturalistic driving studies involve the collection of driving data in an instrumented vehicle without direct experimental manipulation (Stutts et al.,

2005). Data loggers in the vehicle measure a range of performance metrics (e.g., vehicle speed) while driver behaviour is recorded by a video camera to gauge levels of driver inattention and distraction (Klauer et al., 2006). This methodology allows researchers to examine the flow of events (e.g., reading a text message) that occurs directly prior to a safety-critical event in the real world (Craft & Preslopsky, 2012). The analytic approach taken by such studies typically involves formulating odds ratios (ORs), which reveal the odds of an event happening compared to baseline driving (Olson et al., 2009). For the purposes of this chapter, the term 'safety-critical event' serves as an umbrella term for crashes (i.e., vehicle hitting another object) and near-crashes (i.e., a rapid evasive manoeuvre to avoid a crash).

Derived from a recent meta-analysis of naturalistic driving data (Caird, Johnston, Willness, Asbridge, & Steel, 2014), Table 1 shows a number of distracting activities and their associated risk (i.e., ORs) for safety-critical events. The studies examined by Caird et al. (2014) include:

- Klauer et al.'s (2006) study, which is perhaps the most cited naturalistic driving study. This 100-car study was conducted by the Virginia Tech Transportation Institute over a period of 12–13 months between 2003 and 2004. The main purpose of the study was to examine the role of driver inattention in car incidents using a sample of experienced drivers. The authors note that personal grooming (e.g., application of makeup) and glance

Table 1. Selected Distraction Activities and Associated ORs.

Driver Distraction Category	Klauer et al. (2006)	Olson et al. (2009)	Hickman et al. (2010)	Fitch et al. (2013)
Text message on cell phone		**23.24**	**163.6**	1.73
Personal grooming	**3.13**	**4.48**		
Dial cell phone	**2.79**	**5.93**	**3.51**	0.99
Handle CD	**2.25**			
Look outside vehicle	**3.7**	**0.54**		
Reach for object in vehicle	1.38	**3.09**		**3.65**
Drink from container	**1.03**	0.97		
Eating	**1.57**	1.01	**1.11**	
Talk or listen to a hand-held cell phone	**1.29**	1.04	**0.89**	0.79
Talk or listen to a hands-free cell phone		**0.44**	**0.65**	0.73
Smoking (reach, light and extinguish)		**0.60**		
Smoking (cigarette in mouth or hand)		0.97		
Interact with or look at passenger	**0.5**	**0.35**		

Source: Caird et al. (2014).
Note: Bolded ORs were found to be statistically significant in their respective studies.

behaviour directed outside the vehicle had the strongest association with the likelihood of a safety-critical event proceeding.

- Olson et al.'s (2009) study, which used similar methods and analyses to Klauer et al. (2006) to examine the link between driver distraction and likelihood of a safety-critical event using a sample comprised of 203 commercial motor vehicles and 55 trucks. The authors note that, while grooming behaviours and dialling a cell phone were risky behaviours (OR > 4), text messaging posed the greatest risk for a safety-critical event (OR=23).
- Hickman et al.'s (2010) study analysed the role of driver distraction in road safety using a sample of commercial trucks and buses. Text messaging behaviour was highly associated with the risk of safety-critical events (OR = 163.5).
- Fitch et al.'s (2013) study, which examined the impact of hand-held and hands-free cell phone use on driving performance and safety-critical event risk. This study employed a sample of 204 ordinary drivers for 31 days and examined cell phone use (e.g., texting and dialling, reaching for it) in particular. Surprisingly, cell phone engagement was not significantly associated with a higher risk of safety-critical events compared to baseline driving. The authors did not speculate as to what could account for this counterintuitive finding.

With the exception of Fitch et al. (2013), there seems to be a strong link between the interaction with cell phones and safety-critical driving events in the real world, as evidenced by the large ORs summarised in Table 1 (Caird et al., 2014). This association is consistent with prior research highlighting decrements in driving performance associated with concurrent engagement in text messaging behaviour. For example, Caird et al. (2014) found that drivers displayed degraded lateral and longitudinal control, increased reaction times to hazards and increased perceived subjective workload when simultaneously text messaging compared to drivers who were not. It may be these types of driving impairments lead to a higher risk for safety-critical events in real-world scenarios.

One common characteristic of secondary activities, such as cell phone use, personal grooming and reaching for an object, is that the performance of these activities tends to require a diversion of visual attention away from the forward roadway (towards the secondary activity/task). A number of naturalistic driving studies have investigated how visual activity, independent of what the distracting task was, away from the roadway, increases the risk of a safety-critical event. In addition to the increased crash risk associated with glances of 2 seconds or longer away from the forward roadway for experienced drivers noted earlier (i.e., Klauer et al., 2006), Simons-Morton, Guo, Klauer, Ehsani, and Pradhan (2014) suggest that single glances of just 1 second or more off the forward roadway substantially increase the risk of safety-critical events for novice teenage drivers (OR = 1.7). This finding relates back to the potential for driver age and experience to moderate distraction, as these novice drivers tend not to be as adept as experienced drivers at managing secondary tasks during complicated driving manoeuvres.

For data pertaining to naturalistic studies and other paradigms utilised to investigate the link between driver distraction and safety-critical events, the reader should be aware of a number of methodological and theoretical issues.

Firstly, attributions regarding what constitutes a source of distraction and which activities are critical for safe driving are not fixed and are therefore difficult to determine a priori (Regan et al., 2009). This can encourage researchers to adopt different operational definitions for these constructs, which may account for why the prevalence of distraction varies across studies (Gordon, 2009). In addition, these varying operational definitions could entice the use of different coding systems between studies. For example, some researchers may only code distractions as being in-vehicle events or objects and neglect outside sources, or perhaps define behaviours that would be typically categorised as driver inattention (e.g., fatigued state) as a driver distraction (Beanland et al., 2013). Lastly, naturalistic studies often fail to capture certain types of distraction, such as mind wandering, and cognitive distractions, such as daydreaming (Gordon, 2009). However, despite these methodological and theoretical criticisms, the link between driver distraction and the risk of safety-critical events is difficult to refute based on the evidence discussed.

DRIVER DISTRACTION IN HIGHLY AUTOMATED VEHICLES

Automated vehicles are those in which at least some aspects of a safety-critical control function (e.g., steering, throttle or braking) occur without direct driver input (e.g., assisted cruise control, collision warning systems; NHTSA, 2013). One of the market appeals for automated vehicles is that they aim to relieve drivers of the many moment-to-moment demands of driving (Fitch, Bowman, & Llaneras, 2014). However, this significant reduction in driver workload may not be necessarily beneficial for safe driving and may exacerbate driver inattention and distraction. This is a significant issue today as automated vehicle technology is not yet advanced enough to be considered 100% safe and reliable, therefore requiring human input (i.e., take-over of control) when the automated vehicle reaches the limits of its competence (e.g., due to sensory degradation in foggy road conditions or inability of the algorithms driving the vehicle to control the vehicle safely in some traffic scenarios).

Highly automated driving will bring about significant changes in the driver's role in that an active driver will likely become merely a passive monitor of the autonomous driving task (i.e., 'supervising' the automated driving itself). This task may be quite monotonous for some drivers and may render them more prone to experiencing passive fatigue (Saxby et al., 2008), which is associated with diminished levels of driver attentional capacity (Young & Stanton, 2002). Indeed, this passive fatigue induced by high levels of automation has been shown to degrade simulated driving performance when the driver is required to suddenly regain control of the vehicle to avoid an impending collision (e.g., Saxby et al., 2008).

On the other hand, monotonous periods of automated driving may encourage drivers to engage in secondary activities that may be more stimulating or entertaining. For example, simulated driving studies have shown that drivers in automated driving conditions are more likely to partake in activities such as eating (Jamson, Merat, Carsten, & Lai, 2013) or watching a DVD (Carsten, Lai, Barnard, Jamson, & Merat, 2012) than those in conditions of little/no driving automation. Individuals exposed to higher levels of automation also tend to divert more visual attention away from the forward roadway compared to manual drivers (Carsten et al., 2012). Perhaps unsurprisingly, research shows that drivers are poor at regaining control of automated driving when engaged in secondary tasks (e.g., Merat, Jamson, Lai, & Carsten, 2012) or when visual attention is not maintained on the forward roadway (Zeeb, Buchner, & Schrauf, 2015) (Cunningham & Regan, 2015a, 2015b, in press).

The concept of *situational awareness* may help explain why driver inattention and distraction may impair drivers' ability to regain manual control from automated driving. Situational awareness, colloquially, is a driver's comprehension of what is going on around them and use this information to anticipate future events (Endsley, 1995). When a driver is inattentive or distracted they are less likely to gain information about the driving and road context, and their mental model of the driving context will likely be partial. Therefore, in the case of an automated driving system failing, the driver's model of the driving context must be first updated before they can initiate an appropriate response and regain manual control of the vehicle. If a driver has too low a level of situational awareness (due to inattention and distraction), more time – and perhaps too much – will be required for this take-over and could result in a safety-critical incident (Zeeb et al., 2015). In addition, the failure to anticipate critical automation events in the driving context (e.g., due to driver inattention and distraction) will likely lead to the experience of 'automation surprises', which are unexpected or 'surprising' actions performed by the automated systems (Hollnagel & Woods, 2005). Effectively, these 'automation surprises' can erratically increase driver workload and impair the driver's ability to effectively respond to them (Vahidi & Eskandarian, 2003).

While automated driving may diminish the workload associated with manual driving, there is still debate regarding whether or not this is necessarily beneficial for safe driving. Driver inattention and distraction may be especially relevant in future automated vehicles as drivers may over-rely on the less-than-perfect technology, relinquish vigilance of the driving task and, in turn, may feel it is appropriate and safe to misuse it (e.g., sleep or read while in automation mode and effectively leaving little/no attention to the driving task). These human factors issues will undoubtedly impair drivers' readiness and alertness to manually re-engage in the case of automation limitation or failure.

COUNTERMEASURES AND MITIGATION OF DRIVER DISTRACTION

It may not be possible to ever eliminate distraction or driver inattention. At best, it can be well managed. Regan et al. (2009) have estimated that 55% of all known sources of distraction are avoidable, implying that there is ample scope for countermeasure development.

Education and Training

Drivers, particularly those who are young and inexperienced, need to be educated about the risks associated with secondary task engagement while driving and be trained to maintain focus on the forward roadway (Simons-Morton et al., 2014). With the advent of smartphones and other new devices the potential for distraction is constantly there and is even promoted (e.g., frequent Facebook notifications). While government policies and police enforcement regimes restricting cell phone use are proliferating and indeed may have the potential to reduce distraction risk, drivers will often attempt to be 'sneaky' and hide usage, which in itself could be as dangerous (e.g., drivers diverting visual attention to a phone hidden in their lap). Drivers should also be aware that fatigue, intoxication and simply glancing away from the forward roadway for too long could exacerbate the risk of safety-critical events.

Preliminary research into training programmes for novice drivers seems promising, with one study showing that focussed-attention training reduced both the number and length of off-road glances of novice drivers when engaged in driving-irrelevant in-vehicle activity (Pradhan et al., 2011).

An Ergonomically Designed Human–Machine Interface

With the advent of the intelligent information age, an ergonomically well-designed human–machine interface may have the greatest potential to both prevent and mitigate driver distraction. Firstly, and perhaps primarily, active real-time vehicle systems can prevent or limit driver exposure to competing activities when the driver is predicted to be under high mental workload (e.g., a mobile phone call is diverted while the driver is navigating a sharp turn; Victor, 2011). Various items of information, such as that derived from the driving task itself, and how potentially distracting the competing activity is can be combined to assess the driver's 'readiness' to interact with a competing activity without jeopardising the performance of activities critical for safe driving. These vehicle systems are known as 'workload managers', as they are designed to maintain a reasonable driving load – one that is neither too high nor too low (Victor, 2011).

On the other hand, engaging in other activities may be unavoidable (e.g., hearing another vehicle honking) and thus countermeasures also need be aimed at reducing the amount and effect of distraction when it occurs. A range of new vehicles are testing variations of driver state assessment technology (see Dong, Hu, Uchimura, & Murayama, 2011, for a review). These have the ability to provide feedback to help the driver shift attention back to driving when they are judged as being 'too distracted' according to predetermined criteria set by the system, driver or vehicle owner (Rauch, Kaussner, Krüger, Boverie, & Flemisch, 2009). These devices are generally comprised of two measures: direct measures, which monitor the eye-gaze, blink process and head rotation of the occupant in real-time (Boverie & Giralt, 2008; Hargutt, 2000; Wierwille, 1999) and indirect measures, which examine driving- and performance-related metrics which are used to draw conclusions about the driver's state. For example, direct measures aimed at measuring the direction of the head, face or gaze may be used to deduce whether the driver is looking on or off the road (i.e., is visually distracted). On the other hand, indirect measures of lane keeping performance may be utilised to assess the level of driver fatigue. For example, lane keeping performance is heavily impaired with increasing fatigue (Stephan et al., 2006).

The aim of such mitigation strategies would be to alert and help the driver realise that they are glancing away from the roadway for too long or too often and to 'train' the driver to recognise some limits. Preliminary research suggests that this form of driver performance feedback can have beneficial outcomes for road safety (e.g., Donmez, Boyle, & Lee, 2008; Heinzman & Zelinsky, 2010). For example, recent trials of head rotation-monitoring systems have been shown to reduce distraction event frequency by almost 80% through the use of various types of feedback, including distraction alerts and real-time transmitted dispatcher feedback (Croke & Cerneaz, 2009; Heinzman & Zelinsky, 2010).

Other Countermeasures

There are, of course, other countermeasures that have been developed by governments and industry to prevent and mitigate the adverse effects of distraction. These are reviewed in Regan et al. (2009) and include improved data collection methods, for more accurately quantifying the role of driver inattention, including distraction, in crashes and other safety-critical events; improved laws for managing driver exposure to distraction; improved enforcement of these laws; improved graduated licensing systems that systematically limit young driver exposure to sources of distraction until they are capable of managing distraction (see 'self-regulation' earlier in chapter); and improved vehicle and road design to minimise driver workload and provide drivers with more reserve capacity to deal with both anticipated and unexpected distractions.

CONCLUSIONS

Even though they have been the focus of traffic-related research for several decades, driver distraction and other forms of inattention remain as significant road safety problems in Australia and many other countries. Further work is needed to increase our understanding of their prevalence in crashes, their effects on driver safety and how best to manage them. The aim of this chapter was to review the accumulating body of literature on driver distraction and provide the reader with an understanding of the key issues.

The first section of the chapter discussed one way of conceptualising the relationship between driver distraction and other forms of inattention, as well as the mechanisms by which these forms of inattention may operate to compromise the performance of activities critical for safe driving. The second section of the chapter highlighted how driver distraction can derive from a myriad of secondary activities and sources, and how the potential for such competing activities to degrade driving performance and safety may be moderated by a diversity of factors. The third section of the chapter reviewed recent literature on the types of driving performance impairments and crash risks associated with driver engagement in distracting activities and tasks. The fourth section highlighted recent literature on the outlook for driver distraction and inattention in highly automated vehicles of the future. Finally, we reviewed some of the more promising strategies aimed at preventing driver distraction, such as driver education and training, and countermeasures to help mitigate the effects of distraction once it occurs.

Many of the driving tasks that drivers currently perform are becoming automated, and highly automated vehicles are already here. Therefore, future research needs to be devoted to assessing how driver distraction and inattention will operate to influence traffic safety in this domain. For example, further research is needed to help us understand how moderating factors, such as age and experience, which have been shown to moderate the impact of distraction in manual driving conditions, may affect driver distraction and safety in highly automated vehicles; especially when the driver is required to take control of the vehicle in driving conditions in which it can no longer operate autonomously. Until vehicles are able to self-drive with 100% reliability in all driving conditions, driver distraction and other forms of inattention are likely to continue to be significant safety issues for both drivers and other road users.

REFERENCES

Bayly, M., Young, K., & Regan, M. (2008). Sources of distraction inside the vehicle and their effects on driving performance. In M. A. Regan, J. D. Lee, & K. L. Young (Eds.), *Driver distraction: Theory, effects and mitigation*. Boca Raton, FL: CRC Press.

Beanland, V., Fitzharris, M., Young, K., & Lennac, M. (2013). Driver inattention and driver distraction in serious casualty crashes: Data from the Australian National Crash In-depth Study. *Accident Analysis & Prevention*, *54*, 99–107.

Boverie, S., & Giralt, A. (2008). Driver vigilance diagnostic based on eyelid movement observation. *The proceedings of the IFAC world congress,* Seoul, Korea.

Brown, I. (1986). Functional requirements of driving. Paper presented at the Berzelius symposium on cars and causalities, Stockholm, Sweden.

Caird, J. K., & Dewar, R. E. (2007). Driver distraction. In R. E. Dewar & R. Olsen (Eds.), *Human factors in traffic safety* (2nd ed., pp. 195–229). Tucson, AZ: Lawyers & Judges Publishing.

Caird, J. K., Johnston, K. A., Willness, C. R., Asbridge, M., & Steel, P. (2014). A meta-analysis of the effects of texting on driving. *Accident Analysis & Prevention*, *71*, 311–318.

Carsten, O., Lai, F., Barnard, Y., Jamson, A. H., & Merat, N. (2012). Control task substitution in semi-automated driving: Does it matter what aspects are automated? *Human Factors*, *54*, 747–761.

Craft, R. H., & Preslopsky, B., (2012). Distraction and inattention: Top crash causes in the USA (Chapter 8). In M. A. Regan, J. D. Lee, & T. W. Victor (Eds.), *Driver distraction and inattention: Advances in research and countermeasures* (Vol. 1). Surrey: Ashgate.

Croke, D., & Cerneaz, N. (2009). Real time distraction detection and warning system improves safety on public roads: A case study. Paper presented at the first international conference on driver distraction and inattention, Göteborg, Sweden.

Cunningham, M. & Regan, M. (2015a). Autonomous Vehicles: Human Factors and Research Opportunities. In: *The Proceedings of the Australasian Road Safety Conference*, Gold Coast, Australia, October 14–16.

Cunningham, M. & Regan, M. (2015b). Driver distraction and inattention in autonomous vehicles. In: *The Proceedings of the 4th International Driver Distraction and Inattention Conference*, Sydney, Australia, November 9–11.

Cunningham, M., & Regan, M. (in press). Driver distraction and inattention in the realm of automated vehicles, *IET Intelligent Transport Systems*.

Cunningham, M. & Regan, M. (2016). The impact of emotion, life stress and mental health issues on driving performance and safety. *Road and Transport Research "Special Issue – Road User Behaviour"*, *25*(3), 40–50.

Cunningham, M., Regan, M., & Imberger, K. (2017). Understanding driver distraction associated with specific behavioural interactions with in-vehicle and portable technologies. *Journal of the Australasian College of Road Safety*, i(1), 27–40.

Decker, J., Stannard, S., McManus, B., Wittig, S., Sisiopiku, V., & Stavrinos, D. (2014). The impact of billboards on driver visual behavior: A systematic literature review. *Traffic Injury Prevention*, *16*(3), 234–239.

Dingus, T. A., Antin, J. F., Hulse, M. C., & Wierwille, W. W. (1989). Attentional demand requirements of an automobile moving-map navigation system. *Transportation Research A*, *23*, 301–315.

Dingus, T. A., Hulse, M. C., McGehee, D. V., & Manakkal, R. (1994). Driver performance results from the Travtek IVHS camera car evaluation study. *The proceedings of the 38th annual meeting of the Human Factors & Ergonomics Society* (pp. 1118–1122). Santa Monica, CA: Human Factors Society.

Dong, Y., Hu, Z., Uchimura, K., & Murayama, N. (2011). Driver inattention monitoring system for intelligent vehicles: A review. *IEEE Transactions on Intelligent Transportation Systems*, *12*(2), 596–614.

Donmez, B., Boyle, L., & Lee, J. (2008). Mitigating driver distraction with retrospective and concurrent feedback. *Accident Analysis & Prevention*, *40*(2), 776–786.

Dukic, T., Ahlstrom, C., Patten, C., Kettwich, C., & Kircher, K. (2013). Effects of electronic billboards on driver distraction. *Traffic Injury Prevention*, *14*(5), 469–476.

Endsley, M. R. (1995). Toward a theory of situation awareness in dynamic systems. *Human Factors*, *37*(1), 32–64.

Engström, J., Johansson, E., & Östlund, J. (2005). Effects of visual and cognitive load in real and simulated motorway driving. *Transportation Research Part F*, *8*(2), 97–120.

Fillmore, M. T. (2003) Reliability of a computerized assessment of psychomotor performance and its sensitivity to alcohol-induced impairment. *Perceptual and Motor Skills*, *97*(1), 21–34.

Fitch, G., Bowman, D., & Llaneras, R. (2014). Distracted driver performance to multiple alerts in a multiple-conflict scenario. *Human Factors*, *56*(8), 1497–1505.

Fitch, G. M., Soccolich, S. A., Guo, F., McClafferty, J., Fang, Y., Olson, R., … Dingus, T. (2013). *The impact of hand-held and handsfree cell phone use on driving performance and safety-critical event risk.* Report no. DOT HS 811 757. Washington, DC: National Highway Traffic Safety Administration.

Gladstones, W. H., Regan, M. A., & Lee, R. B. (1989). Division of attention: The single-channel hypothesis revisited. *Quarterly Journal of Experimental Psychology*, 41A, 1–17.

Gordon, C. P. (2009). Crash studies of driver distraction. In M. A. Regan, J. D. Lee, & K. L. Young (Eds.), *Driver distraction: Theory, effects and mitigation.* Boca Raton, FL: CRC Press.

Hallett, C., Regan, M. A., & Bruyas, M. P. (2011). Development and validation of a driver distraction impact assessment test. *The electronic proceedings of the 2nd international conference on driver distraction and inattention*, Gothenburg, Sweden, September 5–7.

Hargutt, V. (2000). Eyelid movements and their predictive value of fatigue stages. Paper presented at 3rd international conference of psychophysiology in ergonomics, San Diego, CA, July 30.

Harrison E. L. R., & Fillmore M. T. (2011). Alcohol and distraction interact to impair driving performance. *Drug Alcohol Dependence*, *117*, 31–37.

Hawkins, H. G., Kuo, P.-F., & Lord, D (2014). Statistical analysis of the traffic safety impacts of on-premise digital signs. *Proceedings of the 93rd annual meeting of the TRB*, Washington, DC.

Heinzman, J., & Zelinsky, A. (2010). Automotive safety solutions through technology and human-factors innovation. Paper presented at the 12th international symposium on experimental robotics, Delhi, India.

Hickman, J. S., Hanowski, R. J., & Bocanegra, J. (2010). *Distraction in commercial trucks and buses: Assessing prevalence and risk in conjunction with crashes and near-crashes.* Washington, DC: Federal Motor Carrier Safety Administration.

Hollnagel, E., & Woods, D. D. (2005). *Joint cognitive systems: Foundations of cognitive systems engineering.* Boca Raton, FL: Taylor and Francis.

Horberry, T., & Edquist, J. (2008). Distractions outside the vehicle. In M. A. Regan, J. D. Lee, & K. Young (Eds.) *Driver distraction: Theory, effects, and mitigation* (pp. 215–228). Boca Raton, FL: CRC Press.

Horrey, W. J., & Wickens, C. D. (2004). Driving and side task performance: The effects of display clutter, separation, and modality. *Human Factors: The Journal of the Human Factors and Ergonomics Society*, *46*(4), 611–624.

Horrey, W., & Wickens, C. (2006). The impact of cell phone conversations on driving using meta-analytic techniques. *Human Factors*, *48*, 196–205.

Hosking, S. G., Young, K. L., & Regan, M. A. (2009). The effects of text messaging on young novice driver performance. *Human Factors*, *51*, 582–592.

Izadpanah, P., Reza, O., Koo, S., & Hadayeghi. A. (2014). Effect of static electronic advertising signs on road safety: An experimental case study. *Journal of Orthopaedic Trauma*, *28*, S33–S36.

Jamson, A., Merat, N., Carsten, O., & Lai, F. (2013). Behavioural changes in drivers experiencing highly-automated vehicle control in varying traffic conditions. *Transportation Research Part C: Emerging Technologies*, *30*, 116–125.

Kahneman, D. (1973). *Attention and effect.* Englewood Cliffs, NJ: Prentice Hall.

Klauer, S. G., Dingus, T. A., Neale, V. L., Sudweeks, J. D., & Ramsey, D. J. (2006). *The impact of driver inattention on near-crash/crash risk: An analysis using the 100-car naturalistic driving study data.* Report no. DOT HS 810 594. Washington, DC: National Highway Traffic Safety Administration.

Klauer, S. G., Guo, F., Simons-Morton, B. G., Ouimet, M. C., Lee, S. E., & Dingus, T. A. (2014). Distracted driving and risk of road crashes among novice and experienced drivers. *New England Journal of Medicine*, *370*, 54–59.

Lee, J. D., Roberts, S. C., Hoffman, J. D., & Angell, L. S. (2012). Scrolling and driving: How an MP3 player and its aftermarket controller affect driving performance and visual behavior. *Human Factors*, *54*, 250–263.

Merat, N., Jamson, H., Lai, F., & Carsten, O. (2012). Highly automated driving, secondary task performance and driver state. *Human Factors*, *54*, 762–771.

Michon, J. A. (1985). A critical view of driver behavior models: What do we know, what should we do? In L. Evans & R. Schwing (Eds.), *Human behavior and traffic safety* (pp. 485–520). New York, NY, London: Plenum Press.

National Highway Traffic Safety Administration. (2000). *On DWI laws in other countries* (DOT HS 809 037). Washington, DC: NHTSA

National Highway Traffic Safety Administration. (2013). *Preliminary statement of policy concerning automated vehicles*. Washington, DC: NHTSA.

National Highway Traffic Safety Administration. (2015). *Traffic safety facts research note*. Washington, DC: NHTSA.

Olson, R. L., Hanowski, R. J., Hickman, J. S., & Bocanegra, J. (2009). *Driver distraction in commercial vehicle operations*. Report no. FMCSA-RRR-09-042. Washington, DC: U.S. Department of Transportation.

Pashler, H. (1994). Overlapping mental operations in serial performance with preview. *Quarterly Journal of Experimental Psychology*, *47*, 161–191.

Pradhan, A. K., Divekar, G., Masserang, K., Romoser, M., Zafian, T., Blomberg, R. D., … Fisher, D. L. (2011). The effects of focused attention training (FOCAL) on the duration of novice drivers' glances inside the vehicle. *Ergonomics*, *54*, 917–931.

Rakauskas, M. E., Ward, N. J., Boer, E. R., Bernat, E. M., Cadwallader, M., & Patrick, C. J. (2008). Combined effects of alcohol and distraction on driving performance. *Accident Analysis & Prevention*, *40*, 1742–1749.

Rauch, N., Kaussner, A., Krüger, H. P., Boverie, S., & Flemisch, F. (2009). The importance of driver state assessment within highly automated vehicles. *Conference proceedings of the 16th world congress on ITS*, Stockholm, Sweden.

Recarte, M. A., & Nunes, L. M. (2003). Mental workload while driving: Effects on visual search, discrimination, and decision making. *Journal of Experimental Psychology: Applied*, *9*(2), 119–137.

Regan, M., Hallett, C., & Gordon, C. (2011). Driver distraction and driver inattention: Definition, relationship and taxonomy. *Accident Analysis & Prevention*, *43*(5), 1771–1781.

Regan, M. A., Lee, J. D., & Young, K. L. (2009). Factors moderating the impact of distraction on driving performance and safety. In M. A. Regan, J. D. Lee, & K. Young (Eds.) (2009). *Driver distraction: Theory, effects and mitigation* (Chap. 19). Boca Raton, FL: CRC Press.

Saxby, D., Matthews, G., Hitchcock, E., Warm, J., Funke, G., & Gantzer, T. (2008). Effect of active and passive fatigue on performance using a driving simulator. *Human Factors and Ergonomics Society Annual Meeting Proceedings*, *52*(21), 1751–1755.

Simons-Morton, B. G., Guo, F., Klauer, S. G., Ehsani, J. P., & Pradhan, A. K. (2014). Keep your eyes on the road: Young driver crash risk increases according to duration of distraction. *Journal of Adolescent Health*, *54*, 61–67.

Smiley, A., Smahel, T., & Eizenman, M. (2004). Impact of video advertising on driver fixation patterns. *Transportation Research Record*, *1899*(1), 76–83.

Smiley, A., Persaud, B., Bahar, G., Mollett, C., Lyon, C, Smahel, T., & Kelman, W. (2005). Traffic safety evaluation of video advertising signs. *Transportation Research Record: Journal of the Transportation Research Board, 1937*(1), 105–112.

Stephan, K., Hosking, S., Regan, M., Verdoorn, A., Young, K., & Haworth, N. (2006). *The relationship between driving performance and the Johns drowsiness scale as measured by the Optalert system.* MUARC Report no. 252. Monash University Accident Research Centre, Clayton, Victoria.

Strayer, D. L., & Johnston, W. A. (2001). Driven to distraction: Dual-task studies of simulated driving and conversing on a cellular telephone. *Psychological Science, 12*(6), 462–466.

Strayer, D. L., Drews, F. A., & Johnston, W. A. (2003). Cell phone-induced failures of visual attention during simulated driving. *Journal of Experimental Psychology: Applied, 9*(1), 23–32.

Stutts, J., Feaganes, J., Reinfurt, D., Rodgman, E., Hamlett, C., Gish, K., & Staplin, L. (2005). Driver's exposure to distractions in their natural driving environment. *Accident Analysis & Prevention, 37*, 1093–1101.

Tivesten, E., & Dozza, M. (2014). Driving context influences drivers' decision to engage in visual-manual phone tasks: Evidence from a naturalistic driving study. *Transportation Research Part F: Traffic Psychology and Behaviour, 26*, 258–272.

Törnros, J. E. B., & Bolling, A. K. (2005). Mobile phone use – Effects of handheld and hands-free phones on driving performance. *Accident Analysis & Prevention, 37*, 902–909.

Trick, L. M., & Enns, J. T. (2009). A two-dimensional framework for understanding the role of attention selection in driving. In C. Castro (Ed.), *Human factors of visual and cognitive performance in driving* (pp. 63–72). Boca Raton, FL: CRC Press.

Tsimhoni, O., Smith, D., & Green, P. (2004). Address entry while driving: Speech recognition versus a touch-screen keyboard. *Human Factors, 46*(4), 600–610.

Vahidi, A., & Eskandarian, A. (2003). Research advances in intelligent collision avoidance and adaptive cruise control. *IEEE Transactions on Intelligent Transportation Systems, 4*(3), 143–153.

Verster, J. C., Wester, A. E., Goorden, M., van Wieringen, J., Olivier, B., & Volkerts, E. R. (2009). Novice drivers' performance after different alcohol dosages and placebo in the divided attention steering simulator (DASS). *Psychopharmacology, 204*, 127–133.

Victor, T. (2011). Distraction and inattention countermeasure technologies. *Ergonomics in Design: The Quarterly of Human Factors Applications, 19*(4), 20–22.

Victor, T. W., Engstrom, J., & Harbluk, J. L. (2008). Distraction assessment methods based on visual behaviour and event detection. In M. A. Regan, J. D. Lee, & K. L. Young (Eds.), *Driver distraction: Theory, effects, and mitigation* (pp. 135–165). Boca Raton, FL: CRC Press Taylor & Francis Group.

Wierwille, W. W. (1999). *Historical perspective on slow eyelid closure: When PERCLOS?* (pp. 31–53). Technical conference on ocular measures of driver alertness, Herndon, Virginia, FHWA Technical Report no. MC-99-136. Federal Highway Administration, Office of Motor Carrier and Highway Safety, Washington, DC.

Wikman, A., Nieminen, T., & Summala, H. (1998). Driving experience and time-sharing during in-car tasks on roads of different width. *Ergonomics, 41*(3), 358–372.

Yannis, G., Papadimitriou, E., Papantoniou, P., & Voulgari, C. (2013). A statistical analysis of the impact of advertising signs on road safety. *International Journal of Injury Control and Safety Promotion, 20*(2), 111–120.

Young, K. L., & Lenné, M. G. (2010). Driver engagement in distracting activities and the strategies used to minimise risk. *Safety Science, 48*, 326–332.

Young, K., & Regan, M. (2007). Driver distraction: A review of the literature. In I. J. Faulks, M. Regan, M. Stevenson, J. Brown, A. Porter, & J. D. Irwin (Eds.), *Distracted driving* (pp. 379–405). Sydney: Australasian College of Road Safety.

Young, M., & Stanton, N. (2002). Attention and automation: New perspectives on mental underload and performance. *Theoretical Issues in Ergonomics Science*, *3*(2), 178–194.

Young, K. L., Regan, M. A., & Lee, J. D. (2008). Factors moderating the impact of distraction on driving performance and safety. In M. A. Regan, J. D., Lee, & K. L. Young (Eds.), *Driver distraction: Theory, effects, and mitigation* (pp. 335–352). Boca Raton, FL: CRC Press Taylor & Francis Group.

Zeeb, K., Buchner, A., & Schrauf, M. (2015). What determines the take-over time? An integrated model approach of driver take-over after automated driving. *Accident Analysis & Prevention*, *78*, 212–221.

Zhang, H., Smith, M. R. H., & Witt, G. J. (2006). Identification of real-time diagnostic measures of visual distraction with an automatic eye-tracking system. *Human Factors*, *48*(4), 805–821.

TRANSPORT NETWORK: CHALLENGES AND SOLUTIONS

CHAPTER 5

URBAN AND SUBURBAN ARTERIALS

Tom Brijs and Ali Pirdavani

ABSTRACT

Purpose – *Urban and suburban arterials carry a large share of urban traffic and contend with a relatively large proportion of transport network crashes. Road crashes and their consequent societal costs diminish the sustainability of transportation systems, highlighting the need to identify road safety problems and their corresponding solutions. This chapter briefly outlines problems and solutions associated with crash risk on urban and suburban arterials. In addition, this chapter studies and discusses several safety countermeasures – ranging from local treatments to integral frameworks – and their effectiveness on improving traffic safety of urban and suburban arterials.*

Approach – *Crash occurrence on urban and suburban arterials is affected by numerous contributing factors. This chapter pays attention primarily to the effects of traffic characteristics and road design features. In this regard, several pertinent variables which have been extensively examined in the literature are reviewed and their contributions to the safety of urban and suburban arterials are discussed.*

Findings – *A review of the literature identifies a number of variables as influential factors of crashes on urban and suburban arterials. Although*

Safe Mobility: Challenges, Methodology and Solutions
Transport and Sustainability, Volume 11, 85–106

ISSN: 2044-9941/doi:10.1108/S2044-994120180000011005

the associations of some variables (e.g., traffic volume) are consistent with expectations, others (e.g., lane width and speed) show mixed and sometimes counterintuitive results. These findings signify that additional research is needed to reveal the correct functional form and magnitude of these relationships.

Practical implications – *The results show that while the general direction and magnitude of effects of some engineering and management-related treatments are known, additional research is needed to consolidate the impact and effectiveness of integrated approaches.*

Keywords: Crash contributing factor; explanatory/predictive variables; safety countermeasures; sustainable safety; urban and suburban arterials

INTRODUCTION

This chapter is the first of five chapters focusing on influences of the transport network on crash risk. While this chapter focuses on urban and suburban arterials, it complements the current knowledge regarding the crash risks related to the operation of controlled access facilities (Chapter 6), rural and urban intersections (Chapter 7), roundabouts (Chapter 8) and real-time traffic operations and safety (Chapter 9).

Urban and suburban arterials provide for the majority of urban traffic and experience a relatively large proportion of crashes, including severe crashes. Safety-related concerns on urban and suburban arterials are quite complicated due to the diverse mix of road users and large variety of design and operational features of these facilities, including traffic control devices, access controls and roadway design elements. Hence, it is essential to review the characteristics of these facilities to appropriately address their traffic safety-related challenges.

Urban and suburban arterial roads are major roads that carry large amounts of traffic within, between or through major activity centres. These roads are generally high-capacity and high-speed roads, and mainly serve long-distance traffic ('Federal Highway Administration Planning Glossary' (U.S. Department of Transport, 2008)). According to the concept of functional classification (The Federal Highway Administration (FHWA), 2013), arterial roads tend to be relatively limited access roads with a high degree of throughput. However, slightly different definitions of urban arterials are seen across countries. The FHWA (2008) defines urban arterials as

'Major streets or highways, many of multi-lane or freeway design, serving high-volume traffic corridor movements that connect major generators of travel'. The European Union has its own definition of urban arterials: urban arterials are roads that serve medium- to long-distance movements and may cross through urban areas to provide suburban movements (Saurabh, 2013). According to the system of roads classification in the United Kingdom (UK), urban arterials are major roads intended to 'provide large-scale transport links within or between areas or to feed traffic between A roads and smaller roads on the network' ('Guidance on Road Classification and the Primary Route Network' (GOV.UK, 2012)). As seen from these definitions, common characteristics and similar functionality exist. Considering these definitions, urban and suburban arterials generally (FHWA, 2013):

- Carry high traffic volumes of moderate- to long-distance trips.
- Provide continuity of rural arterials' traffic flow from and towards urban areas.
- Serve large portions of urban traffic on a relatively small share of total urban road network length.

Urban and suburban arterials are often categorised into 'minor' and 'major or principal' arterials. Although the general definition and functionality of these arterials are comparable, among transportation agencies their specifications and designs can vary. For instance, speed limits may vary between 40 kph and 90 kph (30 mph and 60 mph) or geometric properties and traffic volume orders can be significantly diverse depending on the surrounding development. Urban and suburban arterials are diverse and difficult facilities to manage, not only by transportation engineers and designers, but also by road users. Unlike other types of facilities (e.g., motorways) where distinct measures and unique design features adequately convey their functionality to road users, urban and suburban arterials may give rise to misperceptions for road users. For example, imagine driving on a continuous stretch of arterial road that crosses a suburban area and goes through an urban neighbourhood. The speed limit changes from 70 kph to 50 kph; however, other road features remain unchanged, leading to inappropriate driving behaviour and potentially leading to crashes as a result of inappropriate speed adaptation. The abovementioned example illustrates one dimension of the complexity, with facilities having different designs and operational features classified as urban and suburban arterials and, naturally, as a result the traffic safety challenges, solutions and analysis techniques appropriate for these facilities will also be complex.

Moreover, urban and suburban arterial facilities comprise two major components: road segments and intersections. These components may be different from site to site based on neighbouring land uses, degree of access, traffic control features and geometric design. Arterial road segments can be configured as two-lane or multi-lane, divided or undivided, with frequent or limited access points, with or without parking sidewalks, crosswalks, bike lanes, etc. Intersections can be signalised and stop-controlled with a range of attributes, including number of intersection legs, existence of left- and/or right-turn lanes, presence of pedestrian facilities, etc. These differences imply that the constituent components of urban and suburban arterials are intrinsically disparate – confronting road users with a reliably unpredictable set of design and operational features. Accordingly, the management of traffic safety of these facilities is a challenging and complex task.

Crashes on road segments and intersections are inherently different. On road segments the majority of crashes are 'rear-end' collisions, followed by a significantly lower share of 'sideswipe same direction' collisions, while at intersections 'rear-end' and 'angle' collisions are the predominant collision types (American Association of State Highway and Transportation Officials (AASHTO, 2010)). In addition, the severity distributions of crashes are different for road segments and intersections.

As highlighted by the previous discussion, due to their diversity and complexity, urban and suburban arterial road segments and intersections should be studied, analysed and modelled separately. Moreover, crash severity should also be considered when analysing traffic crashes on urban and suburban arterials. Societal impacts and associated costs of injury crashes are known to be much greater than the costs of property-damage-only crashes and, therefore, reducing these serious crashes leads to a more sustainable transport system.

The remainder of this chapter is organised as follows. First, the most frequently employed variables relevant to the safety performance of urban and suburban arterials are summarised. Next, different methodological approaches for evaluating the traffic safety of these facilities are discussed, followed by a summary of potential solutions to alleviate road safety concerns on urban and suburban arterials. Lastly, conclusions wrap up this chapter.

MEASURING CRASH RISK ON URBAN AND SUBURBAN ARTERIALS

Traffic unsafety can be measured using various measures, such as crash frequency, crash rate or surrogate safety measures (Gettman & Head, 2003). Crash frequency and rate can further be categorised based on crash type (e.g., multiple-vehicle crashes, bicycle and pedestrian crashes, etc.) and crash severity (e.g., injury crashes, fatal crashes, total crashes, etc.). Since crashes with higher injury severity and crashes involving vulnerable road users often cause relatively greater social costs, alleviating them will have a greater impact on the sustainability of transportation systems. Hence, it is of great importance to account for different severity levels as well as crash types when evaluating traffic safety performance of urban and suburban arterials. Results of traffic safety analyses of any kind will be undermined if merely the total number of crashes is taken into account, without acknowledging crash severity and involvement of vulnerable road users.

FACTORS KNOWN TO INFLUENCE CRASH RISK ON URBAN AND SUBURBAN ARTERIALS

Road crashes on urban and suburban arterials are influenced by many contributing factors. These factors are generally associated with traffic and road characteristics, driver behaviour and vehicle characteristics. The remainder of this section focuses on the traffic and road characteristics (i.e., geometric design and physical features) of urban and suburban arterials.

The most frequently examined explanatory variables to either explain or predict traffic safety performance of urban and suburban arterials, broken down by road segments and intersections, is provided below (a full list of all potential variables can be found in the final report of the NCHRP Project (Harwood et al., 2007):

- Road segments:
  - traffic volume per day (veh/day) or traffic volume in peak period (veh/h);
  - posted or observed speeds;
  - lane width;
  - access density (number and type of driveways/km);
  - intersection density (intersections/km);
  - number of lanes;
  - horizontal and vertical alignment;
  - median (number and type of openings);
  - shoulder width and type;
  - lighting.
- Intersections:
  - traffic volume per day (veh/day) or traffic volume in peak period (veh/h) by intersection approach;
  - posted or observed speeds on approach;
  - sight distance;

— type of traffic control;
— signal phasing;
— presence and number of left- and right-turn lanes;
— intersection approach angles;
— number of intersection legs;
— lane widths on approaches;
— number of lanes on approaches;
— lighting.

The most important variables, as reflected in the literature, are briefly reviewed next. The challenges regarding their application in road safety management of urban and suburban arterials are then identified.

Traffic Volume

Traffic volume at a site is one of the most important predictors of traffic safety (Saha, Alluri, & Gan, 2015), as it is the only variable that captures the potential exposure of motorists to risk. It is measured in a variety of ways, such as average annual daily traffic, average daily traffic (ADT), hourly traffic volume and vehicle kilometres travelled – all with slightly different interpretations. The inclusion of an exposure variable is essential in conducting a safety performance evaluation, because the potential for single and multiple vehicle crashes are related – in different ways – to the opportunities for collision as captured by exposure. As one might imagine, the relationship between crashes and exposure is quite complex; for example, single vehicle crashes may decrease with increasing volumes, whereas multiple vehicle crashes may increase, such that traffic volumes are often an approximation of a more complex underlying relationship. While it is often difficult to obtain an exposure measure for the entire year, a daily or hourly measure can serve as a reliable replacement (e.g., ADT; Harwood et al., 2007). The primary difference in traffic volume on road segments and intersections is that traffic flow is sufficient to capture exposure on road segments, whereas entering volumes and conflicting traffic volumes at intersections complicate the exposure–crash relationship. It may also be advantageous for each road user group (i.e., vehicles, pedestrians and cyclists) and each vehicle class (e.g., heavy vehicles, passenger cars and motorcyclists) to have their own exposure metrics; however, while desirable, often exposure metrics for these sub-populations are not available. See Chapters 10 and 11 for detailed discussions related to crash risk estimation for pedestrians and cyclists, respectively.

In general, research has shown that increasing traffic volume is associated with increasing crash frequency on both road segments and intersections (C. Wang, Quddus, & Ison, 2013). Traffic volume is also associated with severe crashes (Ma, Yan, Abdel-Aty, Huang, & Wang, 2010; Nevarez, Abdel-Aty, Wang, & Santos, 2009; Pande & Abdel-Aty, 2009). It is often the most influential predictive variable in safety performance evaluations of urban and suburban arterials (Abdel-Aty, Devarasetty, & Pande, 2009; Berhanu, 2004; Chen & Persaud, 2014; Das & Abdel-Aty, 2011; Dixit, Pande, Abdel-Aty, Das, & Radwan, 2011; Ma et al., 2010; P. Y. Park & Abdel-Aty, 2011; J. Park, Abdel-Aty, Lee, & Lee, 2015; Saha et al., 2015; Sultana, Qin, Chitturi, & Noyce, 2014). While exposure is undoubtedly a critical factor in crash explanation and prediction, exploring alternative functional forms of this complex relationship, along with the inclusion of sub-population exposure metrics, is likely to remain the focus of future research.

Access and Intersection Density

Access features, such as driveways, median openings and median treatments, have substantial impacts on the safety performance of urban and suburban arterials (Harwood et al., 2007). The existence of driveways and median openings increases the number of potential conflict points along the arterials and raises the crash occurrence risk. Specifically, the number, type, location and spacing of driveways and median openings have been shown to have significant associations with traffic safety on urban and suburban arterials. Recurring access points to and from arterials, frequent traffic signals and arterials without medians have been found to be positively associated with crash occurrence and severity (Sultana et al., 2014). In a study conducted by Eisele and Frawley (2005), it was concluded that access point density is positively associated with crashes. C. Lee, Xu, and Nguyen (2012) studied midblock crashes on urban arterials and found a positive relation between crashes and the presence of an access point. Similar associations were found by other researchers indicating that access density has a positive relation with crash frequency and severity of crashes on urban and suburban arterials (Avelar, Dixon, Brown, Mecham, & van Schalkwyk, 2013; X. Wang, Song, Yu, & Schultz, 2014). In addition to access density, signal spacing impacts safety on urban and suburban arterials. Signal spacing includes both signal density and uniformity of signal placement. Similar to signal density, studies have revealed that crash frequency is positively associated with intersection density (El-Basyouny & Sayed, 2009; Mauga & Kaseko, 2010).

To summarise, there is a positive association between access control, signal placement and density, and intersection density and traffic safety on urban and suburban arterials in line with expectations, as access points and intersections give rise to increased potential vehicle conflicts. Often, careful decision-making regarding access control features (e.g., median width and type, signing and striping and signage) is a trade-off between ease and convenience of mobility and crash risk.

Speed

Vehicle speed is a traffic-related variable that has been shown to affect traffic safety through influences on crash frequency and severity. Vehicle speed is often measured imprecisely and most commonly through posted speed limits and less commonly through observed speeds (i.e., radar or loop detectors). Several studies have found empirical evidence for an association between speed and crashes on urban and suburban arterials (Abdel-Aty et al., 2009; Das & Abdel-Aty, 2011; Dixit et al., 2011; Ma et al., 2010; Mauga & Kaseko, 2010; X. Wang & Chen, 2012). Other less commonly examined measures of speed include speed variance and roadway design speed.

The existence of a causal relationship between speed and safety has been studied for many years. Accumulation of evidence suggests that the crash risk increases when driving fast due to reduced control of the vehicle, greater reaction time demands and greater distances travelled per unit time when travelling at higher speeds (Li, El-Basyouny, & Kim, 2015). Higher speeds are directly associated with more severe crashes, consistent with Newton's laws of motion, where kinetic energy increases in proportion to the square of velocity, leading to more forceful impacts and subsequently higher injury-severity levels (Tay & Rifaat, 2007; Yasmin, Eluru, Bhat, & Tay, 2014). The results of several studies have shown that both road segments and intersections on urban arterial roads with higher speed limits experience more severe crashes compared with arterials with lower speed limits (Das & Abdel-Aty, 2011; Dixit et al., 2011).

The association between speed and crash frequency is, however, complex. Although a positive relation between speed and crash frequency has been observed, several studies have obtained opposite results, with crash frequency inversely associated with average speed or posted speed limit. A study undertaken by Mauga and Kaseko (2010) showed that posted speed limit was negatively associated with all crash types at all severity levels. Similar results have been obtained by other researchers (Abdel-Aty et al., 2009; Das & Abdel-Aty,

2011; De Pauw, Daniels, Thierie, & Brijs, 2014c; X. Wang & Chen, 2012). A critical determining factor is the way in which speed is measured; that is, average speed, speed variance or posted speed limit. Hence, it is possible that lower average speeds are associated with larger speed variances. Moreover, larger speed variances have been found to be more problematic than average speed (Brijs, Wets, Krimpenfort, & Offermans, 2006; Quddus, 2013). As a result of this complex relationship between safety and speed, statistical associations may change from positive to negative, and vice versa, depending on the way in which speed data are collected. From a statistical modelling point of view, these potentially counterintuitive relationships are attributable to omitted-variable bias, multicollinearity among explanatory variables and inappropriate or mis-specified model functional forms (Lord & Mannering, 2010). Considering these relationships, driving at higher speeds is presumably more risky due to the greater distance travelled for a fixed reaction time and due to the resulting higher crash impact speeds. It is possible that some drivers are more cautious when driving faster due to increased perception of risk. Finally, facilities with higher speed limits may be equipped with higher quality safety features that aim to reduce both the risk and severity of crashes.

Taking prior research into account, speed has been found to have mixed impacts on the traffic safety of urban and suburban arterials. Additional research is needed to determine the precise nature of association between speed and crash frequency on urban and suburban arterials, and which speed measurements best capture crash risk. In this respect, researchers are advised to pursue two research directions: (1) make a clear distinction between crash severity and crash frequency when analysing crashes and (2) examine and explore different speed measures (i.e., average speed, speed limit, speed variance and potentially other speed measures) to the extent possible so as to reveal all possible associations between crash and speed variables. The former is important in the sense that mitigating crashes with higher injury severities contributes substantially more to the sustainability of transport systems in comparison to property-damage-only crashes. The latter will play a key role in gaining an improved understanding of the causal relationship and reveal the nature of association between crashes and speed, perhaps as a function of other factors.

Lane Width

Lane width is one of the most frequently employed variables to explain and predict the safety performance of urban and suburban arterials. The effect this variable has on crash occurrence has revealed inconsistent results. Although wider lanes may provide more space between adjacent travel lanes, and therefore increase the buffer area for lateral drift between adjacent vehicles, they also increase crossing distances for pedestrians (see Chapter 10) and impose potential threats on these particular road users (Potts, Harwood, & Richard, 2007). This is particularly relevant in urban and suburban contexts where pedestrian activity is significant and safe pedestrian movements are needed. Moreover, wider lanes may encourage drivers to travel faster and feel relatively more safe, which could possibly increase crash risk and severity (Hauer, Council, & Mohammedshah, 2004). While some researchers have shown statistically insignificant relationships (J. Park et al., 2015; Strathman, Dueker, Zhang, & Williams, 2003; Yanmaz-Tuzel & Ozbay, 2010) between lane width and crash occurrence, others have concluded that increasing lane widths is expected to decrease crash rates on undivided urban arterials (Hadi, Aruldhas, Chow, & Wattleworth, 1995), or that wider lanes are associated with higher crash frequencies (Hauer et al., 2004). A recent study by Potts et al. (2007) revealed that there was no statistically significant association between lane width and safety conditions of both road segments and approaches towards intersections of urban and suburban arterials. They suggest that there is no evidence of reduced traffic safety as a result of narrowed lanes, with the extra width available to support turning lanes, bicycle lanes or pedestrian sidewalks, or to facilitate median treatments.

METHODOLOGICAL APPROACHES USED TO EVALUATE ROAD SAFETY ON URBAN AND SUBURBAN ARTERIALS

Several crash estimation methodologies are applicable for evaluating the traffic safety performance of urban and suburban arterials. These methods include statistically analysing existing (historical) crash data, road safety audits (site visits and assessments by road safety engineering teams) and evaluating surrogate safety measures (AASHTO, 2010). Often the choice of approach will be suited to motivation, resources and time constraints.

Statistical analysis is the most commonly employed approach for crash estimation. For several years researchers have been establishing and advancing statistical techniques to analyse crashes. Statistical models have been developed to characterise the association between crash occurrence and explanatory/predictive variables. Like for any other road network facility, statistical models for urban and suburban arterials have been evolving from simple methods (Hadi et al., 1995; Hauer et al., 2004; Ma et al., 2010; Retting, Weinstein, Williams, & Preusser, 2001; Sawalha & Sayed,

2001; Strathman et al., 2003; Tay & Rifaat, 2007) to more sophisticated techniques (Das & Abdel-Aty, 2011; P. Y. Park & Abdel-Aty, 2011; Yasmin et al., 2014). There are numerous chapters in this book devoted to addressing the challenges of analysing crash data. Chapter 12 in this book details the estimation of models based on cross-sectional crash data, whereas Chapter 13 examines the use of time-series crash data, Chapter 14 explores the challenges of analysing crash data when samples are small or there are a large number of zeroes and Chapter 15 looks at crash severity modelling. Chapter 16 explores the challenges with detecting crash hot-spots and Chapter 17 the use of crash surrogate safety measures. Several applications of surrogate safety measures have been conducted to assess the traffic safety performance of urban and suburban arterials (Abdel-Aty et al., 2009; Dixit et al., 2011; Eisele & Frawley, 2005; J. Lee, Park, Malakorn, & So, 2013; Pirdavani, Brijs, Bellemans, & Wets, 2010; Polders et al., 2015; Stevanovic, Stevanovic, & Kergaye, 2013; Tao, Foomani, & Alecsandru, 2015; Wen & Luo, 2011).

STRATEGIES FOR IMPROVING ROAD SAFETY ON URBAN AND SUBURBAN ARTERIALS

One of the most significant social costs of transportation is crash occurrence. Therefore, reducing the risk of road crashes will contribute to the sustainability of transportation networks. To this end, a variety of traffic safety strategies can be implemented to improve the traffic safety of road networks. These strategies can be divided into two major categories or groupings: engineering and behavioural change strategies (VTPI, 2012). Engineering strategies involve investing in safer vehicles and roads, while behavioural strategies involve programs, policies and investments focused on improving human interaction with the vehicle (e.g., fatigued driving, impaired driving, licensing and education, speed enforcement, etc.). Behaviour- and engineering-related strategies are often interrelated, as certain measures may influence road users' behaviour through physical changes applied to the roads (i.e., self-enforcing). As an example, in the case of traffic calming, driver behaviours change after a speed hump physically restricts their speeds.

Expected crash frequencies and severities can be reduced by implementing appropriate treatments, which are interchangeably and often referred to as safety countermeasures, interventions or actions. These potential actions include geometric design elements of roads, traffic control systems and devices, as well as operational aspects of road networks (AASHTO, 2010). The *Highway Safety Manual* (AASHTO, 2010) includes an extensive review of safety countermeasures and reports their expected traffic safety effects based on current available evidence. Their safety effects are referred to as crash modification factors and are estimates of the reduction (or increase) in crashes expected as a result of implementing specific countermeasures. Most of these countermeasures are engineering related and involve enhancements to physical attributes of the roads or roadside environments. The magnitude of effects of countermeasures is determined by road type and function, in addition to the surrounding land-use environment. The following section elaborates on the potential safety benefits of several significant, design-related actions appropriate for urban and suburban arterials.

Access Management

Access management describes a set of actions that influence or limit access to roadways and includes features such as median openings, driveway policies and intersection spacing and design. While access management aims to appropriately balance or trade off mobility and accessibility, it can also be applied to improve the safety performance of roadways. Access management is achieved through feature design, such as signal spacing, driveway density, median opening density and median treatments, including raised medians and two-way, left-turn lanes. Several research teams have evaluated the impact of access management features on traffic safety of urban and suburban arterials. The installation of median barriers, providing a raised median and extending the width of existing median openings, has been shown to significantly reduce crash risk and severity (Eisele & Frawley, 2005; C. Lee et al., 2012; Ma et al., 2010; Mauga & Kaseko, 2010; Schultz, Allen, & Boschert, 2010; Schultz, Braley, & Boschert, 2009; Yanmaz-Tuzel & Ozbay, 2010). Moreover, reducing the frequency of driveways, median openings and intersection densities can significantly reduce crash frequencies and injury severities on urban and suburban arterials (Avelar et al., 2013; Sultana et al., 2014; X. Wang et al., 2014; Williams, Stover, Dixon, & Demosthenes, 2014). This finding implies that restricting the number of intersections, driveways and median openings on urban and suburban arterials serve as effective countermeasures for addressing safety issues on these facilities. A practical consideration is that 'accident migration' can arise from access restriction. For example, vehicles that otherwise would turn across an access-restricted median might instead utilise a turning lane at the nearest downstream intersection to make a 'U-turn', and possibly increase potential conflicts at that intersection. These and similar cases where conflict points are moved to other network locations should be considered where access is concerned. Ideally, potential conflicts through access control will be 'migrated' to points with greater control of vehicle movements, lower vehicle speeds and improved sight distances for drivers.

Traffic Calming Strategies

Traffic calming concerns a variety of design features and strategies that are chiefly envisioned to reduce traffic speed on road networks. Among others, speed reduction programs, together with some physical interventions such as road diet and installation of speed humps, are effective strategies to improve the traffic safety level of urban and suburban arterials. A recent study conducted by Li et al. (2015) showed that implementing automated mobile speed enforcement on urban arterial roads can reduce severe crashes up to 20%. It was also confirmed by De Pauw, Daniels, Brijs, Hermans and Wets (2014a) that installing speed and red light cameras at intersections decreases severe crashes by 14%, whereas fixed speed cameras on road segments decrease severe injury crashes by 29% (De Pauw, Daniels, Brijs, Hermans, & Wets, 2014b). Other measures and techniques, such as transversal rumble strips (Ariën et al., 2014), repetition of speed signs (Jongen, Brijs, Mollu, Brijs, & Wets, 2011), additional road marking (Daniels, Vanrie, Dreesen, & Brijs, 2010) or speed enforcement by means of mobile radar with advance warning signs (Wilmots, Hermans, Brijs, & Wets, 2013), are proven to reduce speed that can translate to higher safety. Speed reduction programs in general and speed enforcement in particular are predominantly effective in reducing fatal and severe crashes (Elvik, 2005; Soole, Watson, & Fleiter, 2013). The reason lies in the fact that travelling at high speed is associated with possible collisions with greater impact (discussed above in the speed section).

'Road diet' refers to another traffic calming strategy that is mainly adopted on high-volume arterials (Rosales & Brinckerhoff, 2006). A road diet is actually the conversion of a four-lane arterial into a three-lane arterial: two through-lanes and a centre left-turn lane. In return, the extra width can be dedicated to bicycle lanes or pedestrian sidewalks. In a study undertaken by Stout, Pawlovich, Souleyrette, and Carriquiry (2006), the crash frequency per mile and number of injury crashes reduced by 25% and 34%, respectively, after implementing road diet settings. Harkey et al. (2008) found that road diets are expected to reduce crashes at all severity levels by almost 30% on urban and suburban arterials. Speed humps are most frequently installed on urban or suburban arterials with lower speed regimes and are particularly effective in improving traffic safety near school arterials (AASHTO, 2010).

Treatments for Pedestrians and Cyclists

This section addresses specific treatments for vulnerable road users – pedestrians and cyclists – in relation to urban and suburban arterials, and complements more in-depth discussions provided in this book about crash risks for pedestrians (Chapter 10) and cyclists (Chapter 11).

Road crashes involving vulnerable road users are usually injury crashes at different severity levels. Hence, it is of great importance to pay specific attention to these types of crashes in order to enhance a sustainable transportation system. Several safety countermeasures proven to improve traffic safety of pedestrians and cyclists on urban and suburban arterials are outlined below:

- *Providing a sidewalk or a wide shoulder* seems to reduce 'walking-along-roadway' pedestrian crashes (McMahon et al., 2002). It is also shown that benefits will be substantially greater for residential zones with higher pedestrian exposure levels.
- *Installation of raised pedestrian crosswalks* is also found to be beneficial in reducing frequency of particularly injury crashes. This is mainly due to the fact that these facilities reduce driving speed (Elvik, Høye, Vaa, & Sørensen, 2009).
- *Providing a raised median and refuge island* is known to have safety benefits for pedestrians crossing at both intersections and midblock cross points on urban and suburban arterials (AASHTO, 2010). It should be noted that although some advantages are obtained for pedestrians these installations might result in vehicle–island crashes, which didn't occur before the installation of these new settings.
- *Widening medians* to the extent possible may also reduce pedestrian crashes, mostly in areas with higher pedestrian exposure. This is because pedestrians will have more space to safely stand between two opposite directions (AASHTO, 2010).
- *Providing dedicated bicycle lanes* is widely proven to reduce bicycle–vehicle crashes on urban arterials (AASHTO, 2010). Dedicated bicycle lanes should be visually distinguishable by means of sufficient pavement markings or different pavement colour to maximise effectiveness.
- *Providing separate bicycle facilities* is a desirable alternative, mainly on principal arterials where vehicle exposure and operational speeds are high. As proposed by a Dutch guideline, cycling facilities should be physically separated from motorised traffic movements where the speed limit is equal and higher than 50 kph (Dijkstra, Eenink, & Wegman, 2007). These facilities practically eliminate any possible interaction between cyclists and vehicles

alongside the roads and are, therefore, favourable for roadways with higher traffic exposures (e.g., principal/major urban arterials). Having said that, designers should be cautious about the potential conflict zones between cyclists and other road users, specifically at intersections or crossovers.

Other Measures

In addition to the abovementioned treatments, there are other measures that can improve the traffic safety level of urban and suburban arterials. There exist two major types of parking facilities: off-street parking in parking lots and on-street parking. Reviewing the literature shows that prohibiting on-street parking can reduce injury crashes up to 20% (Elvik et al., 2009). Appropriate lighting of roadways, sidewalks and cycle paths is also known to noticeably reduce fatal and injury night-time crashes. As indicated by the *Handbook of road safety measures* (Elvik et al., 2009), at night fatal pedestrian crashes appear to reduce by almost 80% in urban areas, while all fatal crashes are expected to reduce by nearly 40%. Another effective countermeasure to improve traffic safety on urban and suburban arterials is prohibiting certain movements when they are found to be dangerous. Prohibiting left-turn and/or U-turn movements seem to reduce left-turn, U-turn and intersection crashes by up to 70% (AASHTO, 2010). However, it should be noted that crash migration is a potential downside that needs to be accounted for when considering prohibiting certain movements at certain locations. In addition to these measures, there appear to be other interventions, such as streetscape designs (Harvey & Aultman-Hall, 2015), signal coordination (Stevanovic et al., 2013) and unconventional designs (e.g., superstreet design; Ott, Haley, Hummer, Foyle, & Cunningham, 2012) that can improve traffic safety on urban and suburban arterials.

Sustainable Safety

So far, most of the strategies that have been discussed are local treatments that can be applied individually or in combination with other measures. Alleviating crash risk and its consequent social costs can also be viewed in a broader framework by which several aspects of road safety can be addressed. One applicable comprehensive road safety concept is 'sustainable safety'. Sustainable safety refers to a systematic approach that aims to prevent crashes or reduce their severity by adapting transportation systems' properties to human limitations – including physical and cognitive capacities

(Wegman, Aarts, & Bax, 2008). Since human errors are inevitable, sustainable safety aims at minimising the reliance of traffic safety on road users by creating inherently safe transportation systems. In order to achieve sustainable safety several principles have been defined. Since the whole concept and the principles can be applied on different types of facilities, their application to urban and suburban arterials seems to be beneficial.

Sustainable safety principles advise that roads should be mono-functional and the homogeneity of mass, speed and direction should be guaranteed to the extent possible. In addition, forgiveness of the environment is another key principle. A road's environment and elements should provide road users the opportunity to correct and minimise the consequences of their erroneous actions. Forgiving roads are designed safer and are equipped with elements that reduce the severity of crashes. As an example, changing roadside barriers to less rigid types (e.g., from concrete barriers to steel guardrails or cable barriers) on urban and suburban arterials is expected to reduce fatal and injury run-off-road crashes significantly, or installing more flexible crash cushions appears to considerably reduce the severity of fixed-object crashes (AASHTO, 2010).

Predictability of roads is another important principle of sustainable safety. Road layout and elements should be designed in a distinctive way, so that road users can easily identify the functionality and characteristics of the road. This is commonly referred to as the concept of 'self-explaining' roads. Self-explaining roads provide consistent and clear information to road users concerning their expected behaviour. Unlike some self-enforcing treatments (e.g., traffic calming), which directly influence road users' behaviour by physical restrictions, self-explaining roads try to affect the road users' behaviour psychologically.

Although the self-explaining concept has been predominantly adopted on motorways its application on other road classes, such as urban or suburban arterials, seems to be relevant and useful. One major concern, specifically on suburban arterials, is that the transition zones between rural and built-up areas usually lack sufficient and standard design elements and information to help drivers behave appropriately in response to different road settings, such as speed limits. This problem can be remedied by appropriately implementing the self-explaining concept.

SYNTHESIS AND CONCLUSIONS

Urban and suburban arterials represent a vast and yet diverse variety of road network facilities. These road network features consist of two major elements – road segments and intersections – that accommodate mobility for a mixture of road users. The design traits of urban and suburban arterials can be quite different in terms of traffic control, access features and operational characteristics such as speed. Hence, it is of great importance to distinguish between different road elements when analysing crashes. In addition, due to the presence of different road users (e.g., cyclists, pedestrians and motor vehicle users) it is essential to account for different crash types separately. Moreover, crash severity should also be taken into account when analysing road safety on urban and suburban arterials, due to the greater impact of more severe crashes on the sustainability of the whole road network system. It is evident that the actual traffic safety level of a facility is substantially undermined if the involvement of vulnerable road users or severity of crashes is not acknowledged.

There are several contributing factors for crashes, categorised as traffic and road characteristics, driver behaviour or vehicle properties. This chapter paid special attention to the traffic and road characteristics. In this regard, it reviewed a few of the most important factors that have been extensively used in the literature within the context of urban and suburban arterials. Reviewing the literature showed that the association between crashes and some of these variables (e.g., speed and lane width) is rather disputed, signified by counterintuitive results. This could be attributed to different reasons, such as the omitted-variable bias, multicollinearity among explanatory variables or inappropriate and mis-specified model functions. Here the important lesson to be learnt is that causality should always be considered superior to statistical associations.

Several strategies can be adopted in order to improve traffic safety on urban and suburban arterials. The strategies that have been studied in this chapter are either engineering- or behaviour-related. Although urban and suburban arterials are characterised as facilities on which mobility is the most important function, it is their access feature that contributes more to their safety conditions. Hence, implementing access management strategies is regarded as a useful solution in alleviating crashes and their consequences. Speed is known to be the primary factor in severe crash occurrence. Traffic calming strategies are found to be appropriate strategies through which to improve traffic safety levels of urban and suburban arterials. Due to the importance of the safety of vulnerable road users and their immense contribution to the sustainability of transportation systems, certain solutions should be foreseen to address crashes which involve vulnerable road users. This chapter also discussed a number of solutions that have been proven to be beneficial for pedestrians and cyclists on urban and suburban arterials (see also Chapters 10 and 11).

Besides these rather localised treatments, there appears to be a broader framework that can be applied to improve the safety of road network facilities, named 'sustainable safety'. This road safety concept aims at making roads inherently safe by adapting road design features to human capacities and limitations. This is a proactive approach that works by integrating several principles. Although the implementation of this approach sounds promising in addressing safety issues on urban and suburban arterials more empirical research is needed this area.

REFERENCES

AASHTO (2010). *Highway safety manual* (1st ed.). Washington, D.C.: American Association of State Highway and Transportation Officials.

Abdel-Aty, M., Devarasetty, P. C., & Pande, A. (2009). Safety evaluation of multilane arterials in Florida. *Accident Analysis & Prevention*, *41*(4), 777–788. doi:10.1016/j.aap.2009.03.015.

Ariën, C., Brijs, K., Ceulemans, W., Vanroelen, G., Jongen, E. M. M., Daniels, S., ... Wets, G. (2014). A driving simulator study on the effect of transversal rumble strips located nearby dangerous curves under repeated exposure. In *AHFE conference on advances in human aspects of transportation: Part II*, 557. Jagiellonian University, Krakow, Poland.

Avelar, R., Dixon, K., Brown, L., Mecham, M., & van Schalkwyk, I. (2013). Influence of land use and driveway placement on safety performance of arterial highways. *Transportation Research Record: Journal of the Transportation Research Board, 2398*(December), 101–109. doi:10.3141/2398-12.

Berhanu, G. (2004). Models relating traffic safety with road environment and traffic flows on arterial roads in Addis Ababa. *Accident Analysis & Prevention, 36*(5), 697–704. doi:10.1016/j.aap.2003.05.002.

Brijs, T., Wets, G., Krimpenfort, R., & Offermans, C. (2006). Impact of hourly measured speed on accident risk in the Netherlands: Results from exploratory study using geographic information systems. *Transportation Research Record: Journal of the Transportation Research Board, 1972*(1), 85–93. doi:10.3141/1972-13.

Chen, Y., & Persaud, B. (2014). Methodology to develop crash modification functions for road safety treatments with fully specified and hierarchical models. *Accident Analysis & Prevention, 70*(September), 131–139. doi:10.1016/j.aap.2014.03.012.

Daniels, S., Vanrie, J. Dreesen, A., & Brijs, T. (2010). Additional road markings as an indication of speed limits: Results of a field experiment and a driving simulator study. *Accident Analysis & Prevention, Assessing Safety with Driving Simulators*, *42*(3), 953–960. doi:10.1016/j.aap.2009.06.020.

Das, A., & Abdel-Aty, M. A. (2011). A combined frequency–severity approach for the analysis of rear-end crashes on urban arterials. *Safety Science, 49*(8–9), 1156–1163. doi:10.1016/j.ssci.2011.03.007.

De Pauw, E., Daniels, S., Thierie, M., & Brijs, T. (2014a). Safety effects of reducing the speed limit from 90 km/h to 70 km/h. *Accident Analysis & Prevention, 62*(January), 426–431. doi:10.1016/j.aap.2013.05.003.

De Pauw, E., Daniels, S., Brijs, T., Hermans, E., & Wets, G. (2014b). To brake or to accelerate? Safety effects of combined speed and red light cameras. *Journal of Safety Research, 50*(September), 59–65. doi:10.1016/j.jsr.2014.03.011.

De Pauw, E., Daniels, S., Brijs, T., Hermans, E., & Wets, G. (2014c). An evaluation of the traffic safety effect of fixed speed cameras. *Safety Science*, *62*(February), 168–174. doi:10.1016/j.ssci.2013.07.028.

Dijkstra, A., Eenink, R., & Wegman, F. (2007). Met Een Veilige Snelheid over Wegen: SWOV-Visie Op 'de Grijze Weg.' *Verkeerskunde*, *57*(7), 48–52.

Dixit, V. V., Pande, A., Abdel-Aty, M., Das, A., & Radwan, E. (2011). Quality of traffic flow on urban arterial streets and its relationship with safety. *Accident Analysis & Prevention*, *43*(5), 1610–1616. doi:10.1016/j.aap.2011.01.006.

Eisele, W., & Frawley, W. (2005). Estimating the safety and operational impact of raised medians and driveway density: Experiences from Texas and Oklahoma case studies. *Transportation Research Record: Journal of the Transportation Research Board*, *1931*(January), 108–116. doi:10.3141/1931-14.

El-Basyouny, K., & Sayed, T. (2009). Accident prediction models with random corridor parameters. *Accident Analysis & Prevention*, *41*(5), 1118–1123. doi:10.1016/j.aap.2009.06.025.

Elvik, R. (2005). Speed and road safety: Synthesis of evidence from evaluation studies. *Transportation Research Record: Journal of the Transportation Research Board*, *1908*(January), 59–69. doi:10.3141/1908-08.

Elvik, R., Høye, A., Vaa, T., & Sørensen, M. (2009). *The handbook of road safety measures* (2nd ed.). Bingley, UK: Emerald Group Publishing.

Gettman, D., & Head, L. (2003). *Surrogate safety measures from traffic simulation models*. FHWA-RD-03-050. Tucson, AZ: Siemens Gardner Transportation Systems.

GOV.UK. (2012). Guidance on Road Classification and the Primary Route Network. *Road Classification and the Primary Route Network Publications*. Retrieved from https://www.gov.uk/government/publications/guidance-on-road-classification-and-the-primary-route-network.

Hadi, M. A., Aruldhas, J., Chow, L.-F., & Wattleworth, J. A. (1995). Estimating safety effects of cross-section design for various highway types using negative binomial regression. *Transportation Research Record: Journal of the Transportation Research Board*, *1500*. Retrieved from http://trid.trb.org/view.aspx?id=452743.

Harkey, D. L., Srinivasan, R., Baek, J., Council, F. M., Eccles, K. A., Lefler, N. X., Gross, F., et al. (2008). *Accident modification factors for traffic engineering and ITS improvements*. NCHRP REPORT 617. NCHRP, Transportation Research Board, Washington, D.C. Retrieved from http://trid.trb.org/view/2008/M/864254.

Harvey, C. W., & Aultman-Hall, L. (2015). *Urban streetscape design and crash severity*. Transportation Research Record: Journal of the Transportation Research Board, No. 2500, Transportation Research Board, Washington, D.C., 2015, pp. 1–8. DOI: 10.3141/2500-01.

Harwood, D., Bauer, K., Richard, K., Gilmore, D., Graham, J., Potts, I., … Hauer, E. (2007). *Methodology to predict the safety performance of urban and suburban arterials*. Final Report of Phases I and II. NCHRP Project 17-26. Midwest Research Institute, Kansas City, MO. Retrieved from http://onlinepubs.trb.org/onlinepubs/nchrp/nchrp_w129p1&2.pdf.

Hauer, E., Council, F., & Mohammedshah, Y. (2004). Safety models for urban four-lane undivided road segments. *Transportation Research Record: Journal of the Transportation Research Board*, *1897*(January), 96–105. doi:10.3141/1897-13.

Jongen, E. M. M., Brijs, K., Mollu, K., Brijs, T., & Wets, G. (2011). 70 km/h speed limits on former 90 km/h roads: Effects of sign repetition and distraction on speed. *Human Factors: The Journal of the Human Factors and Ergonomics Society*, *53*(6), 771–785. doi:10.1177/0018720811419848.

Lee, C., Xu, X., & Nguyen, V. (2012). Analysis of midblock crashes in an urban divided arterial road. *Journal of Transportation Safety & Security*, *4*(1), 1–18. doi:10.1080/19439962.2011.607938.

Lee, J., Park, B., Malakorn, K., & So, J. (2013). Sustainability assessments of cooperative vehicle intersection control at an urban corridor. *Transportation Research Part C: Emerging Technologies*, *32*(July), 193–206. doi:10.1016/j.trc.2012.09.004.

Li, R., El-Basyouny, K., & Kim, A. (2015). *A before-and-after empirical Bayes evaluation of automated mobile speed enforcement on urban arterial roads*. Transportation Research Record: Journal of the Transportation Research Board, No. 2516, Transportation Research Board, Washington, D.C., 2015, pp. 44–52. DOI: 10.3141/2516-07.

Lord, D., & Mannering, F. (2010). The statistical analysis of crash-frequency data: A review and assessment of methodological alternatives. *Transportation Research Part A: Policy and Practice*, *44*(5), 291–305. doi:10.1016/j.tra.2010.02.001.

Ma, M., Yan, X., Abdel-Aty, M., Huang, H., & Wang, X. (2010). Safety analysis of urban arterials under mixed-traffic patterns in Beijing. *Transportation Research Record: Journal of the Transportation Research Board*, *2193*(December), 105–115. doi:10.3141/2193-13.

Mauga, T., & Mohamed Kaseko, M. (2010). Modeling and evaluating safety impacts of access management features in the Las Vegas, Nevada, Valley. *Transportation Research Record: Journal of the Transportation Research Board*, *2171*(December), 57–65. doi:10.3141/2171-07.

McMahon, P. J., Zegeer, C. V., Duncan, C., Knoblauch, R. L., Stewart, J. R., & Khattak, A. J. (2002). *An analysis of factors contributing to 'walking along roadway' crashes: Research study and guidelines for sidewalks and walkways*. FHWA-RD-01-101. U.S. Department of Transportation: The Federal Highway Administration (FHWA), McLean, VA. Retrieved from http://trid.trb.org/view.aspx?id=719134.

Nevarez, A., Abdel-Aty, M., Wang, X., & Santos, J. (2009). *Large-scale injury severity analysis for arterial roads: Modeling scheme and contributing factors*. Presented at the 88th Transportation Research Board Annual Meeting, Washington, D.C.

Ott, S. E., Haley, R. L. Hummer, J. E. Foyle, R. S., & Cunningham, C. M.. (2012). Safety effects of unsignalized superstreets in North Carolina. *Accident Analysis & Prevention*, *45*(March), 572–579. doi:10.1016/j.aap.2011.09.010.

Pande, A., & Abdel-Aty. M. A. (2009). A novel approach for analyzing severe crash patterns on multilane highways. *Accident Analysis & Prevention*, *41*(5), 985–994. doi:10.1016/j.aap.2009.06.003.

Park, P. Y., & Abdel-Aty, M. (2011). A stochastic catastrophe model using two-fluid model parameters to investigate traffic safety on urban arterials. *Accident Analysis & Prevention*, *43*(3), 1267–1278. doi:10.1016/j.aap.2011.01.009.

Park, J., Abdel-Aty, M., Lee, J., & Lee, C. (2015). Developing crash modification functions to assess safety effects of adding bike lanes for urban arterials with different roadway and socio-economic characteristics. *Accident Analysis & Prevention*, *74*(January), 179–191. doi:10.1016/j.aap.2014.10.024.

Pirdavani, A., Brijs, T., Bellemans, T., & Wets, G. (2010). Evaluation of traffic safety at un-signalized intersections using microsimulation: A utilization of proximal safety indicators. *Advances in Transportation Studies an International Journal*, *22*(A), 43–52.

Polders, E., Cornu, J., De Ceunynck, T., Daniels, S., Brijs, K., Brijs, T., … Wets, G. (2015). Drivers' behavioral responses to combined speed and red light cameras. *Accident Analysis & Prevention*, *81*(August), 153–166. doi:10.1016/j.aap.2015.05.006.

Potts, I., Harwood, D., & Richard, K. (2007). Relationship of lane width to safety on urban and suburban arterials. *Transportation Research Record: Journal of the Transportation Research Board, 2023*(December), 63–82. doi:10.3141/2023-08.

Quddus, M. (2013). Exploring the relationship between average speed, speed variation, and accident rates using spatial statistical models and GIS. *Journal of Transportation Safety & Security, 5*(1), 27–45. doi:10.1080/19439962.2012.705232.

Retting, R. A, Weinstein, H. B., Williams, A. F., & Preusser, D. F. (2001). A simple method for identifying and correcting crash problems on urban arterial streets. *Accident Analysis & Prevention, 33*(6), 723–734. doi:10.1016/S0001-4575(00)00086-5.

Rosales, J., & Brinckerhoff, P. (2006). *Road diet handbook: Setting trends for livable streets* (2nd ed.). Bingley, UK: Parsons Brinckerhoff.

Saha, D., Alluri, P., & Gan, A. (2015). Prioritizing highway safety manual's crash prediction variables using boosted regression trees. *Accident Analysis & Prevention, 79*(June), 133–144. doi:10.1016/j.aap.2015.03.011.

Saurabh, V. (2013). "CARE DATABASE: Common Accident Data Set (CADaS)." 3.3. Directorate-General for Mobility and Transport of the European Commission.

Sawalha, Z., & Sayed, T. (2001). Evaluating safety of urban arterial roadways. *Journal of Transportation Engineering, 127*(2), 151–158. doi:10.1061/(ASCE)0733-947X(2001)127:2(151).

Schultz, G., Allen, C., & Boschert, T. (2010). Making the most of an existing system through access management at major arterial intersections. *Transportation Research Record: Journal of the Transportation Research Board, 2171*(December), 66–74. doi:10.3141/2171-08.

Schultz, G., Braley, K., & Boschert, T. (2009). Relationship between access management and other physical roadway characteristics and safety. *Journal of Transportation Engineering, 136*(2), 141–148. doi:10.1061/(ASCE)TE.1943-5436.0000085.

Soole, D. W., Watson, B. C., & Fleiter, J. J. (2013). Effects of average speed enforcement on speed compliance and crashes: A review of the literature. *Accident Analysis & Prevention, 54*(May), 46–56. doi:10.1016/j.aap.2013.01.018.

Stevanovic, A., Stevanovic, J., & Kergaye, C. (2013). Optimization of traffic signal timings based on surrogate measures of safety. *Transportation Research Part C: Emerging Technologies, 32*(July), 159–178. doi:10.1016/j.trc.2013.02.009.

Stout, T. B., Pawlovich, M. D., Souleyrette, R. R., & Carriquiry, A. (2006). Safety impacts of 'road diets' in Iowa. *ITE Journal, 76*(12), 24–28.

Strathman, J. G., Dueker, K. J., Zhang, J., & Williams, T. (2003). Analysis of design attributes and crashes on the Oregon highway system. *Applied Health Economics and Health Policy, 2*(4), 201–212.

Sultana, M., Qin, X., Chitturi, M., & Noyce, D. (2014). Analysis of safety effects of traffic, geometric, and access parameters on truck arterial corridors. *Transportation Research Record: Journal of the Transportation Research Board, 2404*(December), 68–76. doi:10.3141/2404-08.

Tao, H., Foomani, M. G., & Alecsandru, C. (2015). *A two-step microscopic traffic safety evaluation model of reserved lanes facilities: An arterial case study*. Presented at the 94th Transportation Research Board Annual Meeting, Washington, D.C.

Tay, R., & Rifaat, S. M. (2007). Factors contributing to the severity of intersection crashes. *Journal of Advanced Transportation, 41*(3), 245–265. doi:10.1002/atr.5670410303.

The Federal Highway Administration (FHWA) *Highway functional classification: Concepts, criteria and procedures*. (2013). FHWA-PL-13-026. U.S. Department of Transportation: Washington D.C.

U.S. Department of Transportation. (2008). *Federal highway administration planning glossary*. Retrieved from http://www.fhwa.dot.gov/planning/glossary/.

VTPI. 2012. "VTPI." *Online TDM Encyclopedia*, Victoria Transport Policy Institute. Retrieved from http://www.vtpi.org/tdm/index.php.

Wang, C., Quddus, M. A., & Ison, S. G. (2013). The effect of traffic and road characteristics on road safety: A review and future research direction. *Safety Science*, *57*(August), 264–275. doi:10.1016/j.ssci.2013.02.012.

Wang, X., & Chen, M. (2012). Safety analysis on urban arterials considering operational conditions in Shanghai. *Procedia Engineering, 2012 International Symposium on Safety Science and Technology*, *45*, 836–840. doi:10.1016/j.proeng.2012.08.247.

Wang, X., Song, Y., Yu, R., & Schultz, G. G. (2014). Safety modeling of suburban arterials in Shanghai, China. *Accident Analysis & Prevention*, *70*(September), 215–224. doi:10.1016/j.aap.2014.04.005.

Wegman, F., Aarts, L., & Bax, C. (2008). Advancing sustainable safety: National road safety outlook for the Netherlands for 2005–2020. *Safety Science*, *46*(2), 323–343. doi:10.1016/j.ssci.2007.06.013.

Wen, H., & Luo, J. (2011). Traffic incident detection for urban arterial road based on data fusion and learning vector quantization. In *ICTIS 2011*, 971–78. American Society of Civil Engineers. Retrieved from http://ascelibrary.org/doi/abs/10.1061/41177%28415%29123.

Williams, K. M., Stover, V. G., Dixon, K. K., & Demosthenes, P. (2014). *Access management manual* (2nd ed.). Washington, D.C.: Transportation Research Board. Retrieved from http://trid.trb.org/view/2014/M/1341995.

Wilmots, B., Hermans, E., Brijs, T., & Wets, G. (2013). *The effect of speed control with a mobile radar with or without an advanced warning sign: A statistical analysis on speeding behavior*. Presented at the 92nd Transportation Research Board Annual Meeting, Washington, D.C..

Yanmaz-Tuzel, O., & Ozbay, K. (2010). A comparative full Bayesian before-and-after analysis and application to urban road safety countermeasures in New Jersey. *Accident Analysis & Prevention*, *42*(6), 2099–2107. doi:10.1016/j.aap.2010.06.023.

Yasmin, S., Eluru, N., Bhat, C. R., & Tay, R. (2014). A latent segmentation based generalized ordered logit model to examine factors influencing driver injury severity. *Analytic Methods in Accident Research*, *1*(January), 23–38. doi:10.1016/j.amar.2013.10.002.

CHAPTER 6

CONTROLLED ACCESS FACILITIES (FREEWAYS)

Francesca La Torre

ABSTRACT

Purpose – *Freeway networks are designed to higher standards and are safer infrastructures as compared to other road types, if properly designed. On the other hand, these facilities are driven at very high speeds and therefore speed and design consistency are essential for achieving safe infrastructure designs. This chapter describes the criteria for speed and design consistency and looks at new tools and criteria for improving freeway safety in new and in existing infrastructures.*

Methodology – *This chapter describes the criteria to evaluate if there are speed, design and human factors inconsistencies, as well as potential solutions for tackling local deficiencies and speeding issues. As one of the critical issues in freeway safety is represented by run-off-road crashes, a specific section in the chapter is devoted to newly developed design and assessment tools for improving roadside safety. The potential implications of Intelligent Transportation System (ITS) technologies on freeways design and management are also presented.*

Safe Mobility: Challenges, Methodology and Solutions
Transport and Sustainability, Volume 11, 107–126

ISSN: 2044-9941/doi:10.1108/S2044-994120180000011006

Findings and Social Implications – *The important crash reduction trends observed in the decade 2001–2010 are now slowing down and new actions are required to be coupled with more traditional design checks. The full implementation of cooperative ITS systems is expected to have a very important impact on road safety, but in the short term several safety improvements can be realised: section speed enforcement techniques and high-friction wearing courses have been proven to be extremely effective, as have perceptual measures accounting for human factors principles.*

Keywords Freeways; road safety; speed management; roadside safety; ITS; human factors

INTRODUCTION

This chapter is the second of five chapters focussing on influences of the transport network on crash risk. While this chapter focusses on controlled access facilities, or freeways, it complements the current knowledge regarding the crash risks related to urban and suburban arterials (Chapter 5), rural and urban intersections (Chapter 7), roundabouts (Chapter 8) and real-time traffic operations and safety (Chapter 9).

Freeway networks are very peculiar road facilities that are access-controlled, with physically divided carriageways and high design and travelling speeds (Fig. 1). Direct access to freeways is not allowed, there are no driveways and all the nodes are realised by means of grade-separated interchanges (Fig. 2), and this leads to very different crash patterns and dominant causes as compared to other types of rural and urban roads.

Even though the higher travelling speeds lead to more severe crash consequences, the overall fatality risk on motorways is consistently lower than the average fatality risk on all roads, as shown in Fig. 3, where the fatality risks for all roads and for freeways from several countries have been plotted based on the IRTAD 2012 database (IRTAD, 2014). This is due to the very peculiar configuration of freeways, with separated carriageways that prevent head-on crashes, grade-separated interchanges that limit side crashes, higher design standards and, typically, higher maintenance standards.

It should be noted, on the other hand, that considerable effort has been made in the decade 2001–2010 to improve freeways' safety with a number of strategic plans. In 1993, the first Communication from the European

Fig. 1. Typical Freeway Segment. *Source:* Autostrade per l'Italia.

Commission to the European Council raised the problem of addressing road safety (EC, 1993) and led to the publication of the first European White Paper in 2001 (EC, 2001) that identified infrastructure safety as one of the issues that had to be tackled to half the number of road fatalities by 2010.

As a first consequence, a Directive aimed at improving tunnels safety in the Trans European Road Network (TERN) was published (EU, 2004) and later adopted by all the European Union (EU) countries. The TERN is the primary European road network, mostly represented by freeways, and several European countries have extended the same safety principles also to tunnels in freeways not part of the TERN (see, for instance, ANAS, 2009).

In 2008 a European Directive on road infrastructure safety management for the TERN network was published (EU, 2008) introducing following mandatory road safety management principles for TERN infrastructures:

- Road safety impact assessment for infrastructure projects.
- Road safety audits for infrastructure projects.
- Safety ranking and management of the road network in operation.
- Safety inspections.

Fig. 2. Freeway Grade-Separated Interchange. *Source:* Autostrade per l'Italia.

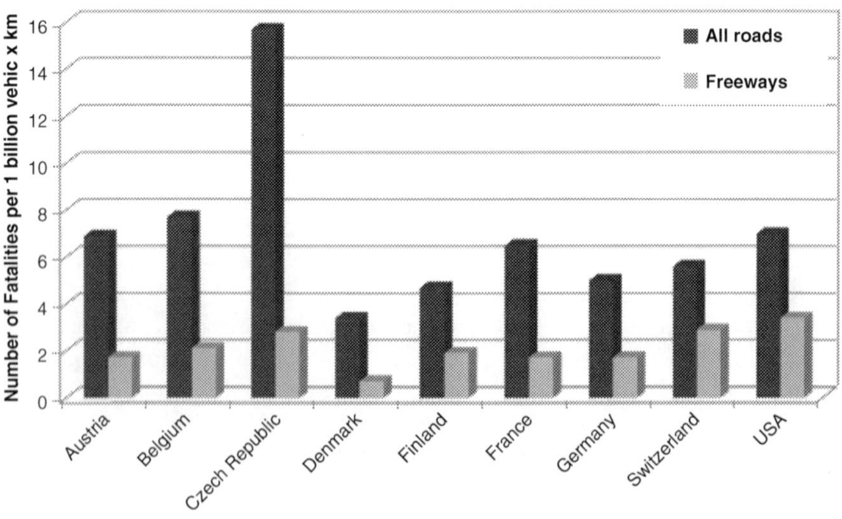

Fig. 3. Comparison of Freeways vs. All Roads Risk for Different Countries, Year 2012. *Data Source:* IRTAD (2014); Own Illustration.

In the United States (US), in 1997, a plan was produced focussing on 22 specific highway safety challenges (AASHTO, 2005). In 2003 a target of not more than one fatality per 100 million vehicle × miles (6 per billion vehicles × km) for the year 2008 was set by the AASHTO Board of Directors, the Governors Highway Safety Association, the American Association of Motor Vehicle Administrators and the US Department of Transportation.

Australia has been very active in the past decade in improving road safety, and in 2011 a National Road Safety Strategy for the period 2011–2020 was published (Australian Transport Council, 2011).

This attention towards safety issues has led to a drastic reduction in fatal and injury crashes as well as in fatalities in the decade 2001–2010. Several countries are now experiencing a stall, as shown in Fig. 4 where the time trends for fatal and injury crash and overall fatality risks on the Italian freeway network (a very extensive network of 6,844 km in 2014) have been plotted. After a reduction of over 60% in fatality risk between the year 2002 and the year 2009, the fatality risk has been essentially stable between 2009 and 2014. The risk of having a severe crash (with consequences to the persons) is still reducing but at a very slow rate. In analysing these results, the economic crisis of the last decade should be considered, as the reduction in vehicle use, and therefore in exposure to risk, has clearly affected the overall number of crashes.

To make a quantum leap in freeway road safety there is a need to ensure that all new designs are made according to the essential principles of design and speed consistency, but following new challenges need to be tackled:

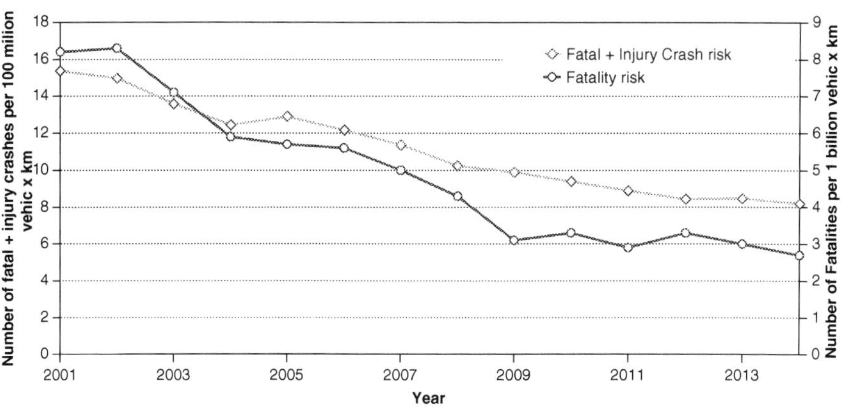

Fig. 4. Freeway Risk Trends in Italy from 2001 to 2014. *Data Source:* AISCAT (2015), Own Illustration.

- Roadside safety.
- Defining criteria to improve safety of existing freeway networks where design and speed consistency principles are not granted or cannot explain recurring crashes.
- Introducing Intelligent Transportation Systems (ITS) in freeway safety management.

DESIGN AND SPEED CONSISTENCY ON FREEWAYS

Freeways are infrastructures typically designed to higher standards, and therefore the gap between travelling speed and design speed is usually lower than for two-lane, two-way rural roads. On the other hand, the design needs to meet the driver's expectations to avoid overloading, fatigue and mistakes that, at high speed, can have severe consequences.

A freeway design that meets drivers' expectations needs to:

- Set a minimum design speed compatible with freeways users' expectations, as an example in Italy (Italian Ministry for Transportation and Infrastructures, 2002) this is set to 90 km/h (approx. 55 mph) in rural freeways and 80 km/h (50 mph) in urban freeways, while in the US (AASHTO, 2011) this is set to 80 km/h (50 mph).
- Consider that the driver will adapt the speed to the perceived geometric design and therefore each curve radius should have a specific design speed (typically given by national standards).
- Consider that the user needs to have the proper time to slow down in approaching a given curve.
- Consider that the freeway user does not expect to be required to make sudden speed reductions and that the attention of the driver is reduced when travelling for a long time at the maximum allowed speed.

To control all these factors several standards have introduced a speed consistency check. The speed diagram from the Italian standard (Italian Ministry for Transportation and Infrastructures, 2002) is shown as an example in Fig. 5. The same approach can be found in the Swiss standard (VSS, 1996).

Based on this type of diagram, the designer should ensure the following:

- Speed drop after a segment travelled at the maximum speed is limited (in the Italian standard this drop is set to a maximum 10 km/h, approx. 6 mph).

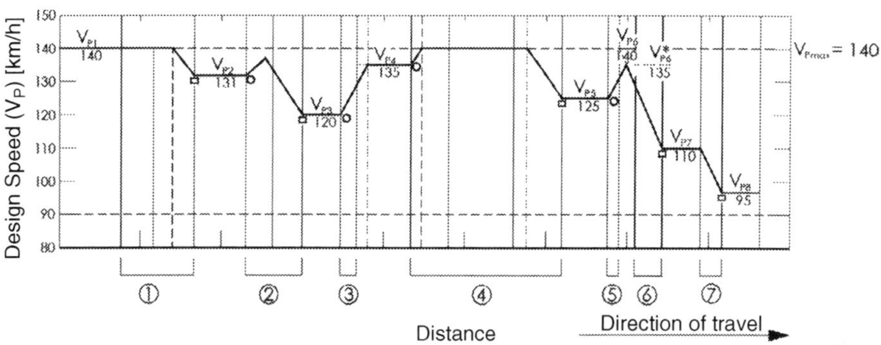

Fig. 5. Speed Diagram According to the Italian Design Standard. *Source:* Italian Ministry for Transportation and Infrastructures (2002); own translation.

- Speed drop between two subsequent curves is limited (in the Italian standard this drop is set to a maximum 20 km/h, 12.5 mph, preferably 15 km/h, approx. 9 mph).
- Sight distance available allows the driver to properly perceive the curve in the point where the expected safe deceleration should start.

In the recently developed Highway Safety Manual (HSM) model for freeway segments (AASHTO, 2014) a Crash Modification Factor (CMF) has been included to account for the segment curvature, but this does not account for design consistency. This means that a curve with a given radius will have the same effect on the number of expected crashes if it is located between other similar curves or at the end of a long, straight segment. Crash patterns on freeways show that there is a need to improve these CMFs to account also for design inconsistencies.

Speeding is a major issue in freeway safety, both in terms of average speed and in terms of speed variance (see Chapter 3). Speed enforcement is an effective strategy for reducing the percentage of speeders, but it is typically a point measure that affects the driving speed only in the enforcement location. For this reason the implementation of section speed control (Fig. 6) – also called point-to-point speed control – is increasing being deployed on freeway sections, although mostly in Europe. Several studies have analysed the effect of this type of enforcement method on driving speed. These studies evaluated average speed, 85th percentile of isolated vehicles' speed (travelling in free-flow conditions), standard deviation of driving speed and the proportion of drivers exceeding the speed limits (Lynch, White, & Napier, 2011; Montella, Punzo, Chiaradonna, Mauriello, & Montanino, 2015; Soole,

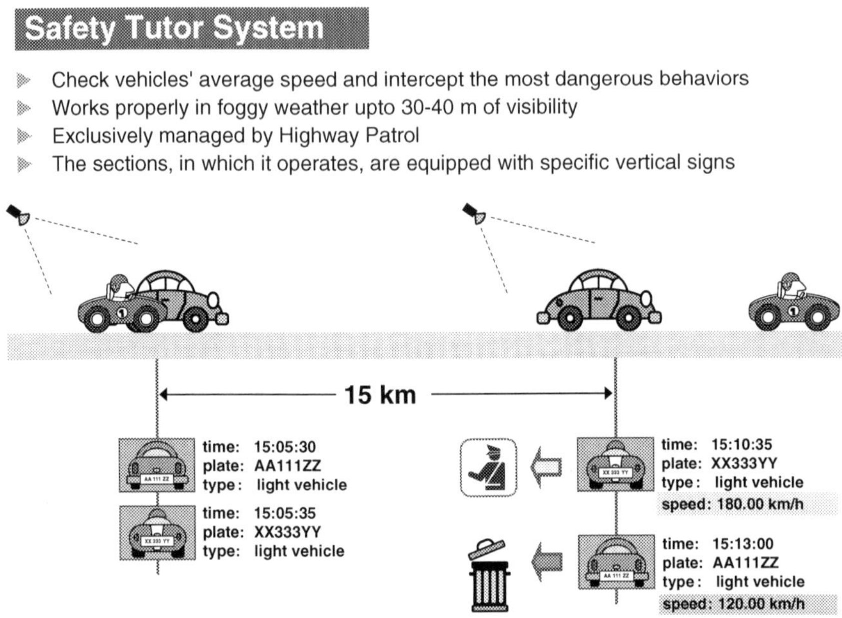

Fig. 6. Section Speed Enforcement Principle. *Source:* Autostrade per l'Italia.

Watson, & Fleiter, 2013). Soole et al. (2013) conducted a very extensive review of all the studies performed on section control enforcement, showing that the most important effect of the section control system is on reducing the percentage of users driving above the speed limit.

A recent study (La Torre, Domenichini, et al., 2016) has shown that enforcing average section speed by means of a clearly visible enforcing system in use on the Italian freeway system (known in Italy as the 'Tutor' device; Fig. 7) leads to an average crash reduction of 2–13% considering different crash types on all sections where the system was implemented. This result is consistent with the safety effect of section control presented in Hoye (2015). Disaggregating the analysis by traffic volumes, La Torre, Domenichini, et al. (2016) found a statistically significant crash reduction for multi-vehicle crashes in medium- to high-volume freeways (above 25,000 vehicles per day) with average crash reductions that can reach 40–50% for very high-volume freeways (above 25,000 vehicles per day). This is likely due to the fact that the effect of the section control enforcement is not only a reduction in the average speed or in the speed of the isolated vehicle but also a reduction in the speed variance that affects multi-vehicle crashes.

Fig. 7. Freeway Section Control Signs in Italy. *Source:* Autostrade per l'Italia.

To understand if speeding is an issue in a specific road segment, speed monitoring campaigns should be conducted to evaluate the operating speed (85th percentile of the isolated vehicle speed travelling in free-flow conditions) to be compared with the design speed. An acceptable speed consistency is achieved when the difference between the operating speed and the design speed is not greater than 10 km/h (approx. 6 mph), while a difference of at least 20 km/h (12.5 mph) is an indicator of poor consistency (Lamm, Psarianos, & Mailaender, 1999).

ROADSIDE SAFETY

Roadside safety is a major safety issue for freeway networks due to the high rate of run-off-road crashes, which are usually high-severity crashes, combined with the high speed of impacts against the hazard located at the roadside or against the roadside protection device (safety barrier, crash cushion or barrier/end terminal). Typical hazards that can result in high-severity crashes are: trees, sign supports, rocks, concrete ditches or high-water reservoirs and stiff structures, with particular attention to the beginning section of the structure (e.g. of a wall or of a tunnel).

In order to reduce the severity of run-off-road crashes the roadside should be made as 'forgiving' as reasonably possible, considering the following design options in decreasing priority order (the second should be applied if the first is unfeasible, etc.):

1. *Remove the hazard*: if possible, a hazard should be removed to completely eliminate any risk for an errant vehicle reaching it.
2. *Redesign the hazard to be safely traversable*: if the hazard cannot be removed, it should be designed flexible enough to be bent by an errant vehicle.
3. *Relocate the hazard further away from the road*: if the hazard cannot be made traversable, it should be moved further away from the road, where it is less likely to be reached by errant vehicles.
4. *Make the hazard passively safe*: if the hazard cannot be relocated, it should be made passively safe to reduce the severity of a possible impact. This can be proven by testing the installed device in accordance with EN 12767 (CEN, 2007) or MASH (AASHTO, 2016).
5. *Install a vehicle restraint system (safety barrier or crash cushion)*: a vehicle restraint system should only be installed if the options above are not possible or unreasonable from a cost-effectiveness perspective.
6. *Delineate the hazard*: if none of the options above are applicable, the hazard should at least be made visible to warn road users of its existence.

Installing a vehicle restraint system is listed only as the fifth option in the abovementioned list, as a collision with a restraint system can also have undesired consequences, and a restraint system should be used only if reaching the hazard is likely to have more severe consequences.

In the recently published supplement to the HSM (AASHTO, 2014) a set of CMFs to account for the presence of a safety barrier and for the width of the available clear zone is given. When a safety barrier is used the number of single-vehicle crashes increase but the severity of the consequences decreases. In the HSM approach it is not possible to distinguish between different type of barriers or containment levels (defined as either based on the European EN 1317-2, CEN, 2010 or on the US MASH, AASHTO, 2016 classification) even though higher containment restraint systems are generally associated with higher crash severities for errant vehicles contained on the carriageway as compared to lower containment devices.

For this aim two parallel studies have been conducted in the US (RSAP, 2013) and in Europe (La Torre, Erginbas, et al., 2015, 2016; SAVeRS, 2015) aimed at providing a quantitative assessment procedure to evaluate the effect of adopting a given barrier as compared to another. The Selection of Appropriate Vehicle Restraint Systems (SAVeRS) tool can be used effectively in planning interventions on existing freeways where a sub-standard barrier is in place. The comparison of the risk associated with the current configuration as compared to the upgraded one allows the road administration to set intervention priorities. All the tools described above are aimed at assessing the effectiveness of passive safety solutions that can reduce the consequences of a run-off-road crash, but active safety solutions are more effective, if feasible, as they aim to reduce the number of expected run-off-road crashes.

Rumble strips are an effective way to reduce the number of expected crashes (La Torre, 2013), but the studies conducted for the development of the HSM model for freeways have shown that this solution is significantly effective only on straight segments and not on bends (AASHTO, 2014)

A recent study has shown that very good results can be achieved in reducing wet pavement fatal and injury crashes in freeway curves, replacing the traditional wearing course with a high-friction wearing course (Karathodorou et al., 2015; Fig. 8). A CMF of 0.27 (a 73% wet pavement run-off-road crashes reduction) was obtained with an Empirical-Bayes before–after assessment, with a very low standard deviation of 0.10 and the 95% confidence interval in the range 0.08–0.47 on 52 treated freeway sections. On the other hand,

Fig. 8. High-Friction Wearing Course. *Source*: Autostrade per l'Italia.

the analysis of several studies on the effectiveness of porous asphalt on wet crashes reduction has shown that this solution, that has the advantage of reducing noise and splash and spray effect, does not lead to a significant crash reduction.

HUMAN FACTORS: CONCEPTS TO IMPROVE SAFETY ON EXISTING FREEWAYS

Human factors are an important component of infrastructure safety and understanding how to evaluate and improve the infrastructure design; considering them allows for the definition of cost-effective solutions, mostly on existing infrastructures where geometric improvements are often technically unfeasible.

In the PIARC (World Road Association) Human Factors Guideline for a Safer Man–Road Interface (PIARC, 2016) three axioms/rules have been introduced to evaluate if the analysed infrastructure has deficiencies that can cause human factors-related crashes:

- *The 6-seconds axiom/rule*: The road should give the driver enough reaction time (estimated as 4–6 seconds prior to a decision needing to be made).
- *The field-of-view axiom/rule*: The road must offer the driver a safe field of view, avoiding monotony, optical guiding lines not parallel to the road edge or optical guiding lines with gaps, or dominant eye-catching objects despite the direction of the road. Improving the guiding elements can also help the driver in tackling situations that do not comply with one of the other two axioms (Fig. 9). Improving the optical density of the field of view has been proven to reduce speed and speed variance in freeway cross-over work zones (Domenichini, La Torre, Branzi, & Nocentini, 2017). This result offers a high potential for safety improvement in freeway work zones that are very critical freeway locations with fatal and injury crashes that can increase over three times as compared to the configuration without the work zone (La Torre, Domenichini, & Nocentini, 2017).
- *The logic axiom/rule*: Roads have to follow the driver's expectation logic as the driver's perception and reactions are affected by the orientation logic formed by their experience and recent perceptions. This is the principle behind the design and speed consistency principles described earlier.

Fig. 9. Field-of-View Improvement for Engaging a Sharp Curve after a Straight Section. *Source*: author.

The PIARC Guide (PIARC, 2016) provides a checklist to evaluate if there are design deficiencies or if any of these axioms/rules are present in a given location.

A study was conducted in Brandenburg, Germany, analysing more than 1,400 crashes on the basis of these principles (PIARC, 2008). The results obtained, summarised in Table 1, show that 68% of the analysed crashes may have been affected by human factors design mistakes in the road.

Crashes in freeways are often associated with fatigue and distraction due to the longer travelled distances (see Chapter 4). To mitigate these issues, four approaches can be used:

- *Education*: Several campaigns are in place all over the world to educate drivers in order to limit the risk of driving distractions or fatigue, including several videos and commercials (see Chapter 2 for more detail).
- *Regulations and enforcement*: In several countries, the use of cell phones without hands-free devices is now prohibited, but enforcement can still be difficult. To reduce fatigue for commercial vehicle drivers, driving time limits and mandatory rest periods have been introduced and several countries have mandatory speed tracking devices recording driving and rest periods.
- *Infrastructure measures*: One of the most common measures to reduce crashes due to fatigue and distraction is the use of shoulder and centreline rumble strips, and these have been proven extremely effective in several studies in the

Table 1. Human Factors Design Mistakes in 1,400 Crashes in Brandenburg, Germany.

	Crashes	%
I. Not influenceable by road design	*383*	*27*
II. Influenceable by road design	**953**	**68**
1. Deficits 6-s axiom	451	32
2. Deficits field-of-view axiom	228	16
3. Deficits logic axiom	274	20
III. Unexplained	*64*	*5*
Sum of analysed crashes	1,400	100

Data source: PIARC (2008).

US and EU. In the current edition of the HSM (AASHTO, 2014) the effectiveness for freeways is limited to straight segments, as the specific research conducted to develop the HSM supplement has shown the effectiveness in bends is statistically non-significant. A recent study conducted in Sweden has shown a 30% reduction of severe injuries and fatalities on freeways with a shoulder rumble (Vadeby, Fors, Kecklund, van Leeuwen, & Åkerstedt, 2015). Even though the concept that a monotonous road surrounding and geometric layout (long, straight segments) can lead to an increase in driver fatigue is well established, there are currently no established principles that link road design (lane width, curvature, visibility of lane markings, etc.) and driver fatigue. Additional research is needed in this area.

- *On-board technologies*: There are several projects devoted to identifying driving fatigue and distraction by means of an on-board monitoring system (including eye-tracking devices). In the US, the Federal Motor Carrier Safety Administration (FMCSA) has an active project to develop a 'Driver Fatigue and Distraction Monitoring and Warning System' (FMCSA, 2017). The advances in the field of automation will lead to a drastic reduction of fatigue and distraction crashes with active lane departure warning or even control systems. A very extensive study on mitigating fatigue in all transportation modes has recently been conducted by VTI in Sweden (Vadeby et al., 2015).

ITS AND FREEWAY SAFETY

Cooperative ITS are completely changing the vehicle-to-vehicle (V2V) and the vehicle-to-infrastructure (V2I; Fig. 10) interaction towards the 'smart road'

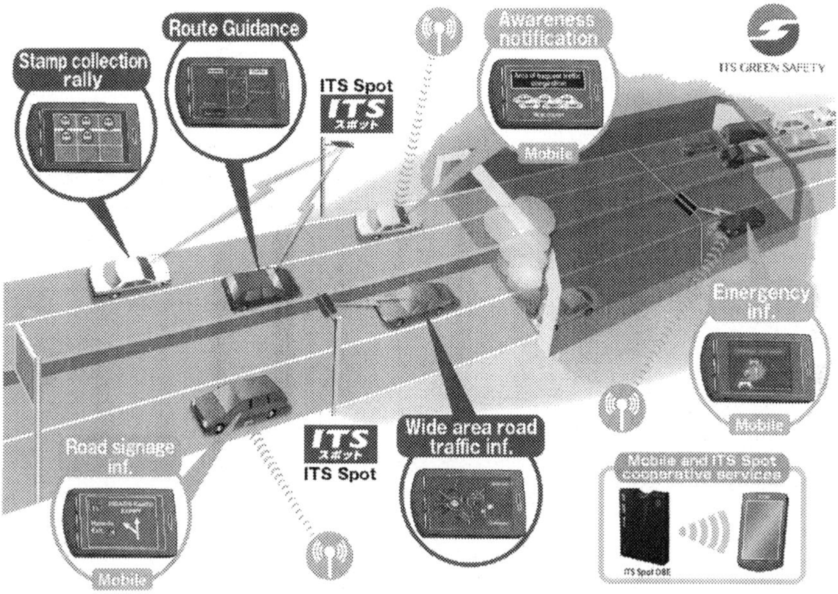

Fig. 10. Vehicle to Infrastructure Cooperative ITS Tools and Applications.
Source: ITS Green Safety – Japan (2013).

concept, but current infrastructure design standards worldwide have been developed independently on the availability of ITS technologies.

In 2010 the Directive 2010/40/EU of the European Parliament and of the Council on the framework for the deployment of ITS in the field of road transport and for interfaces with other modes of transport was published (EU, 2010), identifying safety-related information as priority area 'C'. Furthermore, in 2013 a supplement to the Directive was published specifically dealing with data and procedures for the provision, where possible, of road safety-related minimum universal traffic information free of charge to users (EU, 2010). This supplement indicates that the road safety-related minimum universal traffic information service shall consist of at least one of the following categories:

(a) Temporary slippery road.
(b) Animal, people, obstacles, debris on the road.
(c) Unprotected accident area.
(d) Short-term road works.
(e) Reduced visibility.

(f) Wrong-way driver.
(g) Unmanaged blockage of a road.
(h) Exceptional weather conditions.

The Amsterdam Group (2013) has identified the following set of typical V2V functions:

- Hazardous location warning.
- Slow vehicle warning.
- Traffic jam ahead warning.
- Stationary vehicle warning.
- Emergency brake light.
- Emergency vehicle warning.
- Motorcycle approaching indication.

A set of V2I functions have also been identified by the Amsterdam Group (2013):

- Road works warning.
- In-vehicle signage.
- Signal phase and time.
- Probe vehicle data.

A number of these functions will affect freeway design and management. A recent study has shown that installing a work zone on a freeway segment can lead to a CMF as high as 3.11 (depending on the work zone layout) with an average over all the work zone layouts of 1.33 (La Torre, Domenichini, & Nocentini, 2017). The implementation on a large scale of ITS technologies to warn the driver of work zones and of specific conditions (traffic jams, posted speed, etc.) could drastically reduce this effect.

Furthermore, the common freeway design concepts of sight distance, breaking distance and design speed will have to be redefined, considering also the implementation of on-board safety technologies. Rämä, Kulmala, Sihvola, Luoma, and Schirokoff (2009) estimate that the effect of different on-board technologies on fatalities reduction (Fig. 11) can reach 14–17% for the widespread implementation of electronic stability control systems.

SAFETY ASSESSMENT OF FREEWAY NETWORKS

One of the key issues in road safety management is the availability of models to assess the number of expected crashes in a given location.

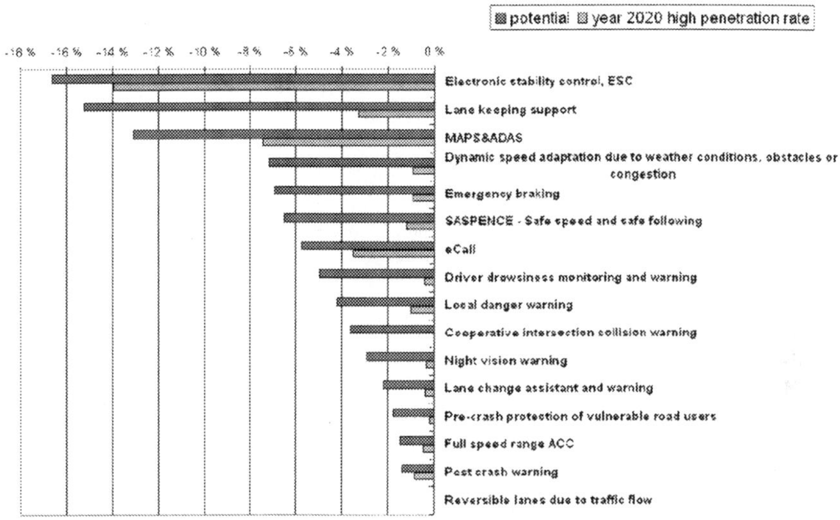

Fig. 11. Estimated Safety Impacts on Fatalities (%) of 16 In-Vehicle Systems.
Source: Rämä et al. (2009).
Note: 'Potential' Assumed 100% Fleet Penetration of Systems; 'Year 2020 High Penetration Rate' Assumed a Promoted Penetration in 2020.

The recently published HSM supplement (AASHTO, 2014) allows the number of predicted crashes in the following infrastructural elements to be estimated:

- Freeway segments.
- Freeway speed change lanes.
- Interchange ramps.
- Collector distributor road segments.

For each of these elements crash prediction models are given for single-vehicle and multi-vehicle crashes as well as for fatal+injury or for property-damage-only crashes.

Even though the HSM is the most up-to-date and complete tool for crash estimation, it should be noted that the models are developed using US data and, to date, only a few studies have been conducted to apply this model to the European road network (La Torre, Domenichini, Corsi, & Fanfani, 2014). A specific project has recently been completed (PRACT – Predicting Road Accidents – a Transferable Methodology across Europe) to develop a European crash prediction model (La Torre, Domenichini et al., 2016; Yannis et al., 2016).

A European trans-national model for estimating fatal+injury run-off-road crashes has been developed in the SAVeRS project, considering data from Austria, Ireland, Italy, Sweden and the UK (La Torre, Erginbas, et al., 2016); it also provides a procedure for adapting the model to different road networks.

FUTURE RESEARCH DIRECTIONS IN FREEWAY SAFETY

Road transport is rapidly moving towards automation, and freeways are expected to be the infrastructures where this new mode of transport will be predominant. Prototype applications are already available for freeways mostly related to 'platooning' vehicles (e.g. DriveMe project in Sweden and Peloton in the US). The core research issue for infrastructure designers and road operators will be the definition of the infrastructure adaptations to meet the automation requirements in the short, medium and long time frame. The major issues are:

- *The design and operation requirements to enable active safety systems to perform properly*. As an example, the Lane Departure Warning system already available in the market works properly only if the road marking is properly designed and maintained (short term).
- *The infrastructure and operational requirements and adaptations to manage the coexistence in a fleet of vehicles with different levels of automation* (from conventional to fully automated vehicles). Several projects have been recently funded in this area in the Horizon 2020 programme in the EU (Trans-Aid, Co-Exist, Brave, Inframix, InterACT and TrustVehicle) and are expected to twin with the US ongoing projects in the same area (medium term).
- *The infrastructure and operational support to full automation* (long term).

REFERENCES

AASHTO. (2005). *Strategic highway safety plan*. American Association of State Highway and Transportation Officials. Retrieved from http://safety.transportation.org/plan.aspx

AASHTO. (2011). *A policy on geometric design of highways and streets* (6th ed.). Washington, DC: American Association of State Highway and Transportation Officials.

AASHTO. (2014). *Highway safety manual supplement.* Washington, DC: American Association of State Highway and Transportation Officials.

AASHTO. (2016). *Manual for assessing safety hardware.* Washington, DC: American Association of State Highway and Transportation Officials.

AISCAT. (2015). *AISCAT informazioni, edizione semestrale* (in Italian, from Vol. II-2002 to Vol. II-2014) Retrieved from http://www.aiscat.it/pubb_semestrali.htm

AMSTERDAM Group. (2013). *Roadmap.* Retrieved from http://www.amsterdamgroup.eu

ANAS. (2009). *Linee Guida per la progettazione della sicurezza nelle Galleria Stradali secondo la normativa vigente* (in Italian). ANAS S.p.A. (Ed.), Direzione Centrale Progettazione.

Anund et al. (2015).

Australian Transport Council. (2011). *National road safety strategy 2011–2020.* Retrieved from https://infrastructure.gov.au/roads/safety/national_road_safety_strategy

CEN. (2007). *EN 12767.* (2007). *Passive safety of support structures for road equipment – Requirements and test methods.* Brussels: European Committee for Standardization.

CEN. (2010). *EN 1317-2.* (2010). *Passive road restraint systems. Part 2: Performance classes, impact test acceptance criteria and test methods for safety barriers including vehicle parapets.* Brussels: European Committee for Standardization.

Domenichini, L., La Torre, F., Branzi, V., & Nocentini, A. (2017). Speed behaviour in work zone crossovers – A driving simulator study. *Accident Analysis and Prevention, 98,* 10–24. doi:10.1016/j.aap.2016.09.018

EC. (1993). *Communication from the commission to the council for an action programme on road safety.* COM(93) 246, Brussels, 9 June 1993.

EC. (2001). *Communication from the Commission to the European Parliament, the Council, the European Economic and Social Committee and the Committee of the regions towards a European road safety area: Policy orientations on road safety 2011-2020.* COM(2010) 389, Brussels, 20 July 2010.

EU. (2004). Directive 2004/54/EC of the European Parliament and of the Council of 29 April 2004 on minimum safety requirements for tunnels in the Trans-European Road Network. *Official Journal of the European Union – L, 167,* 39–91.

EU. (2008). Directive 2008/96/EC of the European Parliament and of the Council of 19 November 2008 on road infrastructure safety management. *Official Journal of the European Union – L, 319,* 59–67.

EU. (2010). Directive 2010/40/EU of the European Parliament and of the Council of 7 July 2010 on the framework for the deployment of Intelligent Transport Systems in the field of road transport and for interfaces with other modes of transport. *Official Journal of the European Union – L, 247,* 6–10.

Hoye, A. (2015). Safety effects of section control – An empirical Bayes evaluation. *Accident Analysis and Prevention, 74,* 169–178.

IRTAD. (2014). *IRTAD database, October2014 – Risk indicators.* Retrieved from http://www.internationaltransportforum.org/irtadpublic/datasets.html

Italian Ministry for Transportation and Infrastructures. (2002). *Norme funzionali e geometriche per la costruzione delle strade* (in Italian). Ministry Decree N. 6792/5.11.2001, Supplemento ordinario alla GAZZETTA UFFICIALE Serie generale - n. 3, 4.1.2002.

ITS Green Safety Japan. (2013). *ITS green safety showcase.* Retrieved from http://www.its-jp.org/

Jenoptik. (2015). *Intelligent speed enforcement with vehicle classification.* Retrieved from http://www.jenoptik.com/en-point-to-point-measurement-switzerland-the-project.

Karathodorou, N., Graham, D., Hu, J., Richter, T., Ruhl, S., Yannis, G., & Domenichini, L. (2015). *Development of new crash modification factors/functions per key safety treatments.* Retrieved from www.practproject.eu

Lamm, R., Psarianos, B., & Mailaender., T. (1999). *Highway design and traffic safety engineering handbook*. New York, NY: McGraw-Hill.

La Torre, F. (2013). *Forgiving roadsides design guide*. Brussels: CEDR.

La Torre, F., Domenichini, L., Corsi, F., & Fanfani, F. (2014). Transferability of the highway safety manual freeway model to the Italian motorway network. *Transportation Research Record: Journal of the Transportation Research Board, 2435*. doi:10.3141/2435-08

La Torre, F., Domenichini, L., Meocci, M., Graham, D., Karathodorou, N., Richter, T., & Laiou, A. (2016). Development of a transnational accident prediction model. *Transportation Research Procedia, 14*, 1772–1781. doi:10.1016/j.trpro.2016.05.143

La Torre, F., Domenichini, L., & Nocentini, A. (2017). Effects of stationary work zones on motorway crashes. *Safety Science, 92*, 148–159. doi:10.1016/j.ssci.2016.10.008

La Torre, F., Erginbas, C., Thomson, R., Amato, G., Pengal, B., Stefan, C., & Hemmings, G. (2016). Selection of the most appropriate Roadside Vehicle Restraint System – The SAVeRS project. *Transportation Research Procedia, 14,* 4237–4246. doi:10.1016/j.trpro.2016.05.395

La Torre, F., Erginbas, C., Williams, G., Thomson, R., Hemmings, G., & Stefan, C. (2015). *Guideline for the selection of the most appropriate roadside vehicle restraint system.* Retrieved from www.saversproject.com

Lynch, M., White, M., & Napier, R. (2011). *Investigation into the use of point-to-point speed cameras.* NZ Transport Agency Research Report 465.

Montella, A., Punzo, V., Chiaradonna, S., Mauriello, F., & Montanino, M. (2015). Point-to-point speed enforcement systems: Speed limits design criteria and analysis of drivers' compliance. *Transportation Research Part C, 53*, 1–18.

PIARC. (2008). *Human factors guideline for safer road infrastructure*. Report 2008R18. Paris: PIARC, ISBN 2-84060-215-6.

PIARC. (2016). *Human factors guideline for a safer man–road interface.* Report 2016 R20EN. Paris: PIARC.

Rämä, P., Kulmala, R., Sihvola, N., Luoma, J., & Schirokoff, A. (2009). Safety effects of intelligent in-vehicle systems. In V. Rouhiainen (Ed.), *Scientific activities in safety & security 2009* (ss. 92–93). VTT. Retrieved from http://www.vtt.fi/inf/julkaisut/muut/2009/SafetySecurityReview09.pdf

RSAP. (2013). *Roadside safety analysis program* version 3.0.1. Retrieved from www.roadsafellc.com

SAVeRS. (2015). *SAVeRS Tool.* Retrieved from www.saversproject.com

Soole, D. W., Watson, B. C., & Fleiter, J. J. (2013). Effects of average speed enforcement on speed compliance and crashes: A review of the literature. *Accident Analysis and Prevention, 54*, 46–56.

Vadeby, A., Fors, C., Kecklund, G., van Leeuwen W., & Åkerstedt, T. (2015). *Countermeasures for fatigue in transportation – A review of existing methods for drivers on road, rail, sea and in aviation.* VTI Rapport 852A.

VSS. (1996). *Tracé – Eléments géométriques du tracé en plan – SN 640100a* (in French). Union des professionnels suisses de la route.

Yannis, G., Dragomanovits, A., Laiou, A., Richter, T., Ruhl, S., La Torre, F., & Li, H. (2016). Use of accident prediction models in road safety management – An international inquiry. *Transportation Research Procedia, 14*, 4257–4266. doi:10.1016/j.trpro.2016.05.397

CHAPTER 7

RURAL AND URBAN INTERSECTIONS

Richard Tay

ABSTRACT

Purpose – *Intersections are hazardous locations and to improve their safety we need to understand the factors contributing to crashes at these locations and provide evidence-based recommendations to reduce them. This chapter provides a summary of the findings on infrastructure-related factors contributing to crashes at urban and rural intersections and some discussions on the implications and potential countermeasures.*

Approach – *A review of the literature on intersection crashes was performed to identify the infrastructure-related crash-contributing factors. Some discussions on the implications and potential countermeasures are then provided.*

Findings – *The factors contributing to road crashes are diverse and complex. While the safety effects of a few factors (e.g., exposure and speed) are relatively consistent, many factors have different impacts on crash frequency and severity (e.g., types of intersection) and different impacts on urban and rural intersections (e.g., bus stops).*

Safe Mobility: Challenges, Methodology and Solutions
Transport and Sustainability, Volume 11, 127–145

ISSN: 2044-9941/doi:10.1108/S2044-994120180000011007

Research Implications – *More studies are needed on developing a stronger theoretical or conceptual foundation on the effects of roadway designs and traffic controls on different dimensions of safety (e.g., exposure, frequency, severity, etc.), types of crashes (e.g., head-on, rear-end, etc.) or road users involved (e.g., drivers, pedestrians, cyclists, etc.).*

Practical Implications – *Transport engineers need to be aware that some treatments may have different effects on different crash types and road users involved. Even though the overall safety may be improved by the treatments designed, they need to consider and mitigate any unintended consequences to satisfy the Pareto improvement principle and the social equity criterion.*

Keywords: Rural intersections; urban intersections; crash-contributing factors

INTRODUCTION

This chapter is the third of five chapters focussing on influences of the transport network on crash risk. While this chapter focusses on rural and urban intersections, it complements the current knowledge regarding the crash risks related to urban and suburban arterials (Chapter 5), controlled access facilities (Chapter 6), roundabouts (Chapter 8) and real-time traffic operations and safety (Chapter 9).

Intersections are recognised as being among the most hazardous locations on the roads. For example, about 43% of all crashes in the United States occur at or near an intersection (Lord, Schalkwyk, Staplin, & Chrysler, 2005) and about 40% of all casualty crashes in Norway occur at junctions (Elvik & Vaa, 2004). Likewise, annual road crash statistics show that more than one-third of crashes (34.31%) occur at intersections in Singapore (Tay & Rifaat, 2007) and about 25% of all road users are killed at intersections in Canada (Transport Canada, 2004). In Australia, intersection crashes comprised 54.2% of total crashes and 20.8% of fatal crashes in the State of New South Wales in 2010 (RTA, 2010) and 48% of police-reported crashes (fatal and injury) in Victoria between 2006 and 2010 (Tay, 2013).

The crash statistics from around the world clearly indicate that intersections are hazardous locations. Hence, to improve road safety significantly we need to prevent and reduce crashes at these locations. Increasing the safety performance of intersections will require us to have a better understanding of the factors contributing to crashes at these locations, especially the roadway design and traffic control variables, in order to develop more targeted countermeasures.

INTERSECTION CRASH ANALYSIS

Intersections are the locations within the road traffic network where two or more traffic flows converge and therefore, by their very nature, provide points of potential conflict. The safety of intersections is dependent upon a number of factors, including:

- *Road users*: knowledge of road rules, familiarity with the road environment, awareness of the environment, walking ability, driving/riding skills and experience, personality, mood, risk perceptions, perceived behavioural control, attitude, intentions, behaviour, age, gender, health, mental and physical conditions, etc.
- *Vehicle and traffic conditions*: traffic volume, traffic speed, vehicle mix, vehicle movement at the time of the crash (types of crash), vehicle's age, size, safety features, occupant protection, aggressiveness of vehicle, etc.
- *Road design and conditions*: roadside furniture, road width, number of lanes, turning lanes, traffic control devices, road alignment, road surface condition, etc.
- *Local environment*: neighbourhood land use (e.g., residential, commercial, industrial and the presence of schools, colleges, emergency services, drinking establishments, etc.), neighbourhood social and demographic characteristics, neighbourhood street patterns, safety corridors, traffic calming, enforcement activities, time of day, weather, etc.
- *Macro-environment*: mobility needs and choices, including public transport and travel distances, especially in rural areas, as well as economic, political, legislative and social environment that determines the road safety culture and society's attitude towards risks (e.g., what is an acceptable level?), risky behaviours, behavioural controls, road and vehicle safety standards, enforcement, etc.

Note that the above crash factors are fairly generic and can be used to analyse most types of crashes. Also, these factors are expected to have an effect on both the likelihood and severity of intersection crashes. In traffic safety analyses the factors contributing to the occurrence of crashes are often separately identified from the factors contributing to the severity of crashes. Moreover, some of the above factors are significantly different in a rural environment compared to an urban environment, and hence it is worthwhile analysing them separately. Although the factors contributing to crashes at signalised and unsignalised intersections are also expected to be different, separate synthesis for this delineation is beyond the scope of this book chapter.

INTERSECTION CRASH LIKELIHOOD OR FREQUENCY

One of the most fundamental analyses that need to be conducted in order to understand the safety issues related to intersections is the estimation of safety performance functions which relate various traffic, roadway, environmental, user, vehicle and crash characteristics to the likelihood or frequency of crashes. The research in this area has been very extensive and diverse, resulting in some mixed results, depending on the types of crashes analysed and the data and methodology used.

The most consistent finding is the positive relationship between exposure (traffic volume) and crash frequency (Dong, Clarke, Yan, Khattak, & Huang, 2014; Dong, Richards, Clarke, Zhou, & Ma, 2014; Wang & Abdel-Aty, 2006, 2008a). As expected, the vast majority of the studies have found that an increase in the amount of vehicular, pedestrian or cyclist traffic will increase the number of traffic collisions. In addition, several studies have found that the percentage of commercial or heavy vehicles is also positively correlated with crash occurrences (Abdel-Aty & Haleem, 2011; Dong, Clarke, et al., 2014; Dong, Richards, et al., 2014).

With regard to roadway characteristics, Table 1 shows that the most consistent finding in the literature is the positive correlation between speed limit and crash frequency. Interestingly, an increase in the number of lanes or lane widths has been associated with an increase in vehicle–vehicle crashes but a decrease in cyclist- or motorcyclist-involved crashes, which may be due to the increased space to accommodate the vulnerable road users, as well as the decrease in exposure by some cyclists due to increased perceived risks associated with cycling on wide roads, which usually also have heavier traffic and higher speed. Hence, it is important to consider the safety of vulnerable road users on multi-lanes and wider roads by separating and protecting them, such as including and separating cycling lanes.

In terms of land use and social influences, the presence of commercial activities around an intersection (Harnen et al., 2003, 2004; Schneider et al., 2010)

Table 1. Roadway Characteristics and Intersection Crash Frequency.

Characteristic	Correlation	Source
Speed limit	+	Abdel-Aty and Haleem (2011), Dong, Richards, et al. (2014), Guo, Wang, and Abdel-Aty (2010), Harnen, Umar, Wong, and Hashim (2003, 2004), Wang and Abdel-Aty (2006, 2008b)
Number of Lane/ lane width	±	Dong, Richards, et al. (2014), Guo et al. (2010), Harnen et al. (2003, 2004), Schepers, Kroeze, Sweers, and Wüst (2011), Wang and Abdel-Aty (2006, 2008b)
Turning lanes	±	Abdel-Aty and Haleem (2011), Dong, Clarke, et al. (2014); Dong, Richards, et al. (2012), Haleem, Abdel-Aty, and Mackie (2010), Schepers et al. (2011), Wang and Abdel-Aty (2006, 2008b)
Median	±	Abdel-Aty and Haleem (2011), Schepers et al. (2011), Schneider et al. (2010), Wang and Abdel-Aty (2006, 2008b)
Road markings	±	Abdel-Aty and Haleem (2011), Schepers et al. (2011)
Road shoulders	±	Haleem et al. (2010), Harnen et al. (2003, 2004)

and the proportion of the population under 18 years of age (Schneider et al., 2010) are positively correlated with crash occurrences.

Urban Intersection Crash Likelihood and Frequency

As expected, increases in pedestrian or vehicular traffic volume (El-Basyouny & Sayed, 2009; M. Kim, Kim, Oh, & Jun, 2012; Lyon & Persaud, 2002; Persaud, Lord, & Palmisano, 2002; Sayed & Rodriguez, 1999) and traffic share of commercial vehicles or motorbikes (Femandes, Miranda-Moreno, & Morency, 2012; Wong, Sze, & Li, 2007; Xu, Wong, & Choi, 2014) are associated with higher crash frequency. One-way traffic is associated with lower crash frequency (Xie, Wang, Chen, & Huang, 2013) due to lower likelihood of potential conflicts, as well as lower mental load for motorists, cyclists and pedestrians. Hence, separating the traffic stream physically may be a way to reduce crashes.

Table 2 shows the findings on the relationship between roadway characteristics and urban intersection crash frequency. While skewed intersections experience lower crash frequency, it is not recommended that engineers introduce skewness in road design. However, installing stop and yield signs, lighting and all red signal phases, as well as reducing visual noise, are relatively cheap measures that can improve safety. Additionally, installing cyclist lanes,

Table 2. Roadway Characteristics and Urban Intersection Crash Frequency.

Characteristic	Correlation	Source
Three-legged intersection	–	Miranda-Moreno, Morency, and El-Geniedy (2011), Xie et al. (2013)
Skew/obtuse/angle	–	Kumara, Chin, and Weerakoon (2003)
Cyclist lane	–	Strauss, Miranda-Moreno, and Morency (2014)
Stop and yield signs	–	Castro, Paletic, and Bhat (2012)
Cross-walk	–	Strauss et al. (2014)
Light rail transit (LRT) stop	–	Pei, Wong, and Sze (2011)
Bus bay	–	Chin and Quddus (2003)
Half-red and all-red phase	–	Femandes et al. (2012)
Countdown timer	–	Anjana and Anjaneyulu (2015)
Lighting	–	Lord, Guikema, and Geedipally (2008)
Curvature	+	Pei et al. (2011), Kumara et al. (2003), Lord et al. (2008), Wong et al. (2007), Xu et al. (2014)
Tram stop	+	Pei et al. (2011), Wong et al. (2007)
Bus stop	+	Chin and Quddus (2003), Miranda-Moreno et al. (2011)
Pedestrian crossings	+	Xu et al. (2014)
Rail crossing	+	Abdel-Aty and Haleem (2011)
Turning/total lane ratio	+	Xie et al. (2013)
Road shoulder	+	Lord et al. (2008)
Flashing light control	+	Bhat, Born, Sidharthan, and Bhat (2014), Castro et al. (2012)
Visual noise	+	Wang and Nihan (2004)
Number of approaches	±	Castro et al. (2012), Chin and Quddus (2003), Pei et al. (2011), Pulugurtha and Samhara (2011), Xu et al. (2014)
Elevated road	±	Wang and Nihan (2004), Xie et al. (2013)
Speed or speed limit	±	Haque, Chin, and Huang (2010), Lord et al. (2008), Mitra and Washington (2012), Mitra, Washington, and van Schalkwyk (2007), Wang and Nihan (2004), Xie et al. (2013)
Number of lanes/land width	±	Femandes et al. (2012), Haque et al. (2010), Pei et al. (2011), Wang and Nihan (2004), Wong et al. (2007), Xu et al. (2014)
Side walk	±	Femandes et al. (2012), M. Kim et al. (2012)
Median	±	Castro et al. (2012), Chin and Quddus (2003), Wang and Nihan (2004)
Turning lanes/pockets	±	Anjana and Anjaneyulu (2015), Chin and Quddus (2003), Femandes et al. (2012), Haque et al. (2010), Kumara and Chin (2005), Lord et al. (2008), Mitra et al. (2007), Mitra and Washington (2012), Pei et al. (2011), Wang and Nihan (2004)

Table 2. (*Continued*)

Characteristic	Correlation	Source
Gradient	±	Bhat et al. (2014), Kumara and Chin (2005), Kumara et al. (2003)
Sight distances	±	Chin and Quddus (2003), Kumara and Chin (2005), Kumara et al. (2003)
Signal cycle time	±	Anjana and Anjaneyu (2015), Pei et al. (2011), Xu et al. (2014)
Number of phases	±	Chin and Quddus (2003), Kumara et al. (20034), Mitra et al. (2007), Mitra and Washington (2012), Pei et al. (2011), Wang and Nihan (2004), Xie et al. (2013)
Safety camera	±	Chin and Quddus (2003), Haque et al. (2010), Kumara and Chin (2005)

cross-walks and bus bays may also be effective at moderate costs. Other site-specific treatments may include on-site road safety messages at tram stops, bus stops, pedestrian crossings and railway crossings.

In terms of the local and macro-environments, commercial land use (Femandes et al., 2012; Miranda-Moreno et al., 2011; Pei et al., 2011; Pulugurtha & Samhara, 2011), increases in the share of a younger population (Mitra et al., 2007; Mitra & Washington, 2012) and the presence of colleges and pubs (Mitra et al., 2007; Mitra & Washington, 2012) are associated with higher crash frequency, whereas the presence of schools (Miranda-Moreno et al., 2011) and commercial centres (Pulugurtha & Samhara, 2011), increases in rainfall, higher proportion of the middle-aged population and increases in number of rainy days are associated with lower crash frequency (Mitra et al., 2007; Mitra & Washington, 2012).

Rural Intersection Crash Likelihood and Frequency

As expected, increases in traffic volume of both major and minor roads (Lord & Park, 2008; Oh, Washington, & Lee, 2009; Ye, Pendyala, Washington, Konduri, & Oh, 2009) and the percentage of turning traffic (Oh, Washington, & Choi, 2004) are associated with higher crash frequency. Whereas the adverse impact of heavy-vehicle traffic in urban intersection crashes is found consistently across several studies, the influence of truck traffic at rural intersections is mixed. For example, Oh et al. (2004) found that the percentage of truck traffic was associated with lower crash frequency but Oh, Lyon, Washington, Persaud, and Bared (2003) found that the percentage of truck traffic was associated with higher crash frequency using the data from the same three American states but from slightly different years. Part of the differences in the results between urban and rural intersections may be attributed to the differences in the built environment, which may create greater sight distance issues. The heavier traffic volume and shorter following distances on the urban roads may also worsen the problem, in addition to greater risks involved in trying to pass a heavy vehicle on urban roads.

Table 3 shows the findings on the relationship between roadway characteristics and rural intersection crash frequency. Unlike urban intersections, both skewed intersections and intersections on a curve have mixed effects on crash frequency. One possible reason for the differences may be that, in a built-up environment, intersections on a curve may create greater challenges for drivers due to poorer sight distance and heavier traffic, which increases the likelihood of a collision more substantially than on rural roads.

Some potential treatments for rural intersections include reducing the speed limit at hazardous intersections, as well as reducing the speed limit difference

Table 3. Roadway Characteristics and Rural Intersection Crash Frequency.

Characteristic	Correlation	Source
Speed limit	+	Oh et al. (2003, 2004, 2009), Vogt and Bared (1998)
Speed limit difference	+	Oh et al. (2009), Ye et al. (2009)
Number of lanes	+	Oh et al. (2009)
Turn lane	±	D. Kim and Washington (2006), Mitra and Washington (2007), Oh et al. (2003, 2009), Ye et al. (2009)
Driveways	+	D. Kim and Washington (2006), D. Kim, Washington, and Oh (2006), Mitra and Washington (2007), Oh et al. (2003, 2004)
Hazard rating	+	Oh et al. (2003), Vogt and Bared (1998)
Crosswalks	−	Oh et al. (2009)
Median	−	Mitra and Washington (2007), Oh et al. (2004)
Lighting	−	D. Kim et al. (2006), D. Kim and Washington (2007), Mitra and Washington (2009), Oh et al. (2004, 2009), Ye et al. (2009)
Bus stops	−	Oh et al. (2009)
Skewed intersections	±	Burchett and Maze (2006), Oh et al. (2003, 2009)
Curvature	+	D. Kim, Lee, Washington, and Choi (2007), Oh et al. (2003, 2004), Vogt and Bared (1998)
Grade	±	Burchett and Maze (2006), D. Kim et al. (2007), Oh et al. (2003, 2004, 2009), Vogt and Bared (1998), Ye et al. (2009)
Signal	±	D. Kim et al. (2007)
Shoulder	±	D. Kim et al. (2007)

between the major and minor roads. Other possible treatments include better delineation of driveways, installing signage to warn drivers of their presence, or building a service road. Additionally, the installation of crosswalks, medians and lighting may be effective in reducing crashes at rural intersections.

In terms of the local and macro-environments, clear weather is associated with a higher frequency of angle crashes and sideswipe crashes in the same direction but a lower frequency of rear-end and sideswipe crashes in opposite directions (D. Kim et al., 2007). Also, a wet road surface is associated with a higher frequency of sideswipe crashes in the same direction but a lower frequency of angle crashes, rear-end crashes and sideswipe crashes in opposite directions (D. Kim et al., 2007). Daylight is associated with a higher frequency of angle crashes and rear-end crashes but a lower frequency of sideswipe crashes (D. Kim et al., 2007). In addition, commercial and residential land uses are positively associated with crashes (Burchett & Maze, 2006).

INTERSECTION CRASH SEVERITY

Besides crash frequency, the other very important dimension of traffic safety is the analysis of the severity of a crash or the injury level of each person involved in a traffic crash. Note that the severity of a crash is usually determined by the road user with the most severe injury. From an economic or health perspective, it is essential to delineate the factors contributing to different severity outcomes because of the substantial differences in social cost and trauma.

One of the most consistent findings in the literature is the positive correlation between speed (e.g., speed limit, traffic speed, impact speed, etc.) and severity outcomes; the higher the speed, the more severe the outcome (Chen, Cao, & Logan, 2012; J. Kim, Ulfarsson, Shankar, & Kim, 2008; Lee & Abdel-Aty, 2005;Moore, Schneider, Savolainen, & Farzneh, 2011). This result is expected because the amount of force and energy involved will increase with the speed of the objects involved in the collisions. In addition, the severity of the outcomes will depend on the types of crashes experienced, with head-on and angle collisions expected to be more severe than rear-end and sideswipe crashes (Abdel-Aty, 2003; Abdel-Aty & Keller, 2005; Moore et al., 2011; Wang & Abdel-Aty, 2008a).

The mass of the objects involved in the collision is the other major factor that determines the amount of force or energy generated (J. Kim et al., 2008; Lee & Abdel-Aty, 2005; Moore et al., 2011). However, since the mass of a vehicle is related to its size and occupation protection, its net effect will be equivocal, although its effect on non-occupants (e.g., people in other vehicles and vulnerable road users including pedestrians and cyclists) will clearly be detrimental. As expected, collisions involving vulnerable road users (see Chapters 10 and 11) will be more severe (Abdel-Aty & Keller, 2005; Chen et al., 2012) due to the lower level of protection.

With respect to road-user characteristics, the demographic risk factors include female (Chen et al., 2012; Lee & Abdel-Aty, 2005; Moore et al., 2011) and ageing drivers (Chen et al., 2012; Moore et al., 2011; Wang & Abdel-Aty, 2008a). This latter result is expected because of the frailty of aging bodies. Also, as expected, risky behaviours, such as not wearing seat-belts or helmets (Chen et al., 2012; Moore et al., 2011; Wang & Abdel-Aty, 2008a), not being insured (Moore et al., 2011), speeding (J. Kim et al., 2008) and alcohol/drug impairment (J. Kim et al., 2008; Lee & Abdel-Aty, 2005; Moore et al., 2011; Wang & Abdel-Aty, 2008a) are also associated with increased injury risks.

Table 4 shows the findings on the relationship between roadway characteristics and intersection crash severity. In general, there are fewer studies on crash severity and fewer mixed results on the effects of roadway characteristics on crash severity compared to crash frequency. To reduce crash severity at urban intersections, transportation engineers and planners may want to consider installing traffic control devices and turning lanes at hazardous intersections and reducing gradient and curvature whenever feasible.

With regard to the local and macro-environment, compared to dark with unlit streets, daylight, dusk/dawn and dark with lit streets are associated with higher crash severity (Haleem & Abdel-Aty, 2010), whereas afternoon

Table 4. Roadway Characteristics and Intersection Crash Severity.

Characteristic	Correlation	Source
Traffic control devices	–	Chen et al. (2012), J. Kim et al. (2008), Lee and Abdel-Aty (2005)
Turning lanes	–	Haleem and Abdel-Aty (2010), Wang and Abdel-Aty (2008a)
Number of lanes	–	Haleem and Abdel-Aty (2010)
Alignment	+	J. Kim et al. (2008), Moore et al. (2011)
Divided roadways	+	J. Kim et al. (2008)
Freeway, interstate and state roads	+	J. Kim et al. (2008)
Dry road surface	+	Moore et al. (2011)
Distance between intersections	+	Haleem and Abdel-Aty (2010)
Road shoulder	+	Haleem and Abdel-Aty (2010)

peak hours and night-time are associated with lower crash severity (J. Kim et al., 2008). Inclement or adverse weather has mixed effects on crash severity (Abdel-Aty, 2003; J. Kim et al., 2008; Lee & Abdel-Aty, 2005). Finally, crashes at rural intersections tend to be more severe than crashes at urban intersections (Abdel-Aty, 2003; Lee & Abdel-Aty, 2005).

Urban Intersection Crash Severity

Drivers who are asleep or fatigued (Obeng, 2007, 2011), alcohol/drug impaired (Chiou, Hwang, Chang, & Fu, 2013; Obeng, 2007, 2011), ill or have medical conditions (Obeng, 2007, 2011) or are younger and older (Chiou et al., 2013; Huang, Chin, & Haque, 2008), as well as an increase in the number of occupants in the vehicle (Obeng, 2011) are associated with higher crash severity, whereas drivers who use a seatbelt (Obeng, 2011) are associated with lower crash severity. In terms of gender, Obeng (2007) and Tay and Rifaat (2007) found male drivers to be associated with higher crash severity but Chiou et al. (2013) found them to be associated with lower severity.

As expected, rear-end and sideswipe crashes are associated with lower crash severity, whereas head-on and run-off-road crashes are associated with higher severity (Obeng, 2007, 2011; Tay & Rifaat, 2007). Compared to vehicles moving straight ahead, vehicles that are turning or overtaking another vehicle experience higher crash severity, whereas vehicles that are stopping or changing lanes experience lower crash severity (Tay & Rifaat, 2007).

In terms of vehicle and traffic conditions, trucks, pick-ups, sport utility vehicles, vans and two wheelers are associated with higher severity (Chiou et al., 2013; Obeng, 2007, 2011; Tay & Rifaat, 2007). Obeng (2007) also found that an increase in the number of vehicles involved and the number of occupants in the vehicles are associated with higher crash severity but Tay and Rifaat (2007) found that single and multiple vehicle crashes experience higher severity than two-vehicle crashes.

Most studies do not investigate the effect of traffic volume in crash severity, because the crash severity outcome can be assumed to be conditioned on the event of a crash and can be assumed to be largely dependent on the energy involved in a crash, the protection of the road users involved and their physical conditions. However, traffic volume may have an indirect effect on the speed of the vehicle under some circumstances. For example, an increase in traffic volume in a congested urban environment may reduce the traffic speed, thereby reducing the speed of the colliding vehicles and the energy generated. Pei et al. (2011) found a negative relationship between traffic volume and crash severity but a positive relationship for commercial and pedestrian traffic. Also, Huang et al. (2008) found that crashes occurring during peak hours tend to be less severe. On the other hand, Obeng (2007) found a positive relationship between volume and crash severity.

Table 5 shows the findings on the relationship between roadway characteristics and urban intersection crash severity. As expected, crashes at intersections with higher speed limits tend to be more severe. Interestingly, while three-legged intersections are associated with lower crash frequency at urban intersections they are associated with higher crash severity. One possible explanation may be the higher prevalence of three-legged intersections on local roads (loops and lollipop street patterns) in urban areas which have lower traffic, and thus lower crash frequency. However, three-legged intersections may have a higher proportion of turning traffic and thus more side-impact crashes, which tend to be more severe.

To reduce the severity of crashes transportation engineers and planners may want to reduce the number of signal stages and amber time, install count-down timers, lighting and merging lanes at intersection hotspots. They may also consider installing site-specific safety message signs or displays, especially near tram or LRT stops where more pedestrian traffic is expected. Additional safety measures that can be considered include

Table 5. Roadway Characteristics and Urban Intersection Crash Severity.

Characteristic	Correlation	Source
Speed limit	+	Tay and Rifaat (2007)
Three-legged, multi-legged	+	Chiou et al. (2013), Huang et al. (2008), Tay and Rifaat (2007)
Tram and LRT stops	+	Pei et al. (2011)
Safety cameras	+	Huang et al. (2008), Tay and Rifaat (2007)
Curvature	+	Pei et al. (2011)
Signal stages and time	+	Pei et al. (2011)
Amber time	+	Obeng (2007)
Count-down timers	−	Anjana and Anjaneyulu (2015)
Turn pockets and lanes	−	Anjana and Anjaneyulu (2015), Pei et al. (2011)
Merging lanes	−	Tay and Rifaat (2007)
Sharp turn	−	Tay and Rifaat (2007)
Lighting	−	Huang et al. (2008)
Lane width	−	Pei et al. (2011)
Number of approaches	−	Pei et al. (2011)
Wet road surface	±	Obeng (2007), Tay and Rifaat (2007)

lowering the speed limit around intersections located at curves and near public transit stops.

In terms of local and macro-environments, crashes occurring at night (Chiou et al., 2013; Huang et al., 2008; Tay & Rifaat, 2007) and in residential locations (Obeng, 2007) experience higher crash severity, whereas crashes occurring in public housing estates, school zones or near train stations, shopping centres and shop houses experience lower crash severity (Tay & Rifaat, 2007).

Rural Intersection Crash Severity

As expected, older drivers and drivers who are fatigued or impaired by alcohol (Barua, Azad, & Tay, 2010) and an increase in the number of occupants in the vehicles (Oh, 2006) are associated with higher crash severity, whereas male and young drivers are associated with lower crash severity (Barua et al., 2010).

With regard to vehicle and traffic conditions, the involvement of tracker-trailers, motorcycles or vans (Barua et al., 2010) and increases in the number of vehicles involved in the crash (Barua et al., 2010) are associated with higher crash severity. However, an increase in traffic volume may have mixed effects on crash severity (Barua et al., 2010; Oh, 2006).

In terms of road design and condition, four-legged intersections and offset intersections (Barua et al., 2010), grade and road curvature (Barua et al., 2010), and increases in the number of commercial driveways (Oh, 2006) or speed limits (Oh, 2006) are associated with higher crash severity, whereas skewed intersections, signalised intersections, crest and wet roads and snow-covered roads are associated with lower crash severity (Barua et al., 2010).

Finding cost-effective measures to reduce the severity of crashes at rural intersections have been and will continue to be challenging due to the higher speed and lower traffic volume involved. Efforts targeted at reducing speed and minimising side-impact crashes should be considered, including installing roundabouts and traffic signals, and reducing grade and curvature (see Chapter 8 for a detailed discussion on roundabouts). Installing site-specific safety message signs or displays should also be considered, as well as tactile warnings (e.g., rumble strips) to alert drivers to the intersection ahead.

With respect to local and macro-environments, summer and fall seasons, night-time and off-peak hours (Barua et al., 2010) are associated with higher crash severities.

DISCUSSION

Many statistical studies have been conducted in the last two decades to identify the factors contributing to crashes at intersections. Although these factors are diverse and complex, they can be categorised into five categories: road user, vehicle and traffic, road design and conditions, local environment and macro-environment. While many of these factors (e.g., speed or speed limit) have the same effect on crash frequency and severity, some factors (e.g., three-legged intersections) have a different effect that may result in possible trade-offs between the two dimensions of safety. The final outcome will depend on how crashes with different severity levels are valued or aggregated (Rifaat, Tay, & de Barros, 2010; Tay, 2002).

Also, some roadway factors may have different safety effects on different types of crashes. For example, D. Kim et al. (2007) found that signalised control is associated with higher frequency of rear-end crashes and sideswipe crashes in the same direction but lower frequency of crashes in angle crashes and sideswipe crashes in the opposite direction. Although rear-end and sideswipe crashes tend to be less severe, they also tend to be more frequent. The net safety effect will depend on how the different types of crashes are valued or aggregated (Rifaat et al., 2010; Tay, 2002).

This review also finds that some contributing factors (e.g., speed limit) are generic to all intersections, while others are more specific to urban or rural intersections. For example, the presence of a bus stop near an urban intersection is associated with an increase in crash frequency (Chin & Quddus, 2003; Mirando-Moreno et al., 2011) but its presence near a rural intersection is associated with a reduction in crash frequency (Oh et al., 2009). However, this comparison is based on cursory observations of the crash factors identified in previous studies which are mostly focussed either exclusively on urban or rural intersections. To date, there are very few studies in the peer reviewed literature that specifically compare crashes at urban and rural intersections (Tay, 2015).

Another interesting observation of the review is that most of the studies assume that the effect of road design on safety is linear, or at least monotonic, which may not be true. For example, Kumara and Chin (2005) found that compared to sight distances between 100 m and 300 m, sight distances of less than 100 m and over 300 m both have a positive correlation with crash frequency. Nonlinearity may therefore contribute in part to the mixed results obtained on some of the roadway design variables if different sites have different base levels. More research needs to be conducted to investigate this potential source of misspecification error.

One clear conclusion of the review is that there have been significant advances in the statistical methodologies applied to analyse crash data. In fact, the main focus of many of the studies is on developing more advanced statistical methods. Although methodological improvements to address estimation issues will advance the skill and knowledge in the field, it also makes it difficult for this review to compare the results obtained across different studies on the effects of roadway characteristics on safety, because part of the differences in the results found may simply be due to the differences in methodologies used.

On the other hand, there have been very few comprehensive conceptual developments (how and why) in modelling the effects of roadway designs and traffic controls on different dimensions of safety (e.g., exposure, frequency, severity, etc.), types of crashes (e.g., head-on, rear-end, etc.) or road users involved (e.g., drivers, pedestrians, cyclists, etc.). Without a better understanding of the different effects some treatments have on crashes at different locations, as well as different crash types and road users involved, it will be more difficult to mitigate any unintended consequences and satisfy the duty of care to all road users, the Pareto improvement principle or the social equity criterion (improving some or all road users without disadvantaging or harming anyone), even though overall safety may be improved by the treatments. Hence, future studies should include more development of the conceptual framework and hypotheses.

This review also found few studies that examine macro-environmental factors, such as political, legislative, economic and social environmental influences, that may determine the safety culture and society's attitudes towards risks, risky behaviour, design standards and behavioural controls. The differences in some findings between urban and rural intersections suggest that the effects of different macro-environmental factors may be very significant across different study locations, and some of these factors may interact with roadway design and traffic control variables, resulting in the mixed results obtained. Hence, future studies should investigate some of the macro-environmental factors using a broader sample, from different jurisdictions, or test the transferability of some of the models and findings.

REFERENCES

Abdel-Aty, M. (2003). Analysis of driver injury severity levels at multiple locations using ordered probit models. *Journal of Safety Research*, *34*, 597–603.

Abdel-Aty, M., & Haleem, K. (2011). Analyzing angle crashes at unsignalized intersections using machine learning techniques. *Accident Analysis and Prevention*, *43*, 461–470.

Abdel-Aty, M., & Keller, J. (2005). Exploring the overall and specific crash severity levels at signalized intersections. *Accident Analysis and Prevention*, *37*, 417–425.

Anjana, S., & Anjaneyulu, M. (2015). Safety analysis of urban signalized intersections under mixed traffic. *Journal of Safety Research*, *52*, 9–14.

Barua, U., Azad, A., & Tay, R. (2010). Fatality risks of intersection crashes in rural undivided highways of Alberta, Canada. *Transportation Research Record*, *2148*, 107–115.

Bhat, C., Born, K., Sidharthan, R., & Bhat, P. (2014). A count data model with endogenous covariates: Formulation and application to roadway crash frequency at intersections. *Analytic Methods in Accident Research*, *1*, 53–71.

Burchett, G., & Maze, T. (2006). Rural expressway intersections characteristics as factors in reducing safety performance. *Transportation Research Record*, *1953*, 71–80.

Castro, M., Paletic, R., & Bhat, C. (2012). A latent variable representation of count data models to accommodate spatial and temporal dependence: An application to predicting crash frequency at intersections. *Transportation Research Part B*, *46*, 253–272.

Chen, H., Cao, L., & Logan, D. (2012). Analysis of risk factors affecting the severity of intersection crashes by logistic regression. *Traffic Injury Prevention*, *13*(3), 300–307.

Chin, H., & Quddus, M. (2003). Applying the random effect negative binomial model to examine traffic accident occurrence at signalized intersections. *Accident Analysis and Prevention*, *35*, 253–259.

Chiou, Y., Hwang, C., Chang, C., & Fu, C. (2013). Modeling two-vehicle crash severity by a bivariate generalized ordered probit model. *Accident Analysis and Prevention*, *51*, 175–184.

Dong, C., Clarke, D., Yan, X., Khattak, A., & Huang, B. (2014). Multivariate random parameters zero inflated negative binomial regression model: An application to estimate crash frequencies at intersections. *Accident Analysis and Prevention*, *70*, 320–329.

Dong, C., Richards, S., Clarke, D., Zhou, X., & Ma, Z. (2014). Examining signalized intersection crash frequency using multivariate zero-inflated Poisson regression. *Safety Science*, *70*, 63–69.

El-Basyouny, K., & Sayed, T. (2009). Collision prediction models using multivariate Poisson-lognormal regression. *Accident Analysis and Prevention*, *41*, 820–828.

Elvik, R., & Vaa, T. (2004). *The handbook of road safety measures*. Amsterdam: Elsevier Science.

Femandes, D., Miranda-Moreno, L., & Morency, P. (2012). Vehicle-pedestrian accidents at signalized intersections: Exposure measures and geometric designs. *92nd annual meeting of the Transportation Research Board*. Washington, DC.

Guo, F., Wang, X., & Abdel-Aty, M. (2010). Modeling signalized intersection safety with corridor-level spatial correlations. *Accident Analysis and Prevention*, *42*, 84–92.

Haleem, K., & Abdel-Aty, M. (2010). Examining traffic crash injury severity at unsignalized intersections. *Journal of Safety Research*, *41*, 347–357.

Haleem, K., Abdel-Aty, M., & Mackie, K. (2010). Using a reliability process to reduce uncertainty in predicting crashes at unsignalized intersections. *Accidents Analysis and Prevention*, *42*, 654–666.

Haque, M., Chin, H., & Huang, H. (2010). Applying Bayesian hierarchical models to examine motorcycle crashes at signalized intersections. *Accident Analysis and Prevention*, *42*, 203–212.

Harnen, S., Umar, R., Wong, S., & Hashim, W. (2003). Motorcycle crash prediction model for non-signalized intersections. *IATSS Research*, *27*(2), 58–65.

Harnen, S., Umar, R., Wong, S., & Hashim, W. (2004). Development of prediction models for motorcycle crashes at signalized intersections on urban roads in Malaysia. *Journal of Transportation and Statistics*, *7*(2), 1–12.

Huang, H., Chin, H., & Haque, M. (2008). Severity of driver injury and vehicle damage in traffic crashes at intersections: A Bayesian hierarchical analysis. *Accident Analysis and Prevention*, *40*, 45–54.

Kim, D., Lee, Y., Washington, S., & Choi, K. (2007). Modeling crash outcome probabilities at rural intersections: Application of hierarchical binomial logistic models. *Accident Analysis and Prevention, 39,* 125–134.

Kim, D., & Washington, S. (2006). The significance of endogeneity problems in crash models: An examination of left-turn lanes in intersection crash models. *Accident Analysis and Prevention, 38,* 1094–1100.

Kim, D., Washington, S., & Oh, J. (2006). Modeling crash types: New insights into the effects of covariates on crashes at rural intersections. *ASCE Journal of Transportation Engineering, 132,* 282–292.

Kim, J., Ulfarsson, G., Shankar, V., & Kim, S. (2008). Age and pedestrian injury severity in motor-vehicle crashes: A heteroskedastic logit analysis. *Accident Analysis and Prevention, 40,* 1695–1702.

Kim, M., Kim, E., Oh, J., & Jun, J. (2012). Critical factors associated with bicycle accidents at 4-legged signalized urban intersections in South Korea. *KSCE Journal of Civil Engineering, 16*(4), 627–632.

Kumara, S., & Chin, H. (2005). Application of Poisson underreporting model to examine crash frequencies at signalized three-legged intersections. *Transportation Research Record, 1908,* 46–50.

Kumara, S., Chin, H., & Weerakoon, W. (2003). Prediction of hazardousness of intersection approaches. *Transportation Research Record, 1840,* 116–122.

Lee, C., & Abdel-Aty, M. (2005). Comprehensive analysis of vehicle–pedestrian crashes at intersections in Florida. *Accident Analysis and Prevention, 37,* 775–786.

Lord, D., Guikema, S., & Geedipally, S. (2008). Application of the Conway–Maxwell–Poisson generalized linear model for analysing vehicle crashes. *Accident Analysis and Prevention, 40,* 1123–1134.

Lord, D., & Park, P. (2008). Investigating the effects of the fixed and varying dispersion parameters of Poisson-gamma models on empirical Bayes estimates. *Accident Analysis and Prevention, 40,* 1441–1457.

Lord, D., Schalkwyk, I., Staplin, L., & Chrysler, S. (2005). *Reducing older driver injuries at intersection using more accommodating design practices.* College Station, TX: Texas Transportation Institute.

Lyon, C., & Persaud, B. (2002). Pedestrian collision prediction models for urban intersections. *Transportation Research Record, 1818,* 102–107.

Miranda-Moreno, L., Morency, P., & El-Geneidy, A. (2011). The link between built environment, pedestrian activity and pedestrian–vehicle collision occurrence at signalised intersections. *Accident Analysis and Prevention, 43,* 1624–1634.

Mitra, S., & Washington, S. (2007). On the nature of over-dispersion in motor vehicle crash prediction models. *Accident Analysis and Prevention, 39,* 439–408.

Mitra, S., & Washington, S. (2012). On the significance of omitted variables in intersection crash modelling. *Accident Analysis and Prevention, 49,* 439–448.

Mitra, S., Washington, S., & van Schalkwyk, I. (2007). Important omitted spatial variables in safety models: Understanding contributing crash causes at intersections. *86th annual meeting of the Transportation Research Board,* Washington, DC.

Moore, D., Schneider, W., Savolainen, P., & Farzneh, M. (2011). Mixed logit analysis of bicyclist injury severity resulting from motor vehicle crashes at intersection and non-intersection locations. *Accident Analysis and Prevention, 43,* 621–630.

Obeng, K. (2007). Some determinants of possible injuries in crashes at signalized intersection. *Journal of Safety Research, 38,* 103–112.

Obeng, K. (2011). Gender differences in injury severity risks in crashes at signalized intersections. *Accident Analysis and Prevention, 43,* 1521–1531.

Oh, J. (2006). Development of severity models for vehicle accident injuries for signalized intersections in rural areas. *KSCE Journal of Civil Engineering, 10*(3), 219–225.

Oh, J., Lyon, C., Washington, S., Persaud, B., & Bared, J. (2003). Validation of FHWA crash model for rural intersections. *Transportation Research Record, 1840*, 41–49.

Oh, J., Washington, S., & Choi, K. (2004). Development of accident prediction models for rural highway intersections. *Transportation Research Record, 1897*, 18–27.

Oh, J., Washington, S., & Lee, D. (2009). Expected safety performance of rural signalized intersections in South Korea. *Transportation Research Record, 2114*, 72–82.

Pei, X., Wong, S., & Sze, N. (2011). A joint-probability approach to crash prediction models. *Accident Analysis and Prevention, 43,* 1160–1166.

Persaud, B., Lord, D., & Palmisano, J. (2002). Calibration and transferability of accident prediction models urban intersections. *Transportation Research Record, 1784*, 57–64.

Pulugurtha, S., & Sambhara, V. (2011). Pedestrian crash estimation models for signalised intersections. *Accident Analysis and Prevention, 43*, 439–446.

Rifaat, S., Tay, R., & de Barros, A. (2010). Effect of street pattern on road safety: Are policy recommendations sensitive to different aggregations of crashes by severity? *Transportation Research Record, 2147*, 58–65.

RTA. (2010). *Road traffic crashes in New South Wales: Statistical statement for the year ended 31 December 2010.* NSW: Road Transport Authority.

Sayed, T., & Rodriguez, F. (1999). Accident prediction models for urban unsignalized intersections in British Columbia. *Transportation Research Record, 1665*, 93–99.

Schepers, J. P., Kroeze, P. A., Sweers, W., & Wüst, J. C. (2011). Road factors and bicycle–motor vehicle crashes at unsignalized priority intersections. *Accident Analysis and Prevention, 43*, 853–861.

Schneider, R., Diogenes, M., Arnold, L., Attaset, V., Griswold, J., & Ragland, D. (2010). Association between roadway intersection characteristics and pedestrian crash risk in Alameda County, California. *Annual meeting of the Transportation Research Board,* Washington, DC.

Strauss, J., Miranda-Moreno, L., & Morency, P. (2014). Multimodel injury risk analysis of road users at signalized and non-signalized intersections. *Accident Analysis and Prevention, 71*, 201–209

Tay, R. (2002). Prioritising road safety countermeasures: Fatalities vs. social costs. *Applied Health Economics and Health Policy, 1*(3), 129–134.

Tay, R. (2013). *Strengthening the effectiveness of intersection safety cameras: Parts 1–3.* Report submitted to the NRMA-ACT Road Safety Trust.

Tay, R. (2015). A random parameters binary probit model of urban and rural intersection crashes. *Accident Analysis and Prevention, 84,* 38–40.

Tay, R., & Rifaat, S. (2007). Factors contributing to the severity of crashes at intersections. *Journal of Advanced Transportation, 41*(3), 245–265.

Transport Canada. (2004). *Vision 2010.* Ottawa: Transport Canada.

Vogt, A., & Bared, J. (1998). Accident models for two-lane rural segments and intersections. *Transportation Research Record, 1635*, 18–29.

Wang, X., & Abdel-Aty, M. (2006). Temporal and spatial analyses of rear-end crashes at signalized intersections. *Accident Analysis and Prevention, 38*, 1137–1150.

Wang, X., & Abdel-Aty, M. (2008a). Analysis of left-turn crash injury severity by conflicting pattern using partial proportional odds models. *Accident Analysis and Prevention, 40*, 1674–1682.

Wang, X., & Abdel-Aty, M. (2008b). Modeling left-turn crash occurrence at signalized intersections by conflicting patterns. *Accident Analysis and Prevention, 40*, 76–88.
Wang, Y., & Nihan, N. (2004). Estimating the risk of collisions between bicycles and motor vehicles at signalized intersections. *Accident Analysis and Prevention, 36*, 313–321.
Wong, S., Sze, N., & Li, Y. (2007). Contributory factors to crashes at signalized intersections in Hong Kong. *Accident Analysis and Prevention, 39*, 1107–1113.
Xie, K., Wang, X., Chen, X., & Huang, H. (2013). Corridor level signalized intersection safety analysis in Shanghai, China using Bayesian hierarchical regression models. *Accident Analysis and Prevention, 50*, 25–33.
Xu, X., Wong, S., & Choi, K. (2014). A two-stage bivariate logistic-tobit model for the safety analysis of signalized intersections. *Analytic Methods in Accident Research, 3*, 1–10.
Ye, X., Pendyala, R., Washington, S., Konduri, K., & Oh, J. (2009). A simultaneous equations model of crash frequency by collision type for rural intersections. *Safety Science, 47*, 443–452.

CHAPTER 8

ROUNDABOUTS

Alfonso Montella

ABSTRACT

Purpose – *Measures aimed at reducing intersection crashes have high potential to be cost effective since intersections constitute only a small part of the overall highway system but intersection-related crashes constitute more than 50% of all crashes in urban areas and over 30% in rural areas. Roundabouts are a proven safety countermeasure, but several issues that significantly affect both crash frequency and severity have been observed at both existing and new roundabouts. This chapter aims to provide guidance on roundabout selection and design criteria.*

Methodology – *The chapter first describes the most relevant criteria to be considered for choosing a roundabout. Then, after the explanation of the roundabout design process and a clear description of the roundabout classification, the chapter provides recommendations for all the steps of the geometric design, highlighting the main design features that contribute to the best safety performances, including speed control and sight distance checks. Finally, the chapter explains traffic control devices and facilities for pedestrians and cyclists.*

Findings – *Roundabout design needs to balance opposing demands and it is important to adopt a performance-based design approach within an iterative process. The most important performance check is the analysis of*

Safe Mobility: Challenges, Methodology and Solutions
Transport and Sustainability, Volume 11, 147–174

ISSN: 2044-9941/doi:10.1108/S2044-994120180000011009

vehicle speeds through the roundabout, since achieving appropriate vehicular speeds has a very positive safety effect.

Keywords: Roundabouts; geometric design; speed control; sight distance; safety

INTRODUCTION

This chapter is the fourth of five chapters focussing on influences of the transport network on crash risk. While this chapter focusses on roundabouts, it complements the current knowledge regarding the crash risks related to urban and suburban arterials (Chapter 5), controlled access facilities (Chapter 6), rural and urban intersections (Chapter 7) and real-time traffic operations and safety (Chapter 9).

Roundabouts are compact, circular intersections in which traffic flows counter-clockwise around a centre island,[1] entering traffic yields to vehicles in the circulating roadway, and approaches are channelised to deflect traffic into a proper entry path. Because of their several safety, operational and environmental advantages roundabouts are hastily spreading around the world (Montella, Turner, Chiaradonna, & Aldridge, 2013; Rodegerdts et al., 2007, 2010).

The use of roundabouts improves intersection safety by eliminating or altering conflict types, reducing crash severity and causing drivers to reduce speeds (Highways Agency, 2007a, 2007b; SETRA, 1998). Indeed, a roundabout can be thought of as operating as a set of right-in, right-out driveways assembled in a ring. The motorist entering the roundabout only conflicts with vehicles coming from the immediate left and crossing conflicts are eliminated at all. In a stop-controlled, four-leg intersection, there are 16 crossing conflicts, 8 diverging conflicts and 8 converging conflicts, while in a four-leg roundabout there are only 4 diverging conflicts and 4 converging conflicts. According to the Highway Safety Manual, conversion of priority intersections into roundabouts (all settings, one or two lanes) produces a statistically significant reduction of 82% and 44%, respectively, for injury crashes and for total crashes, and conversion of signalised intersections into roundabouts produces a statistically significant reduction of 78% for injury crashes and 48% for total crashes (AASHTO, 2010).

Roundabouts increase capacity and improve operation. Roundabouts provide operational benefits, especially when traffic volumes on major and minor roads are comparable and when left-turn volumes from the major roads are relevant. When operating within their capacity, roundabouts typically have lower overall delays than signalised and stop-controlled intersections. The delay reduction is generally most significant during non-peak hours. These performance benefits can often result in reduced lane requirements between intersections. While signalised or stop-controlled intersections can require adding left-turn and/or right-turn lanes, a roundabout may enable maintaining a narrower cross-section in advance of the intersection, although it typically has greater right-of-way needs at the intersection quadrants. Because roundabouts can facilitate U-turns they can be used to reduce or eliminate left-turn movements at driveways between major intersections, thus reducing crashes and delays.

Roundabouts often reduce the number and duration of stops compared with signalised or stop-controlled alternatives. During periods of high traffic flows, vehicles continue to advance slowly in moving queues rather than coming to a complete stop, significantly reducing noise, air quality impacts and fuel consumption because of the lower number of acceleration/deceleration cycles and the shorter time spent idling. The central island and the splitter islands offer the opportunity for more green spaces, thus contributing to reduce the heat island and absorb air pollution caused by vehicular traffic. Further, roundabouts may provide appealing access or centrepieces to communities, provided that they are appropriate for the speed environment in which the roundabout is located. Roundabouts can provide a transition area between high-speed rural and low-speed urban environments and can be used to demarcate commercial areas from residential areas, and can act as the gateway to any kind of neighbourhood, including anchoring the ends of a corridor.

Despite the several advantages and the good safety performance, specific features of roundabouts are very important in determining the way they operate and inadequate design may produce undesirable outcomes (Johnson, 2015; Montella, 2007, 2011; TNZ, 2000). Therefore, it is of paramount importance to understand the roundabout warrant criteria, that is, to understand that the roundabout is not always the best solution and to understand the proper design principles. Actually, in many countries roundabout design standards and/or guidelines do not exist or have only been developed in the last few years.

WARRANT CRITERIA

The most relevant criteria to be considered for choosing a roundabout are: the function of the intersection roads, the capacity needed in relation to the traffic volumes and road users' safety (Dutch Ministry of Transport, Public Works and Water management, 2009; Isebrands & Hallmark, 2008). Generally, roundabouts are a proper design option on highways of several functional classes, and warrant criteria refer to relatively high traffic on the minor road, relatively high left-turn volume from the major road or safety issues. Most standards recommend the use of roundabouts based on the combination of traffic flows on the major and the minor road, highlighting that roundabouts work most efficiently when flows are reasonably balanced between the legs and have greater capacity than priority intersections (Highways Agency, 1995; SETRA, 1998; Spanish Ministry of Development, 1999).

Roundabouts may be useful to: (1) facilitate a change in road category or road type, for example, from dual-carriageway to a single-carriageway road; (2) provide a transition area between high-speed rural and low-speed urban environments; (3) demarcate land use changes; (4) act as a gateway to any kind of neighbourhood, including anchoring the ends of a corridor; (5) eliminate left-turn movements at driveways between major intersections, creating a comprehensive access management strategy with medians, only right-in and right-out intersections, and U-turns at roundabouts; and (6) create traffic-calming effects by reducing vehicle speeds using geometric design rather than relying solely on traffic control devices.

Other roundabout warrant criteria include: (1) poor crash history at stop-controlled or signalised intersections with a high proportion of right-angle crashes, left-turn crashes, red light or stop sign running crashes; (2) where the intersection geometry is skewed, with more than four legs, or with minimal sight distance; (3) where there are stop delays for minor road movements; (4) where there are signal delays; and (5) where there are a high percentage of vehicles turning left.

DESIGN PROCESS

Roundabout design needs to balance opposing demands, such as improving operational performance, reducing crashes and reducing costs (Montella et al., 2013; Rodegerdts et al., 2010). It is important to adopt a performance-based design approach within an iterative process (Fig. 1).

The first step is the operational analysis, combined with the analysis of the constraints (e.g., space availability and land use restrictions). Information from the operational analysis and the analysis of the constraints is used to determine the roundabout type and the required number of lanes, which

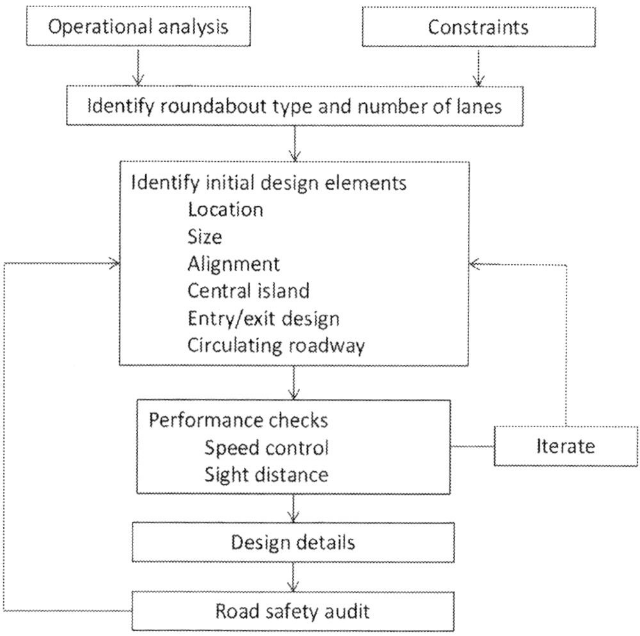

Fig. 1. Recommended Roundabout Design Process.

dictates the required size and many other design details. As a rule of thumb, a roundabout can usually be utilised if the sum of the entering traffic flows is less than 25,000 veh/day for single-lane roundabouts and less than 40,000 veh/day for two-lane roundabouts (Dutch Ministry of Transport, Public Works and Water management, 2009). The sixth edition of the Highway Capacity Manual (HCM; TRB, 2016) presents a procedure to estimate both the capacity of and the level of service of roundabout approaches. The performance measure used to determine the level of service is the control delay, that is, the component of delay that results at each approach because of the yield sign. According to the HCM, the capacity of a single-entry lane conflicted by one circulating lane is based on the conflicting flow and is estimated by the equation $c_{e,pce} = 1{,}380e^{(-1.02\times10^{-3})v_{c,pce}}$, where $c_{e,pce}$ is the lane capacity, adjusted for heavy vehicles (pc/h), and $v_{c,pce}$ is the conflicting flow rate (pc/h). The capacity for two-lane entries conflicted by one circulating lane is estimated as $c_{e,R,pce} = 1{,}420e^{(-0.91\times10^{-3})v_{c,pce}}$, $c_{e,L,pce} = 1{,}420e^{(-0.91\times10^{-3})v_{c,pce}}$, where $c_{e,R,pce}$ and $c_{e,L,pce}$ are, respectively, the right-lane and left-lane capacity.

The capacity of a single-entry lane conflicted by two circulating lanes is estimated as $c_{e,pce} = 1{,}420e^{(-0.85\times10^{-3})v_{c,pce}}$, where all variables are as previously defined ($v_{c,pce}$ is the total of both lanes). The capacity of the right and left lanes of a two-lane roundabout entry conflicted by two circulating lanes is estimated as $c_{e,R,pce} = 1{,}420e^{(-0.85\times10^{-3})v_{c,pce}}$, $c_{e,L,pce} = 1{,}350e^{(-0.92\times10^{-3})v_{c,pce}}$, respectively. At a high level of both entering and conflicting flow, limited priority (in which circulating traffic adjusts its headways to allow entering vehicles to enter), priority reversal (in which entering traffic forces circulating traffic to yield) and other behaviours may occur. In these cases, alternatives tools such as deterministic intersection models and stochastic network models, which are implemented in microsimulation tools, may provide more realistic results. If the operational requirements of single-lane roundabouts with high right-turning traffic volume are not met, it is possible to introduce separate right-turn bypass lanes.

It is then possible to identify the initial design elements: location, size, horizontal and vertical alignment of the approaches, central island, splitter islands, pedestrian crossings and sidewalks, facilities for cyclists, entry width, entry radius, circulatory roadway slope and width, exit width, exit radius and right-turn bypass lanes (if needed).

Most standards and guidelines provide recommendations, which may assist in the roundabout design process, but it is crucial to consider that the interaction of the roundabout geometric elements is more important than their individual impacts. Consequently, performance-based geometric design is the best approach. Performance-based design mainly implies the analysis of vehicle speeds through the roundabout, since achieving appropriate vehicular speeds through the roundabout is the most critical design objective. Furthermore, sight distance has a relevant safety effect and needs to be properly checked before the roundabout design is completed.

In order to guarantee the best safety performance, if design objectives in terms of speed control and sight distance are not satisfied, a change in individual design elements shall be carried out. Thus, the design elements previously discussed are revised and new performance checks are carried out on the revised design. This is an iterative process which can be interrupted only if the performance checks are satisfied. After this step, additional design elements, such as traffic control devices, illumination, landscaping, pavement and drainage facilities, are defined. Finally, yet importantly, at each design stage a road safety audit should be carried out by independent and experienced auditors who set out safety-critical design elements in an audit report (European Parliament and Council of the European Union, 2008). The report should also result in relevant recommendations from a safety point of view.

If relevant changes to the audited design occur it may be appropriate to carry out a new safety audit on the revised design.

ROUNDABOUTS CLASSIFICATION

Generally, roundabouts are classified in three basic categories according to the size and the number of lanes: mini-roundabouts, single-lane roundabouts and multi-lane roundabouts. Furthermore, special types of roundabouts, such as turbo-roundabouts and double-roundabouts, are becoming more popular.

Mini-Roundabouts

Mini-roundabouts are small roundabouts with a fully traversable central island. Mini-roundabouts are a valid design option on local roads in urban areas (Table 1). Undoubtedly, they are very beneficial in low-speed areas where environmental constraints would preclude the use of a larger roundabout. Maximum inscribed circle diameter (ICD, i.e., the diameter of the largest circle that can be fitted into the junction outline) ranges between 24 m (France) and 28 m (UK). The best central island treatment is a fully traversable island, flush for very small mini-roundabouts (ICD $\leq$ 20 m) or slightly domed (h = 40–100 mm) for larger mini-roundabouts. Traffic islands may be provided to separate opposing streams of traffic and, if appropriate, to serve one or more of the following purposes: to provide adequate deflection of the path of vehicles approaching the mini-roundabout, increased conspicuousness to drivers approaching the mini-roundabout, pedestrian use and as a calming feature. Splitter islands are raised, traversable or flush depending on the size of the island and whether trucks will need to track over the top of the splitter island to navigate the intersection.

Single-Lane Roundabouts

This type of roundabout is characterised as having a single-lane entry at all legs and one circulatory lane (Fig. 2). Single-lane roundabouts are allowed in all settings and all types of single-carriageway highways and are considered the most attractive form of roundabouts (Lippold, Lemke, Jaehrig, & Stöckert, 2015). The ICD is not greater than 55 m.

Table 1. Mini-Roundabouts Design Characteristics.

Parameter	France – CERTU	Italy	Spain	Switzerland	UK	US
Highway type	Local	Local	Local	Local	Local	Local
Minimum traffic	–	–	–	–	2-way AADT of all legs $\geq$ 500 v/d	–
Maximum traffic	–	–	–	AADT $\leq$ 15,000 v/d and $V_{ent}+V_{cir} \leq$ 1,200 v/h	–	AADT $\leq$ 15,000 v/d[a]
ICD	$\geq$ 15 m and $\leq$ 24 m	$\geq$ 14 m and $\leq$ 25 m	$\leq$ 28 m	$\geq$ 14 m and $\leq$ 26 m	$\leq$ 28 m	$\geq$ 13 m and $\leq$ 27 m
Central island treatment	Domed $0.10 \leq h \leq$ 0.15 m	ICD < 18 m Fully traversable ICD $\geq$ 18 m Non-traversable + truck apron	Domed $h \leq 0.15$ m $d \leq 4$ m	ICD < 18 m Fully traversable ICD $\geq$ 18 m Non-traversable + truck apron	Flush or slightly domed $h \leq 0.10$ m $d \leq 4$ m	Fully traversable
Splitter islands treatment	Kerbed or flush	–	Kerbed or flush	–	Kerbed or flush	Raised, traversable or flush[b]

[a]Typical daily service volume on a four-leg roundabout which may be expected to operate without requiring a detailed capacity analysis.
[b]Generally discouraged.

A traversable truck apron around the non-traversable part of the central island is typical for most roundabouts to accommodate large vehicles while minimising other roundabout dimensions (Table 2). A truck apron provides an additional paved area to allow the over-tracking of articulated vehicles on the central island without compromising speed control for smaller vehicles. At the same time, the truck apron must be unattractive for use by passenger cars.

Splitter islands should be provided on all single-lane roundabouts. Landscaping and road furniture within splitter islands should not impede visibility of the roundabout or obstruct a driver's sight lines, unless designers decide to reduce visibility with the aim of decreasing approach speeds.

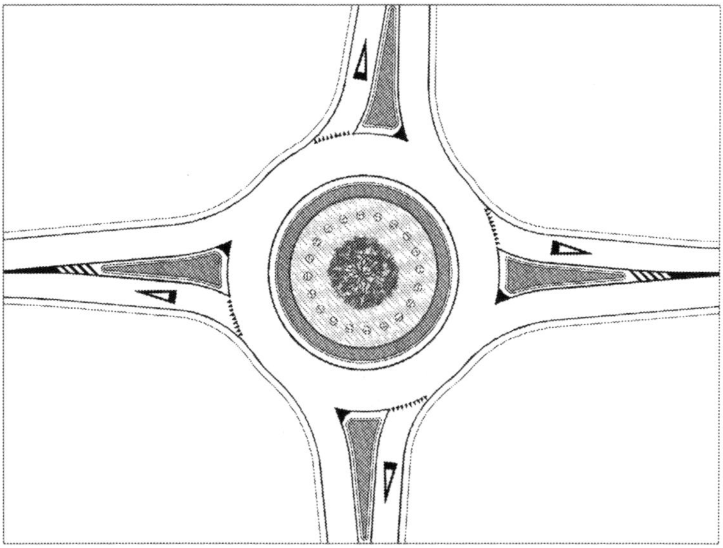

Fig. 2. Example of Single-Lane Roundabout.

Splitter islands should have a reasonably large area and should be long enough to give early warning to drivers that they are approaching an intersection and must slow down. The length of the splitter island may differ depending upon the approach speed. United States (US) guidelines recommend a minimum length equal to 15 m, a desirable length equal to 30 m and a length of 45 m or more on higher speed roadways. In France (SETRA, 1998), the splitter island shape is generated by a construction triangle. The position of the construction triangle is derived from the axis of the leg and the edge of the circulating roadway. The height of the triangle is equal to one-half of the ICD and the base is one-quarter of the height. High vertical curbs should not be used since they may constitute a hazard (Highways Agency, 1993), especially for two-wheeler riders.

The width of the truck apron depends on the swept path of the design vehicle. US guidelines (Rodegerdts et al., 2010) recommend widths between 1.0 and 4.6 m and a cross-slope between 1% and 2% away from the central island. Generally, small widths may be not enough to provide large vehicle tracking and high cross-slopes might facilitate overturning. To discourage use by passenger vehicles, the outer edge of the apron is raised above the circulatory roadway surface. Height generally varies between 40 and 100 mm.

Table 2. Single-Lane Roundabout Design Characteristics.

Parameter	Australia and New Zealand	France – SETRA	Italy	Spain	Switzerland	UK	US
Highway type	All single carriageway	All single carriageway	All single carriageway	All single carriageway	All single carriageway	All single carriageway	All single carriageway
Operating speeds (V_{85})	≤ 80[a] km/h	–	Speed reducing measures	–	–	–	–
Inscribed circle diameter	≥ 26 m and ≤ 54[b] m	≥ 30[c] m and ≤ 50 m	> 25 m and ≤ 50[d] m	> 28 m	> 26 m and ≤ 40 m	> 28 m and ≤ 36 m	≥ 27 m and ≤ 55 m
Central island treatment	Non-traversable + truck apron[e]	Non-traversable + truck apron[f]	Non-traversable	$d \geq 4$ m	Non-traversable + truck apron[e]	Non-traversable + truck apron[e]	Non-traversable + truck apron[e]
Truck apron	$W = f$(vehicle tracking) Cross slope = 2–2.5% $H = 40$–90 mm	$W = 1.50$–2.00 m Cross slope = 4–6% $H = 30$ mm	–	$W = f$(vehicle tracking)	$H = 40$ mm	$W = f$(vehicle tracking)	$W = 1.00$–4.60 m Cross slope = 1–2% $H = 50$–100 mm
Splitter islands treatment	Area ≥ 40[g] m² $L \leq 60$ m	Construction triangle $H = 1/2$ ICD $B = 1/8$ ICD	–	$L_{min} = 20$ m	$W \geq 3$ m	–	$L_{min} = 15$ m $L_{des} = 30$ m $L_{hspeed} \geq 45$ m

[a]If $V_{85} > 80$ km/h consider speed reducing measures.
[b]Greater ICD is allowed if there are more than four legs or if there is a high proportion of heavy vehicles.
[c]On secondary roads with little heavy traffic, ICD between 24 m and 30 m can be considered.
[d]If ICD > 50 m the operational analysis is carried out considering the circulatory roadway as a weaving section.
[e]Truck apron when vehicle tracking indicates this is required.
[f]For ICD ≤ 30 m truck apron is mandatory, for ICD > 30 m truck apron is optional.
[g]At least 10 m² for local streets.

The apron is constructed of a different material than the pavement to differentiate it from the circulatory roadway and to offer contrast against the circulating roadway that is perceptible by day as well as by night (Montella, Turner, Chiaradonna, & Aldridge, 2012).

Multi-Lane Roundabouts

Multi-lane roundabouts have at least one entry with two or more lanes (Table 3) and ICD not smaller than 30 m. Generally, multi-lane roundabouts are a viable design option on several highway types (Montella et al., 2012). The number of lanes can vary from approach to approach. Likewise, the number of lanes within the circulatory roadway may vary depending upon the number of entering and exiting lanes. In general, the number of lanes provided at the roundabout should be the minimum needed for the existing and anticipated demand, as determined by the operational analysis. Irrespective of capacity considerations it is generally important on arterial roads that lane continuity is available through roundabouts; that is, a roundabout serving a two-lane approach should have two entry lanes even if one-lane would have adequate capacity. The number of circulating lanes from any approach must be not smaller than the number of entry lanes on that approach. It is not essential to provide the same number of circulating lanes for the entire length of the circulating carriageway if the appropriate multi-lane exits are provided prior to reducing the number of circulating lanes. In France and in Italy the circulating roadway is a single, wide lane operating without lane-separation markings. Generally, the number of exit lanes should equal the number of circulating lanes prior to the exit.

Design principles are the same for single-lane roundabouts, but the interaction between multiple traffic streams which may enter, circulate through and exit the roundabout side-by-side requires special consideration of some critical issues, such as lane arrangements to allow drivers to select the appropriate lane on entry and navigate through the roundabout without changing lanes, accommodation of side-by-side vehicles through the roundabout and alignment of the legs to prevent exiting–circulating conflicts.

Turbo-Roundabouts

Turbo-roundabouts are multi-lane roundabouts with entries perpendicular to the circulatory roadway, spiral road markings and circulating lanes separated

Table 3. Multi-Lane Roundabout Design Characteristics.

Parameter	Australia and New Zealand	France – SETRA	Italy	Switzerland	UK	US
Highway type	Single and dual carriageways	All single carriageway	Only on single carriageway	–	Single and dual carriageways	Single and dual carriageways
ICD (m)	$\geq$ 34 m and $\leq$ 62[a] m	$\geq$ 30 m and $\leq$ 50 m	> 25 m and $\leq$ 50[b] m	–	> 36 m and $\leq$ 100 m	$\geq$ 46 m and $\leq$ 91 m
Number of entry lanes	Lane continuity Minimum that achieves the desired capacity	2 in the approaches where 1 does not provide the desired capacity	1–2	–	Lane continuity Minimum that achieves the desired capacity	Lane continuity Minimum that achieves the desired capacity
Number of circulating lanes	$\geq$ Number of entry lanes[c]	1 (no lane markings)	1 (no lane markings)	–	$\geq$ Number of entry lanes[c]	$\geq$ Number of entry lanes[c]
Parameter	Australia and New Zealand	France - SETRA	Italy	Switzerland	UK	US
Number of exit lanes	$\leq$ Number of circulating lanes	1 2 if: (a) $V_{exit} \geq$ 1,200 pc/h (b) $V_{exit} \geq$ 900 pc/h and $3 \times V_{circ}$	1	1	$\leq$ Number of circulating lanes	$\leq$ Number of circulating lanes
Central island treatment	Non-traversable + truck apron[d]	Non-traversable + truck apron[d]	Non-traversable	Non-traversable + truck apron[d]	Non-traversable + truck apron[d]	Non-traversable + truck apron[d]
Truck apron	$W = f$(vehicle tracking) Cross slope = 2–2.5% H = 40 – 90 mm	W = 1.50–2.00 m Cross slope = 4–6% H = 30 mm	–	H = 40 mm	$W = f$(vehicle tracking)	W = 1.00 – 4.60 m Cross slope = 1–2% H = 50–100 mm
Splitter islands treatment	Area $\geq$ 40 m^2 $L \leq$ 60 m	Construction triangle H = 1/2 ICD B = 1/8 ICD	–	$W \geq$ 3 m	–	L_{min} = 15 m L_{des} = 30 m $L_{hspeed} \geq$ 45 m

[a]Greater ICD is allowed if there are more than four legs or if there is a high proportion of heavy vehicles.
[b]If ICD > 50 m the operational analysis is carried out considering the circulatory roadway as a weaving section.
[c]Rule for each approach. It is not essential to provide the same number of circulating lanes for the entire length of the circulating carriageway.
[d]Truck apron when vehicle tracking indicates this is required.

by mountable curbs. The peculiar feature is that the addition of a lane is achieved by adjoining the lane on the inner side opposite from an entry. Subtracting a lane can be either performed as a normal lane drop or by a two-lane exit, where the right lane must leave into the exit. Road users must choose the correct lane before entering the roundabout. Within the roundabout itself lane changing is avoided and no weaving or cutting is possible. The roundabout can only be exited via the previously chosen lane. Main benefits are that there are no crashes due to lane changing on the roundabout, drivers entering the roundabout need only give way to traffic in a maximum of two well-demarcated lanes, and driving speed through the roundabout is low because of raised lane dividers. Installation of turbo-roundabouts show satisfactory safety results even though reliable crash modification factors of conversion of multi-lane roundabouts into turbo-roundabouts are not yet available. Turbo-roundabouts are currently used in the Netherlands (Dutch Ministry of Transport, Public Works and Water management, 2009), Germany (Lippold et al., 2015), Spain (Spanish Ministry of Development, 2012) and Slovenia (Tollazzi & Renčelj, 2014). In the US there is an increased use of multi-lane roundabouts with spiral markings which have some similarities with turbo-roundabouts.

Double-Roundabouts

A double-roundabout is a junction comprising two roundabouts separated by a short link (Fig. 3). The main use is: (a) to improve an existing staggered junction where a double-roundabout avoids the need to realign one of the approach roads and can achieve a considerable construction cost-saving compared with a larger, single-island roundabout; (b) for joining two parallel routes separated by a feature, such as a river, a railway line or a motorway; (c) at overloaded single roundabouts where, by reducing the circulating flow past critical entries, a double-roundabout can increase capacity; (d) at junctions with more than four entries, a double-roundabout may achieve better capacity and make more efficient use of space, with better safety characteristics compared with a large roundabout, which may generate high circulatory speeds resulting in a loss of capacity and safety; and (e) at ramp terminals, where it is particularly advantageous if the interchange has a high proportion of left-turn flows from the off-ramps and to the on-ramps. Roundabouts at ramp terminals may also reduce the required width and/or length of bridges, providing a significant cost-benefit. A double-roundabout should be designed as a single system rather than as two individual roundabouts.

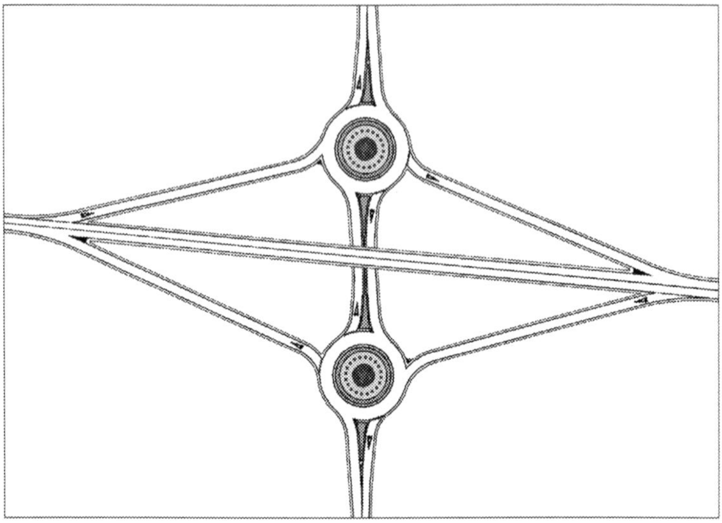

Fig. 3. Example of a Double-Roundabout.

The link joining the two roundabouts is usually short, and there is often an insufficient distance to change lanes. The lane use on the link should be established from the turning volumes and should be checked so that lane balance is produced.

GEOMETRIC DESIGN

Central Island

Generally, the optimal alignment is when all the axes of the legs cross through the centre of the roundabout. An offset of the approach alignment to the left of the roadway centreline allows for increased deflection and reduction of the impact to the right side of the road but may create greater impacts to the left side. An offset to the right is not commonly used because it generates problems in achieving speed control objectives and makes the perception of the central island less conspicuous, with a detrimental safety effect (Montella, 2007; SETRA, 1998), but may be appropriate in some instances, provided that speed requirements and other design considerations are met to minimise impacts to the adjacent properties and/or to the landscape.

The best shape of the central island is circular, as changes in curvature of the circulating carriageway result in differential speeds and increase driver workload. However, elliptical, oblong or other shapes may need to be used to suit unusual and/or constrained site conditions. The central island presents an obstruction to traffic, and therefore it should be recognisable at the required stopping sight distance. Most countries recommend deliberately obstructing forward visibility through the central island. A slightly elevated central island is recommended, possibly with slopes that do not exceed 15%, since recent studies (Daniels, Brijs, Nuyts, & Wets, 2011; Jensen, 2015) show that higher central islands have a positive safety effect. Vegetation and landscaping can be used to achieve this objective, allowing also to: improve the perception of the roundabout from a distance, differentiate the roundabout from the road environment and the surroundings, improve the aesthetic quality, reduce the heat island, absorb the air pollution caused by the vehicular traffic, beautify the road environment and discourage pedestrian traffic through the central island. To ensure stopping sight distance for circulating vehicles a perimeter portion of the central island of width not smaller than 2 m should be landscaped with low-level grass or shrubs (Italian Ministry of Infrastructures and Transports, 2006; Rodegerdts et al., 2010; SETRA, 1998). An obstacle that is likely to cause the sudden stop of an out-of-control vehicle should not be located on the central island.

Entry Width

Entry width, in conjunction with length and sharpness of flare, significantly impacts on capacity and safety. The entry width should be able to accommodate the swept path of the design vehicle. However, it is important that the entry is not any wider than necessary, as too large entry widths can make it difficult to achieve enough speed reduction at the entries to roundabouts, with detrimental safety consequences (Montella et al., 2012). For single-lane entrances, typical entry widths range from 4.0 m to 5.5 m; these are often flared from upstream approach widths. Care should be taken with entry widths greater than 5.5 m, as drivers may wrongly interpret the wide entry to be two lanes when there is only one circulatory lane. Smaller entry widths, in the range from 3.0 m to 4.0 m, are required in mini-roundabouts. In multi-lane roundabouts, the required entry width is dependent upon the number of lanes and the design vehicle. Approach flaring may provide an effective means of increasing capacity without requiring as much right-of-way as a full lane addition. Research from the UK suggests that length of flare affects capacity without a detrimental effect on safety (Rodegerdts et al., 2010). Widths for individual lanes at entry range from 3.0 m to 4.6 m.

Entry Radius

Similarly to the entry width, the entry radius affects both capacity and safety. A proper selection of the entry radius is needed since both too large and too small entry radii may produce undesirable effects. Excessively large entry radii have a higher potential to produce fast entry speeds. On the other hand, too abrupt entry radii may lead to single-vehicle crashes. The outside line of the entry is commonly designed curvilinearly tangential to the outside edge of the circulatory roadway. Likewise, the projection of the inside edge of the entry is commonly curvilinearly tangential to the central island. Some road authorities prefer that this projection passes through a point in the circulating roadway about 1.5 m from the central island to ensure that the vehicle tracks on the pavement rather than mounting the island.

At multi-lane roundabouts, the design of the entry curvature should balance the competing objectives of speed control and adequate alignment of the natural path. Small entry radii may produce low entry speeds but often lead to path overlap on the entry since they tend to lead vehicles of the outside lane into the inside circulatory lane. Greater entry radii reduce the potential for path overlap. US guidelines recommend using a compound curve or tangent along the outside curb. The design consists of an initial small-radius (20–35 m) entry curve set at least 6 m back from the edge of the circulatory roadway. A short section of a large-radius curve (> 45 m) or tangent is then provided between the entry curve and the circulatory roadway to align vehicles into the proper circulatory lane at the entrance line.

French and Swiss guidelines require the smaller entry radii (8–15 m at smaller roundabouts), while guidelines which control vehicle speeds by the limitation of the entry path radius are more flexible allowing greater radii.

Entry Angle

The entry angle is the conflict angle between the entering and the circulating traffic. Usually, too low entry angles produce poor visibility to the left, requiring drivers to strain to look over their shoulders, and may encourage merging behaviour similar to freeway on-ramps. Meanwhile, too high entry angles may not provide enough positive alignment to discourage wrong-way movements, reduce capacity and may produce excessive entry deflection which can lead to sharp braking at entries accompanied by rear-end crashes. Generally, higher entry angles are preferred in low speed environments and are more common in European countries.

Circulatory Roadway Width

The required circulatory roadway width is determined from the number of entering lanes and the turning requirements of the design vehicle. In general, the circulating width should be at least as wide as the maximum entry width and up to 120% of the maximum entry width (Highways Agency, 2007b; Rodegerdts et al., 2010). For single-lane roundabouts, the circulatory roadway width usually remains constant throughout the roundabout. Typical circulatory roadway widths range from 5 m to 7 m. Care should be taken to avoid making the circulatory roadway width too wide within a single-lane roundabout because drivers may think that two vehicles are allowed to circulate side-by-side. The circulatory roadway width should accommodate all but the largest vehicle (in the US only cars and buses). A truck apron will often need to be provided within the central island to accommodate the largest design vehicles. Usually, the left-turn movement is the critical path for determining circulatory roadway width. At mini-roundabouts, larger circulatory widths are needed (typically from 7 m to 8 m). At multi-lane roundabouts, the circulatory roadway width may be variable depending upon the number of lanes and the design vehicle turning requirements. A constant width is not required throughout the entire circulatory roadway, and it is desirable to provide only the minimum width necessary to serve the required lane configurations within that specific portion of the roundabout. A common combination is two entering and exiting lanes along the major roadway but only single entering and exiting lanes on the minor street. Multi-lane circulatory roadway lane widths typically range from 3.5 m to 4.9 m per lane.

Exit Width

Exit lane widths need to be checked for vehicle-swept paths to ensure that the design vehicle is properly accommodated. For single-lane exits typical widths range from 4.0 m to 7.5 m. Smaller exit widths, in the range from 2.5 m to 4.5 m, are required in mini-roundabouts. For double-lane exits typical widths range from 7.0 m to 11.0 m.

Exit Radius

In general, standards and guidelines require the use of a relatively large radius so that it is comfortable for drivers to exit roundabouts. In urban area, this is balanced by the need to maintain slow speeds through the pedestrian crossing on exit. The exit right edge is commonly designed to be curvilinearly tangential to the outside edge of the circulatory roadway. Likewise, the projection of the inside edge of the exit roadway is commonly curvilinearly tangential to the central island (or to an offset of about 1.5 m from the central island).

PERFORMANCE CHECKS

Speed Control

Achieving appropriate vehicular speeds through the roundabout has a very positive safety effect (Montella et al., 2013). A well-designed roundabout reduces the relative speeds between conflicting traffic flows by requiring vehicles to negotiate the roundabout along a curved path. Indeed, several studies show that a geometric design allowing high speeds entering and negotiating a roundabout is associated with angle crashes due to failure to give way to vehicles already on the roundabout and rear-end crashes when vehicles brake suddenly (Arndt & Troutbeck, 2005; Montella, 2007, 2011; Turner, Wood, & Roozenberg, 2006). Main geometric design parameters to control speeds through the roundabout are the radius of deflection, the entry path radius and the deviation angle.

Radius of Deflection

According to French standards, a trajectory's deflection is the radius of the arc that passes at 1.5 m away from the edge of the central island and at 2 m from the edges of the entry and exit lanes (1.0 m from the painted lines). The radius of such an arc should be less than 100 m—the recommended value is 30 m. Generally, the fastest path is the trajectory traced by two opposing arms; in particular circumstances, the fastest path is the right-turn manoeuvre. An analysis of roundabout crash contributory factors in Italy showed: (a) that the most common geometric design contributory factor was an excessive radius of deflection and (b) that the excessive radius of deflection was associated with angle crashes at entry (Montella, 2011).

Entry Path Radius

Recently, there has been a move away from a focus on deflecting to controlling vehicle speeds through geometry of the roundabout entry. This has meant a focus on the entry radius and the maximum central circulating radius. The entry path radius is a measure of the deflection imposed on vehicles entering a roundabout. It is an important determinant of safety at roundabouts because it governs the speed of vehicles through the junction and whether drivers are likely to give way to circulating vehicles. To determine the entry path radius, the fastest path allowed by the geometry is drawn. This is the smoothest, flattest path that a vehicle can take through the entry, round the central island and through the exit (in the absence of other traffic). In the UK (Highways Agency, 2007b), the entry path is assumed to be 2 m wide so that the vehicle following it would maintain a distance of 1 m between its centreline and any kerb or edge marking. The path starts 50 m in advance of the give-way line. The smallest radius of this path on entry that occurs as it bends before joining the circulatory carriageway is called the entry path radius (Fig. 4(a)). The entry path radius should be measured over the smallest best-fit circular curve over a distance of 25 m occurring along the approach entry path in the vicinity of the give-way line, but not more than 50 m in advance of it. The entry path radius must be checked for all turning movements. It must not exceed 70 m at compact roundabouts in urban areas and 100 m at all other roundabout types.

In Australasia (Austroads, 2015), different procedures to construct the entry path are defined for single-lane entries (Fig. 4(b)); two-lane entries, staying in correct lane (Fig. 4(c)); and two-lane entries, cutting across lanes (Fig. 4(d)). For single-lane entries and two-lane entries (staying in correct lane), the maximum value of the entry path radius is 55 m. If the desired driver speed on the leg prior to the roundabout is less than 90 km/h, this value is increased (up to 100 m for $V \leq 40$ km/h). For two-lane entries (cutting across lanes) the maximum radius is 1.5 times (1.9 for $V \leq 40$ km/h) the actual entry path radius when staying in the correct lane. We caution the reader that, differently from the UK and US, in Australasia there is not specific research behind the speed versus geometry parameters.

US guidelines (Rodegerdts et al., 2010) state that the fastest paths must be drawn for all approaches and all movements, including left-turn movements and right-turn movements. Five critical path radii must be checked for each approach (Fig. 4(e)). R1, the entry path radius, is the minimum radius on the fastest through-path prior to the entrance line. R2, the circulating path radius, is the minimum radius on the fastest through-path around the central island. R3, the exit path radius, is the minimum radius on the fastest through-path into the exit. R4, the left-turn path radius, is the minimum radius on the path of the conflicting left-turn movement. R5, the right-turn path radius, is the minimum radius on the fastest path of a right-turning vehicle. The R1–R5 radii represent the vehicle centreline in its path through the roundabout.

A vehicle is assumed to be 2 m wide and maintains a minimum clearance of 0.5 m from a roadway centreline or concrete curb and flush with a painted edge line (Figs. 4(f)–(h)). When drawing the path, a short length of tangent should be drawn between consecutive curves to account for the time it takes for a driver to turn the steering wheel. US guidelines do not directly provide a maximum value of the entry path radius but recommend maximum entry design speed for mini-roundabouts (30 km/h), single-lane roundabouts (40 km/h) and multi-lane roundabouts (40–50 km/h) and provide the radius–speed relationship. As a result, maximum recommended entry path radii are 30 m for mini-roundabouts, 55 m for single-lane roundabouts and 85 m for multi-lane roundabouts. Furthermore, relative speeds between conflicting traffic streams and between consecutive geometric elements should be minimised, such that the maximum speed differential between movements is no more than approximately 15–25 km/h.

Deviation Angle

The Swiss and the Italian standards require a deviation angle imposed by the central islands between two opposite legs greater than 45° (Italian Ministry of Infrastructures and Transports, 2006; VSS, 1999), although this value of the deviation angle is not attainable for mini-roundabouts. The rationale of limiting the deflection angle is that if the vehicle path is not deflected sufficiently from the straight direction of travel by the central island, this will lead to failures to give way, increased pass-through speeds and underestimations of these speeds by conflicting parties. Experimental studies in Switzerland (Spacek, 2004) show a correlation between smaller deviation angles and higher crash rates. Furthermore, an analysis of roundabout crash contributory factors in Italy (Montella, 2011) shows: (a) that the second most common geometric design contributory factor is a small deviation angle and (b) that the small deviation angle is associated with angle crashes at entry.

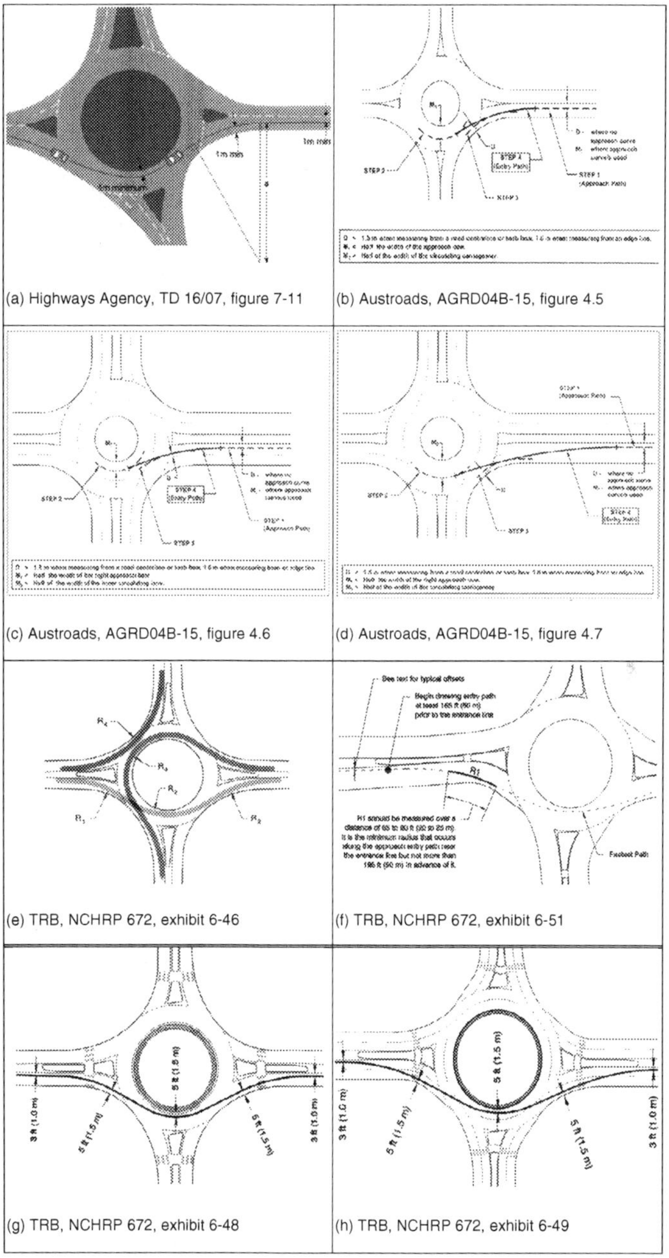

(a) Highways Agency, TD 16/07, figure 7-11

(b) Austroads, AGRD04B-15, figure 4.5

(c) Austroads, AGRD04B-15, figure 4.6

(d) Austroads, AGRD04B-15, figure 4.7

(e) TRB, NCHRP 672, exhibit 6-46

(f) TRB, NCHRP 672, exhibit 6-51

(g) TRB, NCHRP 672, exhibit 6-48

(h) TRB, NCHRP 672, exhibit 6-49

Fig. 4. Procedures to Construct the Fastest Path Radius.

Sight Distance

Inadequate sight distance is a major safety concern. First, the roundabout should be seen well in advance to allow the road user to understand the need to slow down and the geometry of the roundabout must be legible. After perceiving the presence of a roundabout, the user must be able to quickly recognise its various components. According to the French standard (SETRA, 1998), the splitter islands and the central island should become visible at a distance of at least 250 m.

Italian studies show that an incomplete view over the left quadrant of the circulating roadway close to the entrance was a contributory factor in 10% of the investigated roundabout crashes (Montella, 2011). On the other hand, the British, French and US guidelines indicate that it is advantageous to provide no more than the minimum sight distance and that excessive visibility leads to higher vehicle speeds and lower safety for all road users, and advise to add landscaping to restrict sight distance to the minimum requirements. Sight distance at roundabouts refers both to stopping sight distance and to intersection sight distance.

In Australasia (Austroads, 2015), three sight distance criteria must be applied. Criterion 1 refers to approach sight distance: the alignment on the approach should be such that the driver has a good view of the splitter island, the central island and desirably the circulating carriageway. Criterion 2 relates to a car driver entering a roundabout having adequate sight distance to two potentially conflicting movements within the roundabout, namely: (1) a vehicle entering from the approach immediately to the right and (2) a vehicle travelling on the circulating roadway. Criterion 3 refers to a driver approaching a roundabout being able to see other entering vehicles in time to stop before the yield line and avoid a vehicle driving through the roundabout. Criteria 1 and 2 are both mandatory requirements, whereas criterion 3 is optional.

In the UK (Highways Agency, 2007b), five sight distances must be provided: (1) stopping sight distance on the approach, (2) forward visibility at entry, (3) visibility to the right[2], (4) circulatory visibility and (5) pedestrian visibility. On the approach, stopping sight distance must be provided. At entry, drivers must be able to see forward and to the right, from the centre of the nearside lane at a distance of 15 m from the give-way line, the full width of the circulatory carriageway for a visibility distance depending on the ICD. Visibility to the right must be provided also from the give-way line. The visibility distance must be provided also on the circulatory roadway. Drivers approaching a roundabout with a zebra crossing across the entry must be able to see the full width of the crossing so that they can see whether there are pedestrians wishing to cross. At the give-way line, drivers must be able to see the full width of a pedestrian crossing across the next exit if it is within 20 m of the give-way line on that leg.

In the US (Rodegerdts et al., 2010), stopping sight distance should be checked at three critical types of locations: (1) approach, (2) circulatory roadway and (3) crosswalk on exit. At the entry, sight distance to perceive and react to the presence of conflicting vehicles must be provided. The length of the approach leg of the sight triangle is limited to 15 m. If the approach leg of the sight triangle is greater than 15 m it may be advisable to add landscaping to restrict sight distance to the minimum requirements.

In France (SETRA, 1998) and in Italy (Ministry of Infrastructures and Transports, 2006), a complete view over the left quadrant of the circulating roadway at a distance of 15 m from the entrance is required. In Denmark (Jensen, 2015), a complete view over the left quadrant at a distance of 3 m from the entrance is considered adequate.

FACILITIES FOR PEDESTRIANS AND CYCLISTS

This section discusses specific treatments for pedestrians and cyclists at roundabouts and complements more in-depth discussion about pedestrian and cyclist risk provided in Chapters 10 and 11, respectively.

Pedestrians

Due to the reduction of vehicle speeds, roundabouts improve pedestrian crossing opportunities and safety. The pedestrian crossings must be accessible with detectable warnings and appropriate slopes in accordance with requirements for people with disabilities. A typical and minimum crosswalk setback of 6 m from the yield line is recommended. This is the length of one vehicle without any additional distance to account for the gap between vehicles, since ideally the crosswalk is placed within this gap. As far as pedestrian crosswalk alignment options, it is generally recommended to place each leg of the crosswalk perpendicular to the outside curb of the entry and exit lanes. This option creates the shortest crossing distance and makes it easier to build accessible ramps to the sidewalk, since the crossing is perpendicular to the curb.

A key feature is the splitter island refuge area which allows pedestrians to cross the road in two stages, thus providing the ability for pedestrians to focus on one traffic stream at a time while crossing. To adequately provide shelter for pedestrians, the minimum raised splitter island width at the crosswalk should be 1.2 m (preferably 2.5 m) according to the UK standard (Highways Agency, 2007b), 1.8 m according to the US guidelines (Rodegerdts et al., 2010) and 2.4 m according to the Australasian guidelines (Austroads, 2015). The walkway through the splitter island should be cut-through to help wheelchair users and should be aligned with the crosswalks to provide guidance for all pedestrians, particularly for those who are blind or who have low vision. One challenge is that pedestrians with visual impairments may not receive the same level of information at a roundabout as at a typical signalised intersection and may require additional treatments such as pedestrian signalisation.

Wherever possible, sidewalks should be separated by the carriageway with a landscape strip. Landscape strips provide increased comfort for pedestrians, room for street furniture, discourage pedestrians from crossing to the central island or cutting across the circulatory roadway of the roundabout and provide guidance to pedestrians with visibility impairments to the designated crosswalks.

Cyclists

At single-lane roundabouts with low-traffic and low-approach speeds, cyclists can navigate the roundabout as vehicles (Austroads, 2015; Highways Agency, 2007b; Rodegerdts et al., 2010).

At larger single-lane or multi-lane roundabouts, where speeds are higher, off-road peripheral cycle tracks are the safest solution (Austroads, 2015; Highways Agency, 2007b; Jensen, 2015; Ministry of Transport, Public Works and Water management, 2009). A bicycle path physically separated from the circulating carriageway can be a shared path of sufficient width and appropriately marked to accommodate both cyclists and pedestrians around the perimeter of the roundabout. Dutch standards recommend a rectangular shape of the cycle path, so that the path approach entry and exit legs are at right angles, to emphasise that cyclists must give way when crossing a leg of the roundabout.

TRAFFIC CONTROL DEVICES

Road users require clear, unambiguous guidance to select the correct travel path and to identify who has priority when the travel paths of various road users intersect. The provision of good signage allows motorists to plan their route in advance of the roundabout, adjust their speed and select the appropriate lane from which to enter the roundabout. Signs at all roundabouts should aid in detecting the presence of the roundabout ahead, deciding on a destination or exit leg and slowing to an appropriate speed. Signs at multilane roundabouts should also aid in deciding on the correct entry lane. Signs that have the greatest safety effect are yield signs at the entrance, map-type signs in advance of the entrance and direction signs at the exit (Montella, 2011). Yield signs mainly reduce entry-circulating crashes due to failure to give way. Map-type signs create faster reading and display the configuration of the upcoming roundabouts. Direction signs on splitter islands are useful at roundabouts to confirm exit leg choices. It is also very important to not install any advertisement signs that create distractions to road users in the roundabout.

Although signs are important, properly designed markings should provide the primary means of guidance. Poor or insufficient markings can confuse users and provide reduced guidance for the driver approaching and travelling through the roundabout. Permanent raised pavement markers along the splitter islands and the central island significantly improve roundabout perception and delineation at night. Especially at existing roundabouts, significant benefits are expected following markings improvement. Indeed, Jensen (2015) found safety improvement associated with the better signs and markings, and Montella (2011) found that missing or very faded yield lines or symbols were a contributory factor in 30% of the investigated roundabout crashes.

CONCLUSIONS

Roundabouts contribute to creating a sustainable road network by improving aesthetics, environment, traffic operation and safety of all road users. However, caution must be exercised in selecting and designing roundabouts. Indeed, roundabouts are not the best design solution in all circumstances and are not recommended in cases of low traffic volumes on minor roads. Specific features of roundabouts are very important in determining the way they operate, particularly with respect to pedestrian and cyclist safety.

Proper geometric design criteria are essential to achieve the best performance of roundabouts in terms of both capacity and safety. It is noteworthy to observe that the interaction of the roundabout geometric elements is more important than their individual impacts. Consequently, performance-based geometric design is the best approach. Indeed, rigid design procedures which do not take into account safety and operational consequences of the design decisions and the need to balance opposing demands might produce undesirable outcomes.

Performance-based design mainly implies the analysis of vehicle speeds through the roundabout, since achieving appropriate vehicular speeds through the roundabout is the most critical design objective and has a very positive safety effect. A well-designed roundabout reduces the relative speeds between conflicting traffic flows by requiring vehicles to negotiate the roundabout along a curved path. Main geometric design parameters to control speeds through the roundabout are the radius of deflection, the entry path radius and the deviation angle. Sight distance has a relevant safety effect and needs to be properly checked before the roundabout design is completed. Furthermore, proper facilities for pedestrians and cyclists, such as splitter island refuge areas and off-road peripheral cycle tracks, may considerably improve roundabout safety.

Finally, it is noteworthy to highlight the need for more research on the relationships between roundabout geometric design criteria, drivers' behaviour and safety to further improve roundabout operation and safety.

NOTES

1. This chapter is written from the perspective of countries with right-hand traffic. For readers in countries with left-hand traffic, read the opposite in instances referring to the direction or side of traffic flow; that is, 'traffic flows clockwise', 'high right-turning traffic', etc.
2. UK has left-hand traffic.

REFERENCES

AASHTO. (2010). *Highway safety manual* (1st ed.). Washington, DC.

Arndt, O. K., & Troutbeck, R. (2005). Relationship between unsignalised intersection geometry and accident rates. *3rd TRB international symposium on highway geometric design*, Chicago, IL.

Austroads. (2015). *Guide to road design, Part 4B Roundabouts*. Report AGRD04B-15. Sydney, Australia: Austroads.

Daniels, S., Brijs, T., Nuyts, E., & Wets, G. (2011). Extended prediction models for crashes at roundabouts. *Safety Science*, *49*, 198–207.

Dutch Ministry of Transport, Public Works and Water Management. (2009). *Roundabouts – Application and design: A practical manual.* The Hague.

European Parliament and Council of the European Union. (2008). Directive 2008/96/EC of the European Parliament and of the Council of 19 November 2008 on road infrastructure safety management. *Official Journal of the European Union, L, 319*, 59.

Highways Agency. (1993). *Geometric design of roundabouts. Design manual of roads and bridges* (Vol. 6, Section 2, Part 3, TD 16/93). London: Highways Agency.

Highways Agency. (1995). *Geometric design of major/minor priority junctions. Design manual of roads and bridges* (Vol. 6, Section 2, Part 6, TD 42/95). London: Highways Agency.

Highways Agency. (2007a). *Design of mini-roundabouts. Design manual of roads and bridges* (Vol. 6, Section 2, Part 2, TD 54/07). London: Highways Agency.

Highways Agency. (2007b). *Geometric design of roundabouts. Design manual of roads and bridges* (Vol. 6, Section 2, Part 3, TD 16/07). London: Highways Agency.

Isebrands, H., & Hallmark, S. (2008). *Planning-level guidelines for modern roundabouts*. Report CTRE Project 06-255.

Italian Ministry of Infrastructures and Transports. (2006). Decree 19 April 2006: Guidelines for the design of road intersections. *Official Journal of Italian Republic, 170*, 24–26.

Jensen, S. U. (2015). Safety effects of height of central islands, sight distances, markings and signage at single-lane roundabouts. *5th TRB international symposium on highway geometric design*, Vancouver, BC, Canada.

Johnson, M. T. (2015). A synthesis of roundabout design optimization for safety and operations. *5th TRB international symposium on highway geometric design*, Vancouver, BC, Canada.

Lippold, C., Lemke, K., Jaehrig, T., & Stöckert, R. (2015). Country report Germany: The new generation of design guidelines for roads and motorways in Germany. *5th TRB international symposium on highway geometric design*, Vancouver, BC, Canada.

Montella, A. (2007). Roundabout in-service safety reviews: Safety assessment procedure. *Transportation Research Record, 2019*, 40–50.

Montella, A. (2011). Identifying crash contributory factors at urban roundabouts and using association rules to explore their relationships to different crash types. *Accident Analysis and Prevention, 43*, 1451–1463.

Montella, A., Turner, S., Chiaradonna, S., & Aldridge, D. (2012). Proposals for improvement of the Italian roundabout geometric design standard. *Procedia – Social and Behavioral Sciences, 53*, 189–202.

Montella, A., Turner, S., Chiaradonna, S., & Aldridge, D. (2013). International overview of roundabout design practices and insights for improvement of the Italian standard. *Canadian Journal of Civil Engineering, 40*, 1215–1226.

Rodegerdts, L., Blogg, M., Wemple, E., Myers, E., Kyte, M., Dixon, M., … Carter, D. (2007). *Roundabouts in the United States*. NCHRP Report 572, Washington, DC.

Rodegerdts, L., Bansen, J., Tiesler, C., Knudsen, J., Myers, E., Johnson, M., … O'Brien, A. (2010). *Roundabouts: An informational guide* (2nd ed.). NCHRP Report 672, Washington, DC.

SETRA. (1998). *The design of interurban intersections on major roads: At-grade intersections.* Bagneux Cedex, France.

Spacek, P. (2004). Basis of the Swiss design standard for roundabout. *Transportation Research Record, 1881*, 27–35.

Spanish Ministry of Development, General Direction of Highways. (1999). *Recommendations for roundabouts*. Madrid, Spain.

Spanish Ministry of Development, General Direction of Highways. (2012). *Guide for highway intersections*. Madrid, Spain.

TNZ. (2000). *The ins and out of roundabouts: A safety auditors' perspective*. Wellington, New Zealand.

Tollazzi, T., & Renčelj, M. (2014). Comparative analyse of the two new alternative types of roundabouts – Turbo and flower roundabout. *Baltic Journal of Road and Bridge Engineering*, *9*(3), 164–170.

TRB. (2016). *The highway capacity manual: A guide for multimodal mobility analysis* (6th ed.). Washington, DC.

Turner, S., Wood, G., & Roozenberg, A. (2006). Accident prediction models for roundabouts. *22nd ARRB conference*, Canberra, Australia.

VSS. (1999). *Swiss standard SN 640 263: Carrefours, carrefours giratoires*. Zurich.

CHAPTER 9

REAL-TIME TRAFFIC SAFETY AND OPERATION

Mohamed Abdel-Aty, Qi Shi, Anurag Pande and Rongjie Yu

ABSTRACT

Purpose – *This chapter provides details of research that attempts to relate traffic operational conditions on uninterrupted flow facilities (e.g., freeways and expressways) with real-time crash likelihood. Unlike incident detection, the purpose of this line of work is to proactively assess crash likelihood and potentially reduce the likelihood through proactive traffic management techniques, including variable speed limit and ramp metering among others.*

Methodology – *The chapter distinguishes between the traditional aggregate crash frequency-based approach to safety evaluation and the approach needed for real-time crash risk estimation. Key references from the literature are summarised in terms of the reported effect of different traffic characteristics that can be derived in near real-time, including average speed, temporal variation in speed, volume and lane-occupancy, on crash occurrence.*

Safe Mobility: Challenges, Methodology and Solutions
Transport and Sustainability, Volume 11, 175–204

ISSN: 2044-9941/doi:10.1108/S2044-994120180000011010

Findings – *Traffic and weather parameters are among the real-time crash-contributing factors. Among the most significant traffic parameters is speed particularly in the form of coefficient of variation of speed.*

Research implications – *In the traffic safety field, traditional data sources are infrastructure-based traffic detection systems. In the future, if automatic traffic detection systems could provide reliable data at the vehicle level, new variables such as headway could be introduced. Transferability of real-time crash prediction models is also of interest. Also, the potential effects of different management strategies to reduce real-time crash risk could be evaluated in a simulation environment.*

Practical implications – *This line of research has been at the forefront of bringing data mining and other machine-learning techniques into the traffic management arena. We expect these analysis techniques to play a more important role in real-time traffic management, not just for safety evaluation but also for congestion pricing and alternate routing.*

Keywords: Real-time traffic data; crash likelihood; ITS; freeway safety

INTRODUCTION

This chapter is the final of five chapters focusing on influences of the transport network on crash risk. While this chapter focuses on real-time traffic operations and safety of the transport network, it complements the current knowledge regarding the crash risks related to urban and suburban arterials (Chapter 5), controlled access facilities (Chapter 6), rural and urban intersections (Chapter 7) and roundabouts (Chapter 8).

Freeways/expressways in urban and rural areas are high-speed roadway facilities that provide motorists with efficient through movement to their destinations (see Chapter 6). These access-controlled facilities are free of intersections and traffic signals that interrupt traffic flow. However, growing demand and high speeds on freeways/expressways still raise safety concerns. According to a recent Traffic Safety Facts (NHTSA, 2012), 16.1% of all fatal crashes occur on freeways/expressways (including Interstates). Not only do the crashes on freeways/expressways result in a tremendous loss of human lives, they also lead to congestion that greatly increases time and fuel wastage. Consequently, traffic operators are urgently in need of effective freeway/expressway safety management.

Recent decades have seen rapid popularisation of intelligent transportation systems (ITS) in the transportation arena. These systems are designed to aid roadway surveillance and control, route advisories and traffic control centres, etc., while the integration of different ITS technologies for advanced traveller information systems and advanced traffic management systems are also important ITS themes (Roess, Prassas, & McShane, 2004). The expected benefits of ITS technologies include safer vehicles and roadways, enhanced mobility, reduced environmental impact, promotion of technology innovation and improved transportation system information sharing (Barbaresso et al., 2014). Currently, for freeways/expressways, many segments are heavily instrumented with ITS capabilities. Electronic toll collection, travel time estimation, active traffic management and traffic incident management are common ITS practices on freeways/expressways. The functioning of these systems, however, all build on precise automatic traffic detection.

Automatic traffic detection systems can continuously monitor the traffic flow at the installed location, or on a specific segment, and archive the information at short time intervals (30 s or 1 min). Traffic volume, speed, density in terms of occupancy and simple vehicle classification are the most common types of parameters collected by these detection systems. As for the quality and usefulness of the collected data, on one hand they depend on the specific detection technology used, while on the other they rely on the deployment density and scale of the systems over the roadway network. With real-time traffic information, traffic operators can effectively manage these systems and make informed decisions. At the same time, the availability of real-time data has also opened a new frontier for traffic safety researchers. Crashes have long been regarded as a sophisticated phenomenon resulting from the combined effects of human behaviour, traffic conditions, weather conditions and roadway geometric characteristics (Huang & Abdel-Aty, 2010). Traditional safety analyses focusing on crash frequency often use average traffic flow parameters to represent the overall traffic conditions on the studied units (roadway segments or areas). However, researchers have argued that crashes are more affected by traffic states near the crash locations immediately prior to crashes occurring (Mensah & Hauer, 1998). For example, a crash occurring at 9:00 am is more likely to be affected by the traffic at 8:55 am instead of the traffic at 6:00 am near the crash site. As a result, real-time traffic data can help restore traffic conditions before crashes occur. By investigating how the traffic flow parameters impact crash occurrence, safety researchers and practitioners have the potential to prevent crashes or reduce their likelihood and lower their severity in real-time. These types of studies are generally referred to as real-time crash prediction or real-time safety evaluation studies, and their purpose is to have a safer traffic system with proactive safety strategies.

DATA

To conduct real-time crash prediction, the analyses need to combine real-time traffic and crash data and then explore the relationship between them. Based on the research objectives and data availability, many studies also incorporate geometric characteristics and real-time weather data. Still, real-time traffic and crash data are the most fundamental types of data used in real-time crash prediction studies.

Real-Time Traffic Data

Real-time traffic data are normally provided via ITS traffic detection systems. According to Martin, Feng, and Wang (2003), state-of-the-art detection technologies fall into three categories: in-roadway detectors, over-roadway detectors, and off-roadway technologies. The in-roadway detectors are also known as intrusive detectors, and over-roadway detectors are commonly known as non-intrusive detectors. Off-roadway detection technologies, such as probe vehicles and remote sensing, are emerging techniques that are becoming more popular in the transportation field. Since only a small fraction of the time period in traffic data prior to crashes are used in real-time crash prediction, the quality of data heavily depends on the reliability and accuracy of the traffic detection system. Existing studies have utilised real-time traffic data from these three sources.

The in-roadway sensor is one that is embedded in the roadway's pavement, embedded in the roadway's subgrade or taped or otherwise attached to the roadway's surface (Mimbela & Klein, 2000). One of the most representative and widely used in-roadway detectors in real-time crash analysis is the inductive loop detector. Loop detectors have been implemented since the early stages of automatic traffic surveillance, thus they are applications of relatively mature technologies. However, they have several drawbacks, such as disruption of traffic for installation and repair, and high failure rates in certain conditions (Martin et al., 2003), such as poor road surface conditions and adverse weather. Due to their long history of implementation most early real-time safety studies adopted loop detectors as their data source.

An over-roadway sensor is one that is mounted above the roadway itself or alongside the roadway, offset from the nearest traffic lane by some distance. Existing over-roadway sensors range from video image processors to more up-to-date microwave radar sensors (Mimbela & Klein, 2000). Compared with the in-roadway sensors, the over-roadway sensors have a significant advantage in that they minimise the disruption of traffic during installation and maintenance.

A probe vehicle is an off-roadway detection technology that is developing fast. Compared with the other two types of sensors mentioned above, probe vehicles require in-vehicle devices in addition to fixed infrastructure. Current probe vehicle technologies include global positioning systems (GPS), Bluetooth, mobile phones, automatic vehicle identification (AVI) and automatic vehicle location (Martin et al., 2003). With sufficient probe vehicles, real-time traffic information can be provided at the individual vehicle level. Nevertheless, since only a portion of vehicles are equipped with in-vehicle devices, some traffic indicators might not be accurate.

As for the traffic flow parameters used for real-time safety evaluation, the variables vary across detection systems. Loop detectors and microwave radar sensors are able to archive traffic volume, speed, occupancy and vehicle classification per lane at the detection locations and return the information every 20–30 s or every minute. The AVI sensors detect individual vehicles equipped with electronic transponders and record the timestamps when they pass the sensor, from which average speed over segments can be calculated. GPS data reflect the location of a vehicle and its corresponding speed and travelling direction at the time of detection. For both AVI and GPS data the complete traffic volume, occupancy and vehicle classification are generally unavailable.

Crash Data

In real-time crash prediction, crash occurrence is often converted to a categorical variable, which is crash and non-crash, or different crash severity levels. To match the real-time traffic data to each crash case, information from two aspects is critical: (1) the crash location and (2) crash date and time.

The accurate recording of locations is necessary to assign detectors near the crash sites since in practice it is unlikely that there will be a traffic sensor installed exactly at the crash location. Crash locations can be expressed in different ways, such as coordinates and mileposts. Once the crash location is determined, traffic information from multiple detectors that are upstream and downstream to the crash location can be used to represent the traffic conditions at the crash site.

Meanwhile, the temporal property of a crash is even more crucial for real-time safety analysis. While it is rare that a crash date is wrongly recorded, the accuracy of the crash time data could have a major impact on real-time crash prediction results. Although mobile phone use has greatly enhanced the accuracy of reported crash times, there is still a possibility that the crash times are not correctly collected, as shown in Fig. 1. Imprecise crash time data may mix the pre-crash traffic flow with post-crash traffic flow, which would lead to biased real-time crash risk evaluation models. To cope with the issue, Abdel-Aty, Uddin, Pande, Abdalla, and Hsia (2004) propose a rule-based shockwave methodology to verify the actual crash time. It is also suggested to visually check the speed profiles before and after the recorded crash occurrence time to ensure its accuracy.

Besides the crash locations and times, other crash-related information could also be of use for real-time crash prediction at a more specific level. Crash types, injury severities and number of vehicles involved have all been explored in the existing literature (Abdel-Aty, Uddin, & Pande, 2005; Christoforou, Cohen, & Karlaftis, 2011; Xu, Tarko, Wang, & Liu, 2013; Yu & Abdel-Aty, 2013b).

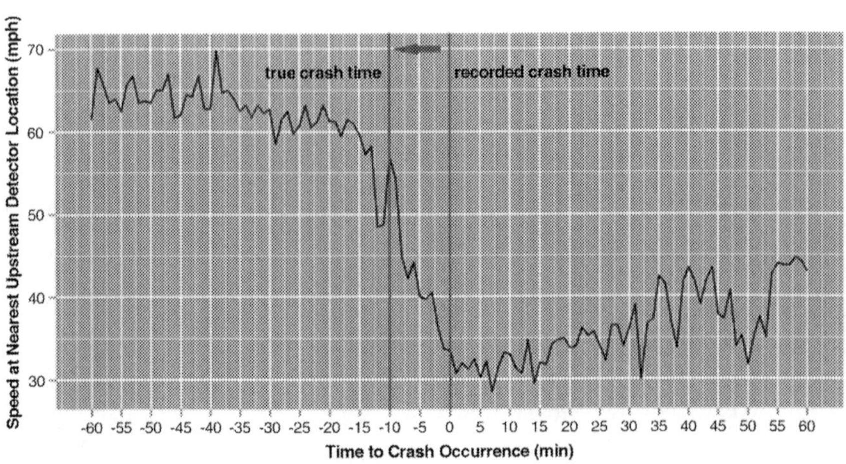

Fig. 1. Speed Profile Before and After Crash Occurrence.

Geometric and Weather Data

Geometric characteristics at crash locations and corresponding weather conditions at the time of crash are also key factors contributing to crashes. The inclusion of geometric information, however, is often affected by the sampling method of non-crash cases. When the matched sampling scheme is applied, geometric factors are controlled for and therefore implicitly included. If a random sampling scheme is applied, geometric elements then become random variables. Their effects on crash occurrence would also be studied in this case. The most widely used geometric characteristics include the number of lanes, shoulder width, median width, pavement conditions, horizontal and vertical curves and proximity to ramps.

Different from the motivation of incorporating geometric variables in real-time crash prediction, the use of weather data often relies on its availability. According to existing studies that have implemented real-time weather data in crash prediction, multiple weather variables have been proven to significantly contribute to crash occurrence. Sources of weather data include road weather information systems, weather stations near the freeway of interest and sometimes airport weather stations. Most of these weather systems provide basic meteorological information, such as temperature, humidity, wind speed, precipitation, and visibility.

Data Processing

When data from different sources are prepared, they need to be integrated into a single set for real-time crash prediction. The process consists of two major steps: data aggregation and combination and selection of non-crash cases.

Data Aggregation and Combination

The real-time traffic data are often archived at 30-s or 1-min time intervals for each travelling lane, which means they contain random noise and are difficult to work with in a modelling framework. Therefore, the most common solution is to aggregate them into 5-min intervals. Researchers have tried to calibrate which 5-min interval prior to crashes performs the best in crash prediction. Traffic data during a 30-min time period before crashes are broken into six time slices, as illustrated in Fig. 2. Variables such as total volume, average speed, average lane occupancy and standard deviation of

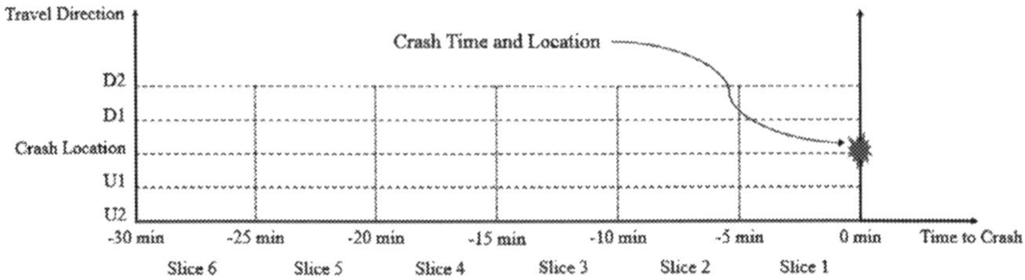

Fig. 2. Time and Location of Data Collection Stations.

these parameters during each time slice are created to represent the traffic operation conditions and its turbulence prior to crash occurrence. In addition to the basic traffic flow variables, several indices have been calculated to represent certain traffic flow characteristics. For example, Oh, Park, and Ritchie (2006) calculated the rear-end crash collision risk index with 5-min level data. Lee, Abdel-Aty, and Hsia (2006) proposed an overall average flow ratio to represent the total number of lane changes in all lanes. Dias, Miska, Kuwahara, and Warita (2009) introduced the congestion index to represent the congestion conditions on the roadway. With the real-time traffic data aggregated at each detector location, usually only the upstream (U1 and U2) and downstream (D1 and D2) stations near the crash sites are selected to represent the traffic conditions.

Matching real-time weather data is different from matching traffic data. Most weather reporting systems do not report weather conditions at fixed time intervals. A new reading will only be created when a specific weather parameter changes. Thus, in most of the real-time crash prediction studies, the last weather reading from the nearest weather sensor/station before the crash occurrence is used as the weather condition for that crash case. Corresponding geometric information to a crash is relatively straightforward; as long as it is determined within which segment the crash occurred, the geometric characteristics can be matched.

Non-Crash Case Selection

The main objective of real-time crash prediction is to identify under what real-time conditions a crash is more likely to occur by comparing the pre-crash traffic conditions with the 'normal' conditions prevailing on the freeway. As a result, an appropriate selection of non-crash cases is needed. Two sampling schemes are extensively deployed: (1) within stratum-matched non-crash data (Abdel-Aty & Pande, 2007) and (2) randomly sampled non-crash data.

For the stratum-matched sampling scheme, traffic data are extracted from the day of the crash and on corresponding (non-crash) days without a crash. 'Corresponding' here means that, for example, if traffic data for 30-min prior to a crash occurring on a specific Monday at 6:00 pm are extracted from the nearest two detectors upstream (U1 and U2) and two detectors downstream (D1 and D2) to represent traffic conditions for the crash case, then traffic data during the same time period from the same detectors on other Mondays without crashes are extracted as the corresponding non-crash cases. Hence, the crash will have detector data consisting of the speed, volume and lane occupancy values for all the lanes from the above four loop stations, with $m+1$ (1 crash and m non-crash cases) observations together forming one stratum. These data collection steps are repeated for N crash cases to form N strata. The idea in this sampling scheme is to control implicitly for parameters other than the traffic conditions (e.g., geometry, driving population, etc.) and use only the traffic conditions measured by the detectors as explicit explanatory variables in the models. The same mechanism can also be applied to other data types. More details of this sampling technique can be found in an earlier study by Abdel-Aty et al. (2004).

Random sampling is another scheme used to select non-crash cases. To generate a random non-crash sample, the study time period is divided into m 1-min intervals (60 min *24 h*n days = m 1-min intervals), which is the number of options available to choose the 'time of non-crash'. In addition, all of the stations on the freeways/expressways serve as 'station of non-crash'. In all, we can choose from p (m 1-min periods * x stations on both travelling directions) options to draw a random combination of time and station to assign as random non-crash cases. These random combinations may be used to extract sets of 30-min data prior to the assigned time of the non-crash from two stations upstream and two stations downstream of the assigned station of non-crash cases, to create a random non-crash sample. Out of these available random non-crash cases a sample of appropriate size may be drawn depending on the sample size requirements of the methodology used for analysis.

In both sampling schemes above, one needs to deal with the issue of how many non-crash cases need to be selected, which is a key step in the data preparation procedure. Abdel-Aty et al. (2004) tested different settings of the number of non-crash cases m, where 1:1, 1:2, 1:3, 1:4 and 1:5 datasets were created and tested. However, no significant differences were observed when changing m. In other studies, different m:1 ratios have been examined; $m=4$

was found to give relatively lower standard errors for the estimated odds ratio (Ahmed, Abdel-Aty, & Yu, 2012b). Kuhn, Egert, Neumann, and Steinbeck (2008) show that negligible power is gained through adding controls beyond 3:1 matching. Therefore, the matched set (stratum) usually comprises m (4)+1 observations in later studies.

MODELLING METHODS

Real-time data are collected for crash and non-crash cases, and pre-crash traffic conditions were compared with normal traffic flow conditions to identify the crash precursors at the individual crash level. In contrast to crash frequency models, real-time crash predictions unveil one important component in traffic safety that aggregated analysis sometimes fails to explore: traffic dynamics – the sudden formation of disrupted traffic conditions (Hossain & Muromachi, 2012) and weather conditions.

For the dependent variable, since the data are split into crash and non-crash cases, a dichotomy variable is utilised (crash as 1 and non-crash as 0). Some studies, meanwhile, have only focused on real-time information for crash cases to identify their effects on different levels of crash severity or crash types. Overall, real-time traffic safety analysis can be regarded as a classification issue. So far, there are generally two categories of modelling methods that are employed in real-time crash risk evaluation studies: the statistical analysis approach and the data mining (machine-learning) approach.

Statistical methods include logistic regression models based on a random sampling scheme (Hourdos, Garg, Michalopoulos, & Davis, 2006; Xu, Tarko, et al., 2013; Xu, Wang, et al., 2013), matched-case control logistic regression (Abdel-Aty et al., 2004, 2005; Zheng, Ahn, & Monsere, 2010), probit regression (Abdel-Aty & Abdalla, 2004; Christoforou et al., 2011; Lee & Abdel-Aty, 2008a; Yu & Abdel-Aty, 2014c), aggregate log linear model (Lee, Hellinga, & Saccomanno, 2003) and Bayesian statistics (Abdel-Aty, Hassan, Ahmed, & Al-ghamdi, 2012; Ahmed, Abdel-Aty, & Yu, 2012a; Oh, Oh, Ritchie, & Chang, 2005; Yu & Abdel-Aty, 2013b). Data mining-based methods include neural networks (Abdel-Aty & Pande, 2005; Pande & Abdel-Aty, 2006a, 2006b), classification trees (Pande & Abdel-Aty, 2006a; Pirdavani et al., 2015; Yu & Abdel-Aty, 2013c), random forests (Ahmed & Abdel-Aty, 2012; Pande, Das, Abdel-Aty, & Hassan, 2011; Shi & Abdel-Aty, 2015), support vector machines (SVMs) (Yu & Abdel-Aty, 2013c), fuzzy c-means

(Oh et al., 2006) and Bayesian belief net (Hossain & Muromachi, 2012). Statistical methods present the effects of variables in an interpretable way through parameter estimates and confidence intervals. However, they assume a linear relationship between the dependent and independent variables. Data mining methods often make predictions with very high accuracy and have promising applications if the purpose of modelling is real-time crash prediction of crash occurrence. However, the drawback is that the predictions are sometimes questioned because the analysis process is seen as a 'black box' and would not help identify crash-contributing factors.

In research with logistic regression based on a random sampling scheme, the effects of geometric characteristics and time of day on crashes have not been controlled for. Therefore, geometric information (number of lanes, road surface width, shoulder width, median width, curvature, ramps and pavement conditions, etc.), weather conditions (rain, fog and visibility, etc.) and time-of-day factors (peak hour, weekend, positions of the Sun, etc.) can all be explicitly incorporated into the analysis. Depending on whether the response variable is binary (mostly crash and non-crash) or multinomial (often seen in severity analysis), the model specification can be expressed as

(1) Binary case:

$$\pi(x) = P(Y=1|\mathbf{x}) = 1 - P(Y=0|\mathbf{x}) \tag{1}$$

$$\log\frac{\pi(x)}{1-\pi(x)} = \text{logit}[\pi(x)] = \beta_0 + \boldsymbol{X\beta} \tag{2}$$

(2) Multinomial case:

$$\pi_j(x) = P(Y=j|\mathbf{x}) \text{ with } \sum\nolimits_j \pi_j(\mathbf{x}) = 1 \tag{3}$$

$$\log\frac{\pi_j(x)}{\pi_1(x)} = \beta_0 + \boldsymbol{X}\beta, \quad j = 2,3,\ldots,J \tag{4}$$

where Y is a categorical response with J categories and $\boldsymbol{X}$ stands for the matrix of explanatory variables. The probability of each category sums to one. β_0 and β are intercept and vector for parameter coefficients. Based on the structure and inter-relationships between the response variables, extensions based on logit models can also be applied in real-time crash prediction studies, such as sequential logistic models (Xu, Tarko, et al., 2013), ordered logit models, nested logit models (Lee & Abdel-Aty, 2008b) and mixed logit models.

Matched case-control logistic models were first introduced in this field by Abdel-Aty et al. (2004). It is by far the most frequently adopted approach in real-time crash prediction analysis. By implementing a matched-case control scheme for non-crash cases, researchers are able to focus on the effects of traffic variables on crash occurrence while controlling for the geometric characteristics (and other factors) that may affect traffic flow features and crash likelihood. Non-crash cases are matched to the crash cases with the same location, day of week and time of day. Using this sampling scheme, the data structure shows N strata equal to the number of crashes with m controls in each stratum. Model specification can be displayed as

$$\log\frac{\pi_j\left(x_{ij}\right)}{1-\pi_j\left(x_{ij}\right)}=\mathrm{logit}\left[\pi_j\left(x_{ij}\right)\right]=\beta_j+\beta_1 x_{1ij}+\beta_2 x_{2ij}+\ldots+\beta_k x_{kij} \tag{5}$$

where $\pi_j\left(x_{ij}\right)$ is the probability that the ith observation in the jth stratum is a crash and $x_{ij}=\left(x_{1ij},x_{2ij},\ldots,x_{kij}\right)$ is a vector of k traffic flow variables $x_1,x_2,\ldots,x_k$, i is the observation indicator in that stratum from 0 to m and j is the stratum number from 0 to N.

Recently, with the rising interest in Bayesian inference, Bayesian models have also been employed in real-time crash risk analysis. Compared with the frequentist inference, which assumes the parameters to be fixed and the data to be random, Bayesian inference considers the data to be fixed and the parameters to be random. One key advantage of the Bayesian inference method is that extra knowledge and experience about the data can be used as prior information in the analyses (Yu & Abdel-Aty, 2013c). Another advantage of using Bayesian models is that they have the capability to deal with sophisticated data structures (Huang & Abdel-Aty, 2010). Bayesian models can incorporate random effects, random parameters and multilevel structures to address the heterogeneity among events. Some possible drawbacks of Bayesian inference are that the model specification is more complicated and the computation takes significantly longer.

Among the various data mining techniques, a neural network may be defined as a massively parallel-distributed processor made up of simple processing units, having a natural propensity for storing experimental knowledge and making it available for use (Christodoulou & Georgiopoulos, 2000). Neural network models generally consist of three entities, namely the model of processing elements, model of interconnections and structures and the learning rules (Pande & Abdel-Aty, 2006a). Multilayer perceptron architecture with feed-forward connections is one of the most widely used neural network architectures. It means that the only connections allowed between nodes are from a layer of a certain index to the next layer with a higher index. The net input to hidden layer neurons is determined through the inner product between the vector of connection weights and the inputs. The activation function is then applied to this net input of hidden neurons. The weights from the hidden to output layers are then used to estimate the output of the network. These weights are the parameters recursively estimated during the supervised learning (i.e., training) process and are used to 'score' unseen observations following calibration. The activation function of hidden neurons is non-linear in nature and is critical in the functioning of the neural network. It allows the network to 'learn' any underlying relationship of interest between inputs and outputs (Pande & Abdel-Aty, 2006a).

A classification tree represents the segmentation of data created by applying a series of simple rules. Each rule assigns an observation to a group based on the value of an input. One rule is applied after another, resulting in a hierarchy of groups within groups. The hierarchy is called a tree and each group is called a node. The final or terminal nodes are called leaves. For each leaf, a decision is made and applied to all observations in that leaf (Breiman, Friedman, Stone, & Olshen, 1984). Classification trees can be used to automatically rank the input variables based on the strength of their contribution to the tree (i.e., variable importance measure or VIM). This ranking may act as the basis for variable selection for subsequent modelling procedures in traffic safety studies.

Nevertheless a single classification tree suffers from high variance or bias. Consequently, the random forest, as an ensemble classifier using many decision tree models to vote for the most popular class, is introduced to generate unbiased estimates of the classification error as trees are added to the forest (Breiman, 2001). Also, the strong law of large numbers guarantees the random forest is robust against over-fitting. One of the basic practices of the random forest approach is to estimate variables' importance. The random forest algorithm estimates the importance of a variable by looking at how much prediction error increases (or decreases) in accuracy when out-of-bag data for that single variable are permuted (Liaw & Wiener, 2002). Another measure is the total decrease in node impurities denoted by the Gini coefficient from splitting on the variable, averaged over all trees (Breiman, 2006). The former measure has the drawback that it overestimates the variable importance of highly correlated variables (Strobl, Boulesteix, Kneib, Augustin, & Zeileis, 2008). The latter, on the other hand, does not perform fairly with predictors of many categories (Strobl, Boulesteix, Zeileis, & Hothorn, 2007).

The SVM is a recently developed pattern classifier based on statistical learning theory and structural risk minimisation (Vladimir & Vapnik, 1995). The algorithm tries to find a separating hyperplane by minimising the distance of misclassified points to the decision boundary (Yu & Abdel-Aty, 2013c). In many cases the data cannot be linearly separated, in which case the SVM becomes an optimisation problem (Cortes & Vapnik, 1995).

A Bayesian belief network (BBN) is a directed graph, together with an associated set of probability tables (Fenton & Neil, 2012). It emerges from the artificial intelligence (AI) field with multifarious usage, including reasoning under uncertainty and making predictions of highly uncertain phenomena (Hossain & Muromachi, 2012). Compared with models involving mathematical formulas and prose, the advantage of BBN is that the graphic model represents the structure of the probabilistic argument in an intuitive format. The BBN graph consists of nodes and arcs. The node represents variables and the arcs denote the causal/influential relationships between variables (Fenton & Neil, 2012). As for the probability tables, Bayes theorem helps update the values of the probabilities of the events of interest. One disadvantage of BBN is the complexity of the model; however, with the fast development of AI technologies, researchers have recently implemented this technique in real-time transportation studies (Dias et al., 2009; Hossain & Muromachi, 2012).

FINDINGS

Crash-prone conditions have been identified using real-time crash risk analysis models. Real-time crash-contributing factors related to traffic flow conditions in addition to weather and geometric factors are the most widely researched topics in existing studies. Table 1 presents the conclusions of previous real-time traffic safety evaluations, from 2000 until 2015. Existing studies were selected using search engines and academic databases such as Google Scholar, ScienceDirect, TRIS, Transportation Research Record, using the key words: 'real-time', 'crash prediction' and 'traffic safety'. In addition, real-time crash prediction studies not identified via searching but referenced in existing papers were also included. The study was included in the review if its objective included freeways or expressways and if explicit conclusions about the effects of the involved factors were stated. The effects of various crash-contributing factors are presented, with '+' indicating a positive effect, '−' indicating a negative effect, '±' implying that both positive and negative effects were found in the same study and '·' indicating that although the variable of interest was significant, whether it would increase or decrease crash likelihood was not explicitly provided. The last condition exists in studies that only used data mining technologies. In real-time safety evaluation employing statistical inference, the effects of variables were available in most cases.

Traffic-related factors in most studies include speed, standard deviation of speed, coefficient of variation of speed, speed difference between upstream and downstream detection locations, lane occupancy, standard deviation of occupancy, volume (or logarithmic volume), standard deviation of volume, volume difference between upstream and downstream detection locations, coefficient of variation of volume and congestion. Time- and weather-related factors are peak hours, adverse weather indicator, visibility, wet pavement, night-time indicator and precipitation. Common geometric-related factors are vertical curve, horizontal curve, median width and shoulder width.

Findings regarding speed are inconsistent. In some studies higher speed contributes to crash occurrence, while in others lower speed leads to higher crash likelihood. The inconsistency might be affected by the study objective (whether total or specific types of crashes are modelled) and the location of the roadway segments (rural/urban). Moreover, in real-time safety studies, speed information from multiple detectors might be extracted. The effects of speed at different detection locations for the same crash can be different. For example, low-speed downstream of crash locations and high-speed upstream of crash locations are more likely to cause rear-end crashes. In summary, as Abdel-Aty et al. (2005) point out, the crash mechanisms under high-speed and low-speed operation conditions are distinct; consequently, whether speed has a positive or negative effect on crash likelihood really depends on the specific roadway under evaluation and the study objectives. Compared to the findings about the effects of speed, different studies have reached unanimous conclusions about standard deviation and coefficient of variation of speed. According to the listed studies in Table 1, higher speed variation increases the crash probability and severity level. In multiple studies it is the speed variation instead of speed that affects crash occurrence. Therefore, reaching a smoother traffic flow helps reduce crash risk.

For lane occupancy most of the existing real-time studies found that occupancy, as a surrogate measure of traffic density, contributes to crash occurrence. In other studies occupancy at different locations (upstream and downstream detection points) or different time slices (10–15 min and 5–10 min prior to crash occurrence) show distinct effects on traffic crashes, thus proving that the dynamics of traffic within short space and time intervals could pose a considerable influence on safety in real-time. In the studies that involve the standard deviation of occupancy, all found that the turbulence in traffic density contributes to crash occurrence.

Table 1. Existing Real-time Safety Studies and Corresponding Findings.

ID	Authors	Country	Study Area	Research Objects		Speed	SD of Speed	Coefficient of Variation of Speed	Speed Difference between Upstream and Downstream	Lane Occupancy	SD of Lane Occupanc
1	Lee, Saccomanno, and Hellinga (2002)	Canada	Gardiner Expressway, Toronto	Total crashes				+		+	
2	Lee et al. (2003)	Canada	Gardiner Expressway, Toronto	Total crashes				+	+	+	
3	Abdel-Aty et al. (2004)	USA	I-4, Orlando, Florida	Total crashes				+		+	
4	Abdel-Aty and Pande (2005)	USA	I-4, Orlando, Florida	Total crashes				+			
5	Abdel-Aty et al. (2005)	USA	I-4, Orlando, Florida	Multi-vehicle crashes	Low speed			+		+	
					High speed					+/–	
6	Hourdos et al. (2006)	USA	I-94, Minnesota	Total crashes		+		+			
7	Abdel-Aty and Pemmanaboina (2006)	USA	I-4, Orlando, Florida	Total crashes				+		+	
8	Abdel-Aty and Pande (2006)	USA	I-4, Orlando, Florida	Total crashes				+		+	
9	Pande and Abdel-Aty (2006b)	USA	I-4, Orlando, Florida	Rear-end crashes	Low speed			+		+	
					High speed	+				+	
10	Abdel-Aty, Pande, Lee, Gayah, and Santos (2007)	USA	I-4, Orlando, Florida	Total crashes	Low speed			+		+	
					High speed					+/–	
11	Park and Oh (2009)	Korea	Seohaean Freeway, Seoul	Total Crashes	Rear-End Collisions					±	
					Night-time crashes towards Seoul	–				+/–	
12	Haj-Salem and Lebacque (2009)	France	Ring way of Paris	Total crashes						±	
13	Zheng et al. (2010)	USA	I-5 Portland, Oregon	Total crashes			+				

olume	Volume Difference between Upstream and Down-stream	SD of Volume	Coefficient of Variation of Volume	Congestion	Peak Hours	Adverse Weather	Visibility	Wet Pave-ment	Night-Time	Rain	Vertical Curve	Hori-zontal Curve	Median Width	Shoul-der Width
					+	−								
					+									
		−												
+/−		−												
							−	+	−					
		−								+				
		−												
		−												
+/−		−												
±														
+/−														

Table 1. *(Continued)*

ID	Authors	Country	Study Area	Research Objects	Speed	SD of Speed	Coefficient of Variation of Speed	Speed Difference between Upstream and Downstream	Lane Occupancy	SD of Lane Occupan
14	Hossain and Muromachi (2010)	Japan	Shinjuku 4 Tokyo Metropolitan Expressway	Total crashes				–		
15	Oh and Kim (2010)	Korea	Seohaean freeway, Seoul	Rear-end crashes	+					
16	Christoforou et al. (2011)	France	A4-A86, Paris	Rear-end crashes	–				–	
				Sideswipe crashes						
				Rear-end crashes with more than 2 vehicles					+	
				Multiple collisions	+					
				Single-vehicle crash						
17	Hassan and Abdel-Aty (2011)	USA	I-4 and I-95, Florida	Visibility-related crashes	+				+	
18	Pande et al. (2011)	USA	I-4 & I-95, Florida	Rear-end crashes	·				·	
19	Ahmed et al. (2012b)	USA	SR 417, Orlando, Florida	Total crashes	–	+				
				Rear-end crashes	–	+				
20	Ahmed and Abdel-Aty (2012)	USA	SR 408 and SR 417, Orlando, Florida	Total crashes	–	+	+			
21	Ahmed et al. (2012a)	USA	I-70, Colorado	Total crashes – Dry season	–					
				Total crashes – Snow season			+			
22	Hossain and Muromachi (2012)	Japan	Shibuya 3 and Shinjuku 4 Tokyo Metropolitan Expressway	Total crashes	+				–	
23	Xu, Liu, Wang, and Li (2012)	USA	I-880, California	Total crashes	–	+			+/–	
24	Christoforou et al. (2012)	France	A4–A86, Paris	Severe crashes	+/–					
25	Yu and Abdel-Aty (2013a)	USA	I-70, Colorado	Total crashes – Weekday	–				+	+
				Total crashes – Weekend	–					+
26	Yu and Abdel-Aty (2013c)	USA	I-70, Colorado	Total crashes	–					+

Volume	Volume Difference between Upstream and Downstream	SD of Volume	Coefficient of Variation of Volume	Congestion	Peak Hours	Adverse Weather	Visibility	Wet Pavement	Night-Time	Rain	Vertical Curve	Horizontal Curve	Median Width	Shoulder Width
	−													
									−					
+											+			
									−		−			
											−	−		
							−				+	−	−	
							−			+	+	−	−	
				+										
−														
+														
		+												

Table 1. *(Continued)*

ID	Authors	Country	Study Area	Research Objects	Speed	SD of Speed	Coefficient of Variation of Speed	Speed Difference between Upstream and Downstream	Lane Occupancy	SD of Lane Occupan
27	Yu and Abdel-Aty (2013b)	USA	I-70, Colorado	Single-vehicle crashes	–					+
				Multi-vehicle crashes	–					+
28	Xu et al. (2013b)	USA	I-880, California	Clear weather crashes	+	+			+	
				Rainy weather crashes	+				+	
				Reduced visibility crashes	+					+
29	Xu et al. (2013a)	USA	I-880, California	Total crashes		+			+	
30	Li, Wang, Chen, Liu, and Xu (2013)	USA	I-880, California	Total crashes		+	+			
31	Hassan and Abdel-Aty (2013)	USA	I-4 and I-95, Florida	Visibility-related crashes	+				+	
				Clear visibility crashes			+		+	
32	Yu and Abdel-Aty (2014c)	USA	I-70, Colorado SR 408, Florida	Crash severity I-70		+				
				Crash severity SR 408		+				
33	Yu and Abdel-Aty (2014a)	USA	I-70, Colorado	Crash severity		+				
34	Xu, Wang, Liu, Guo, and Li (2014)	USA	I-880 and I-5, California	Total crashes			+		+	+
35	Shi and Abdel-Aty (2015)	USA	SR 408, SR 417 and SR 528, Florida	Rear-end crashes	–					

Volume (or a logarithmic of volume), standard deviation of volume and the difference in volume between upstream and downstream stations are also employed in several studies. In most of these studies volume is a significant factor that results in a higher probability of crashes. This is understandable: as volume increases, the exposure to a crash for individual motorists increases as well. The standard deviation of volume has a negative impact on crash likelihood, implying that if the volume of different lanes varies greatly motorists will have larger gaps to make lane changes. Likewise, if the variation of volume between lanes is smaller, then the gap for lane changing becomes smaller and the risk for sideswipe crashes increases.

Volume	Volume Difference between Upstream and Downstream	SD of Volume	Coefficient of Variation of Volume	Congestion	Peak Hours	Adverse Weather	Visibility	Wet Pavement	Night-Time	Rain	Vertical Curve	Horizontal Curve	Median Width	Shoulder Width
+														
							−							
										+				
							−							
+						+						+		−
							−				+			
									−					−
											+			
+														
+				+	+									

In a few studies (Hossain & Muromachi, 2012; Shi & Abdel-Aty, 2015), real-time congestion intensity was evaluated using a specific indicator for a better understanding of the relationship between congestion and safety in real-time. Existing studies in Table 1 demonstrate that congestion contributes to crashes – more specifically rear-end crashes.

Time and real-time weather variables are another important group of factors in real-time safety. Peak hours significantly increase crash likelihood. Morning and evening peak hours usually imply higher traffic demand and congestion, especially in urban areas. Studies that incorporate a night-time indicator (Hourdos et al., 2006; Park & Oh, 2009) also found that crashes are less likely to happen during this time period, which could be interpreted in the same way as the effects of peak hours. The effects of adverse weather in the two studies above do not agree with each other (Lee et al., 2002; Xu, Tarko, et al., 2013). The effects of weather conditions should be interpreted from two aspects: their effects on driving and motorists' reaction to adverse weather. In one sense, adverse weather increases the difficulty of driving; in another sense, drivers might be more cautious under such weather conditions in order to neutralise the effect of the weather. Reduced visibility, based on results from various studies, consistently increases safety risks. This is expected as visibility is among the most important factors determining motorists' driving capability. Wet pavements and rain (precipitation) can alter friction between tyres and the pavement surface and decrease visibility and thus have a significant impact on safety in real-time. Currently, few studies have incorporated real-time weather information due to the lack of data. However, in regions where drastic changes in weather are expected (Ahmed et al., 2012a; Yu & Abdel-Aty, 2014a) studies have proved the importance of including real-time weather data in crash prediction.

As stated earlier, the inclusion of geometric characteristics in real-time crash prediction models mostly depends on the sampling scheme adopted. If matched-case control studies are implemented, the effects of geometric characteristics are implicitly controlled for. In cases where random sampling is in use the effects of geometric variables need to be explicitly considered. Vertical curve, horizontal curve, median width and shoulder width are common variables found to be significant in previous studies. Vertical curves influence drivers' sight distance as well as stopping distance. Nevertheless, they won't increase the crash likelihood for all types of crashes, as seen in Table 1. With regard to the effects of horizontal curves, several studies have concluded that they have a negative effect on crash occurrence. Similar to the effects of adverse weather, the effects of horizontal curvature should be viewed from an engineering and a human behaviour point of view. Wider shoulders and medians could provide drivers with some leeway in the face of incidents and help avoid crashes; thus, they have a beneficial effect on traffic safety.

The above findings are based on a wide range of studies across the world. From the above discussion it can be seen that real-time traffic flow parameters, real-time weather conditions and road geometric characteristics have been extensively studied. Regarding another important aspect in safety investigation, driver behaviour data are currently unavailable in most studies. The lack of driver behaviour data in existing research might also contribute to the inconsistent effects of some variables.

APPLICATIONS

Abundant traffic data are currently available in real-time on many of the high-speed roadway facilities. Many traffic authorities are using these data for traffic operation and management. Travel time estimation, congestion warnings and detour suggestions based on ITS traffic data are common strategies on different systems. However, while safety is an utmost indicator for freeway/expressway performance, real-time safety management has not been adequately addressed so far. Both incident detection and emergency response are reactive strategies to reduce safety issues. In contrast, real-time crash prediction, as discussed above, shows great potential for proactive traffic safety management.

One of the most direct applications is to construct a framework including both real-time traffic operation and safety monitoring, as illustrated in Fig. 3. Real-time ITS traffic data can come from loop detectors, radar sensors, AVI sensors and probe vehicles. Given these data, traffic operation conditions, such as average flow (speed, volume and density), variation of flow (standard deviation or coefficient of variation of above variables) and congestion could

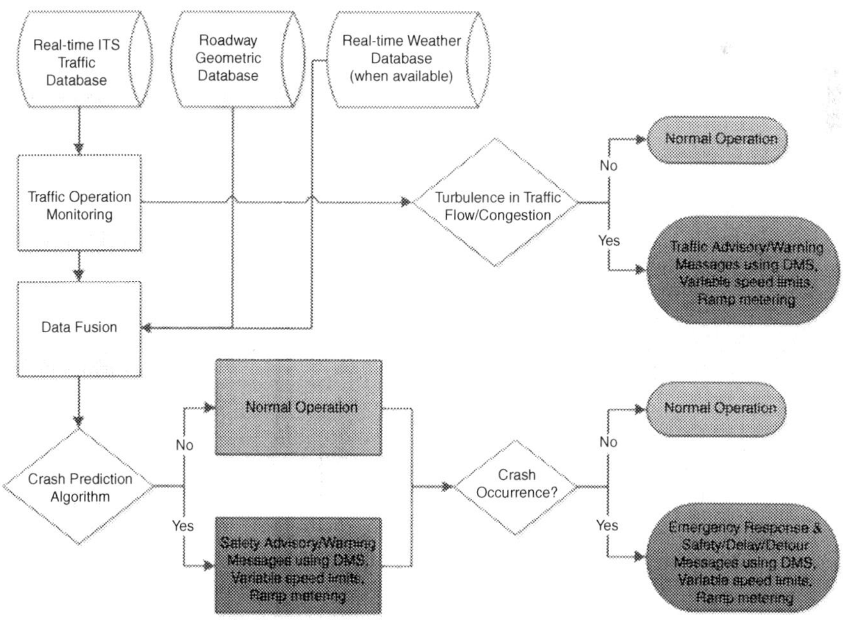

Fig. 3. Framework for Real-time Traffic Operation and Safety Monitoring.

all be monitored in real-time. Both traditional aggregate crash frequency studies and disaggregate real-time crash prediction studies have demonstrated the significant impacts of traffic parameters on safety. Therefore, detecting traffic turbulence and improvement strategies would also be beneficial for traffic safety.

In addition to traffic operation monitoring, real-time traffic safety evaluation is also necessary. Real-time safety management builds on a data fusion process from different sources and the selection of appropriate crash prediction algorithms. Traffic data and geometric characteristics are easier to access compared with weather data. Nevertheless the more data, especially in real-time, the more accurate a prediction could be made. Statistical approaches, data mining techniques or the combination of the two could serve as possible prediction algorithms to suggest potential risks. More specifically, algorithms for different types of crashes and different crash severity levels could all be included. To ensure their feasibility in the field, frequent updates and evaluation of the algorithms are critical. In the case of a crash occurring, safety advice and warning messages could be provided for the flagged sections. As needed, variable speed limits (VSL) and ramp metering could also be implemented. If an actual crash is observed, quick emergency response and other related countermeasures need to be carried out efficiently. To communicate information with road users, many studies have suggested the use of dynamic message signs is a useful tool to display different types of messages.

Other indirect approaches to evaluate real-time crash prediction are also promising. One field is microscopic traffic simulation. Based on two crash risk evaluation models developed by Abdel-Aty et al. (2005) and Abdel-Aty, Dilmore, and Hsia (2006), the effects of a VSL system on real-time freeway safety improvement were evaluated. The simulation study was done in PARAMICS (Quastone Paramics LTD, 2009) while focusing on four key components of the VSL control strategies: (1) speed change patterns (abrupt or gradual), (2) upstream lowering and downstream raising distances, (3) rate of change of speed limits (time step for change and speed step for change) and (4) gap distances. After numerous test scenarios and comparing the crash risk reductions, the best control strategy was identified. In another attempt to incorporate real-time crash prediction for VSL application, Yu and Abdel-Aty (2014b) identified the best VSL control strategies through solving optimisation problems of minimising crash risk for a controlled freeway section. The VSL algorithm contains two major parts: (1) an extension of METANET traffic flow model (Messner & Papageorgiou, 1990) and (2) a real-time crash risk evaluation model. The effects of the control algorithm were evaluated using the micro-simulation software VISSIM (PTV, 2008).

From the simulation results, it can be concluded that the VSL control strategy, by taking account of real-time traffic safety, would effectively improve traffic safety under high and moderate compliance levels.

FUTURE STUDIES

Compared with traditional aggregate crash frequency studies, real-time crash prediction is a more precise approach to improving traffic safety. This field has drawn great attention in the past 15 years, with most existing work occurring after the year 2000. However, there is still room for future exploration and improvement. Future work focusing on, but not limited to, three major aspects can be anticipated: the data sources, prediction methodology and applications.

In the era of Big Data, a gigantic amount of information is generated continuously in almost every aspect of social life. Big Data also opens new frontiers for real-time traffic safety evaluation studies. In the traffic safety field, traditional data sources are infrastructure-based traffic detection systems. The data are mostly aggregated and processed at 30-s or 1-min intervals. However, current technologies have already been able to detect and archive individual vehicle parameters. In the future, if automatic traffic detection systems can provide data at the vehicle level, new variables such as headway could be introduced, and speed variation would also be better captured. Probe vehicles are another promising data source that have not been widely used in real-time crash prediction to date due to the low deployment rate. With sufficient probe vehicles in the future, vehicle trajectories and detailed information about acceleration and deceleration would all shed new light on our understanding of crash mechanisms in real-time. Naturalistic driving study data and microscopic driver behaviour information can also be utilised to develop and improve the model. Last but not least, the Big Data era is characterised by data generated from multiple social network media. Some initial efforts are already underway and it is expected in the future that data from these sources could be utilised in crash prediction studies.

With Big Data from more sources, data fusion and analysis techniques also need updating. Data mining and other machine-learning techniques will play an increasingly important role in real-time crash prediction. Also, the issue of how to produce interpretable results from data mining tools will need to be addressed. In addition, models with the capability of addressing the heterogeneity and spatial correlations from crash and traffic data need to be explored to better describe crash precursors.

As for the application of real-time traffic safety evaluation, the transferability of real-time crash prediction models is of research interest. If the models could be successfully transferred other roadway segments could benefit from the findings on existing facilities. Also, the potential effects of different management strategies to reduce real-time crash risks could be evaluated in a simulation environment. Beyond simulation, whether there is a possibility to incorporate real-time traffic safety evaluation in driving simulators is another field that might show potential in the future.

ACKNOWLEDGEMENTS

The authors wish to thank several former researchers and students in the University of Central Florida who contributed to this work. To name a few, Dr. Chris Lee of the University of Windsor, Dr. Mohamed Ahmed of the University of Wyoming, Dr. Vikash Gayah of PennState, Dr. Hany Hassan, Dr. Kirolos Haleem, Jeremy Dilmore, Albinder Dhindsa, Ryan Cunningham and Rajashekar Pemmanaboina. The authors also acknowledge the funding of several agencies, including FDOT, CDOT, CFX and the UTC STC consortium led by the University of Tennessee. All opinions are those of the authors.

REFERENCES

Abdel-Aty, M., & Abdalla, F. M. (2004). Linking roadway geometrics and real-time traffic characteristics to model daytime freeway crashes: Generalized estimating equations for correlated data. *Transportation Research Record: Journal of the Transportation Research Board, 1897*(1), 106–115.

Abdel-Aty, M., Dilmore, J., & Hsia, L. (2006). Applying variable speed limits and the potential for crash migration. *Transportation Research Record: Journal of the Transportation Research Board, 1953*(1), 21–30.

Abdel-Aty, M., Hassan, H. M., Ahmed, M., & Al-Ghamdi, A. S. (2012). Real-time prediction of visibility related crashes. *Transportation Research Part C: Emerging Technologies, 24*, 288–298.

Abdel-Aty, M., & Pande, A. (2005). Identifying crash propensity using specific traffic speed conditions. *Journal of Safety Research, 36*(1), 97–108.

Abdel-Aty, M., & Pande, A. (2006). ATMS implementation system for identifying traffic conditions leading to potential crashes. *IEEE Transactions on Intelligent Transportation Systems, 7*(1), 78–91.

Abdel-Aty, M., & Pande, A. (2007). Crash data analysis: Collective vs. individual crash level approach. *Journal of Safety Research, 38*(5), 581–587.

Abdel-Aty, M., Pande, A., Lee, C., Gayah, V., & Santos, C. D. (2007). Crash risk assessment using intelligent transportation systems data and real-time intervention strategies to improve safety on freeways. *Journal of Intelligent Transportation Systems, 11*(3), 107–120.

Abdel-Aty, M., & Pemmanaboina, R. (2006). Calibrating a real-time traffic crash-prediction model using archived weather and its traffic data. *IEEE Transactions on Intelligent Transportation Systems, 7*(2), 167–174.

Abdel-Aty, M., Uddin, N., & Pande, A. (2005). Split models for predicting multivehicle crashes during high-speed and low-speed operating conditions on freeways. *Transportation Research Record: Journal of the Transportation Research Board, 1908*(1), 51–58.

Abdel-Aty, M., Uddin, N., Pande, A., Abdalla, F. M., & Hsia, L. (2004). Predicting freeway crashes from loop detector data by matched case-control logistic regression. *Transportation Research Record: Journal of the Transportation Research Board, 1897*(1), 88–95.

Ahmed, M., & Abdel-Aty, M. (2012). The viability of using automatic vehicle identification data for real-time crash prediction. *IEEE Transactions on Intelligent Transportation Systems, 13*(2), 459–468.

Ahmed, M., Abdel-Aty, M., & Yu, R. (2012a). Assessment of interaction of crash occurrence, mountainous freeway geometry, real-time weather, and traffic data. *Transportation Research Record: Journal of the Transportation Research Board, 2280*(1), 51–59.

Ahmed, M., Abdel-Aty, M., & Yu, R. (2012b). Bayesian updating approach for real-time safety evaluation with automatic vehicle identification data. *Transportation Research Record: Journal of the Transportation Research Board, 2280*(1), 60–67.

Barbaresso, J., Cordahi, G., Garcia, D., Hill, C., Jendzejec, A., & Wright, K. (2014). *USDOT's intelligent transportation system (ITS) ITS strategic plan 2015–2019*. Washington, DC. Retrieved from https://www.its.dot.gov/strategicplan.pdf.

Breiman, L. (2001). Random forests. *Machine Learning, 45*(1), 5–32.

Breiman, L. (2006). *Random forest: Breiman and Cutler's random forests for classification and regression*. Retrieved from URL http://stat-www.berkeley.edu/users/breiman/RandomForests, R package version.

Breiman, L., Friedman, J., Stone, C. J., & Olshen, R. A. (1984). *Classification and regression trees*. Boca Raton, FL: CRC Press.

Christodoulou, C., & Georgiopoulos, M. (2000). *Applications of neural networks in electromagnetics*. Norwood, MA: Artech House, Inc.

Christoforou, Z., Cohen, S., & Karlaftis, M. G. (2011). Identifying crash type propensity using real-time traffic data on freeways. *Journal of Safety Research, 42*(1), 43–50.

Christoforou, Z., Cohen, S., & Karlaftis, M. G. (2012). Integrating real-time traffic data in road safety analysis. *Procedia-Social and Behavioral Sciences, 48*, 2454–2463.

Cortes, C., & Vapnik, V. (1995). Support-vector networks. *Machine Learning, 20*(3), 273–297.

Dias, C., Miska, M., Kuwahara, M., & Warita, H. (2009). Relationship between congestion and traffic accidents on expressways: An investigation with Bayesian belief networks. In *Proceedings of the 40th annual meeting of infrastructure planning (JSCE)*, Japan.

Fenton, N., & Neil, M. (2012). *Risk assessment and decision analysis with Bayesian networks*. Boca Raton, FL: CRC Press.

Haj-Salem, H., & Lebacque, J.-P. (2009). Risk index modeling for real-time motorway traffic crash prediction. *Traffic and granular flow '07* (pp. 55–64). Springer: Springer International Publishing Switzerland .

Hassan, H., & Abdel-Aty, M. (2011). Exploring visibility-related crashes on freeways based on real-time traffic flow data. In *Proceedings of the transportation research board 90th annual meeting*.

Hassan, H., & Abdel-Aty, M. (2013). Predicting reduced visibility related crashes on freeways using real-time traffic flow data. *Journal of Safety Research*, *45*, 29–36.

Hossain, M., & Muromachi, Y. (2010). Evaluating location of placement and spacing of detectors for real-time crash prediction on urban expressways. In *Proceedings of the 89th annual meeting of the transportation research board*.

Hossain, M., & Muromachi, Y. (2012). A Bayesian network based framework for real-time crash prediction on the basic freeway segments of urban expressways. *Accident Analysis & Prevention 45*, 373–381.

Hourdos, J. N., Garg, V., Michalopoulos, P. G., & Davis, G. A. (2006). Real-time detection of crash-prone conditions at freeway high-crash locations. *Transportation Research Record: Journal of the Transportation Research Board*, *1968*, (1), 83–91.

Huang, H., & Abdel-Aty, M. (2010). Multilevel data and Bayesian analysis in traffic safety. *Accident Analysis & Prevention*, *42*(6), 1556–1565.

Kuhn, S., Egert, B., Neumann, S., & Steinbeck, C. (2008). Building blocks for automated elucidation of metabolites: Machine learning methods for NMR prediction. *BMC Bioinformatics*, *9*, 400.

Lee, C., & Abdel-Aty, M. (2008a). Presence of passengers: Does it increase or reduce driver's crash potential? *Accident Analysis & Prevention*, *40*(5), 1703–1712.

Lee, C., & Abdel-Aty, M. (2008b). Two-level nested logit model to identify traffic flow parameters affecting crash occurrence on freeway ramps. *Transportation Research Record: Journal of the Transportation Research Board*, *2083*(1), 145–152.

Lee, C., Abdel-Aty, M., & Hsia, L. (2006). Potential real-time indicators of sideswipe crashes on freeways. *Transportation Research Record: Journal of the Transportation Research Board*, *1953*(1), 41–49.

Lee, C., Hellinga, B., & Saccomanno, F. (2003). Real-time crash prediction model for application to crash prevention in freeway traffic. *Transportation Research Record: Journal of the Transportation Research Board*, *1840* (1), 67–77.

Lee, C., Saccomanno, F., & Hellinga, B. (2002). Analysis of crash precursors on instrumented freeways. *Transportation Research Record: Journal of the Transportation Research Board*, *1784*(1), 1–8.

Li, Z., Wang, W., Chen, R., Liu, P., & Xu, C. (2013). Evaluation of the impacts of speed variation on freeway traffic collisions in various traffic states. *Traffic Injury Prevention*, *14*(8), 861–866.

Liaw, A., & Wiener, M. (2002). Classification and regression by randomforest. *R News*, *2*(3), 18–22.

Martin, P. T., Feng, Y., & Wang, X. (2003). *Detector technology evaluation*. Mountain-Plains Consortium. Retrieved from https://www.mountain-plains.org/pubs/pdf/MPC03-154.pdf.

Mensah, A., & Hauer, E. (1998). Two problems of averaging arising in the estimation of the relationship between accidents and traffic flow. *Transportation Research Record: Journal of the Transportation Research Board*, *1635* (1), 37–43.

Messner, A., & Papageorgiou, M. (1990). Metanet: A macroscopic simulation program for motorway networks. *Traffic Engineering & Control*, *31* (8–9), 466–470.

Mimbela, L. E. Y., & Klein, L. A. (2000). *Summary of vehicle detection and surveillance technologies used in intelligent transportation systems*. Las Cruces, NM: The Vehicle Detector Clearinghouse, New Mexico State University.

NHTSA. (2012). Traffic safety facts 2012. In National Highway Traffic Safety Administration (Ed.), *Transportation, U.S.D.O.* Washington, DC.

Oh, C., & Kim, T. (2010). Estimation of rear-end crash potential using vehicle trajectory data. *Accident Analysis & Prevention*, *42*(6), 1888–1893.

Oh, C., Park, S., & Ritchie, S. G. (2006). A method for identifying rear-end collision risks using inductive loop detectors. *Accident Analysis & Prevention*, *38* (2), 295–301.

Oh, J.-S., Oh, C., Ritchie, S. G., & Chang, M. (2005). Real-time estimation of accident likelihood for safety enhancement. *Journal of Transportation Engineering*, *131*(5), 358–363.

Pande, A., & Abdel-Aty, M. (2006a). Assessment of freeway traffic parameters leading to lane-change related collisions. *Accident Analysis & Prevention*, *38*(5), 936–948.

Pande, A., & Abdel-Aty, M. (2006b). Comprehensive analysis of the relationship between real-time traffic surveillance data and rear-end crashes on freeways. *Transportation Research Record: Journal of the Transportation Research Board*, *1953* (1), 31–40.

Pande, A., Das, A., Abdel-Aty, M., & Hassan, H. (2011). Estimation of real-time crash risk. *Transportation Research Record: Journal of the Transportation Research Board*, *2237*(1), 60–66.

Park, J., & Oh, C. (2009). Relating freeway traffic accidents to inductive loop detector data using logistic regression. In *Proceedings of the 4th IRTAD conference*, 16–17 September 2009, Seoul, Korea (pp. 223–231).

Pirdavani, A., De Pauw, E., Brijs, T., Daniels, S., Magis, M., Bellemans T., & Wets, G. (2015). Application of a rule-based approach in real-time crash risk prediction model development using loop detector data. *Traffic Injury Prevention*, *16*(8), 786–791.

PTV. (2008). *VISSIM 5.10 user manual*. PTV Vision, July 18.

Quastone Paramics LTD. (2009). *The paramics manuals* (version 6.6.1). Edinburgh: Quastone Paramics LTD.

Roess, R. P., Prassas, E. S., & McShane, W. R. (2004). *Traffic engineering.* Upper Saddle River, NJ: Prentice Hall.

Shi, Q., & Abdel-Aty, M. (2015). Big data applications in real-time traffic operation and safety monitoring and improvement on urban expressways. *Transportation Research Part C: Emerging Technologies*, *58*(B), 380–394 .

Strobl, C., Boulesteix, A.-L., Kneib, T., Augustin, T., & Zeileis, A. (2008). Conditional variable importance for random forests. *BMC Bioinformatics*, *9*(1), 307.

Strobl, C., Boulesteix, A.-L., Zeileis, A., & Hothorn, T. (2007). Bias in random forest variable importance measures: Illustrations, sources and a solution. *BMC Bioinformatics*, *8*(1), 25.

Vladimir, V. N., & Vapnik, V. (1995). *The nature of statistical learning theory.* New York: Springer.

Xu, C., Liu, P., Wang, W., & Li, Z. (2012). Evaluation of the impacts of traffic states on crash risks on freeways. *Accident Analysis & Prevention*, *47*, 162–171.

Xu, C., Tarko, A. P., Wang, W., & Liu, P. (2013). Predicting crash likelihood and severity on freeways with real-time loop detector data. *Accident Analysis & Prevention*, *57*, 30–39.

Xu, C., Wang, W., & Liu, P. (2013). Identifying crash-prone traffic conditions under different weather on freeways. *Journal of Safety Research*, *46*, 135–144.

Xu, C., Wang, W., Liu, P., Guo, R., Li, Z. (2014). Using the Bayesian updating approach to improve the spatial and temporal transferability of real-time crash risk prediction models. *Transportation Research Part C: Emerging Technologies*, *38*, 167–176.

Yu, R., & Abdel-Aty, M. (2013a). Investigating the different characteristics of weekday and weekend crashes. *Journal of Safety Research*, *46*, 91–97.

Yu, R., & Abdel-Aty, M. (2013b). Multi-level Bayesian analyses for single- and multi-vehicle freeway crashes. *Accident Analysis & Prevention*, *58*, 97–105.

Yu, R., & Abdel-Aty, M. (2013c). Utilizing support vector machine in real-time crash risk evaluation. *Accident Analysis & Prevention*, *51*, 252–259.

Yu, R., & Abdel-Aty, M. (2014a). Analyzing crash injury severity for a mountainous freeway incorporating real-time traffic and weather data. *Safety Science*, *63*, 50–56.

Yu, R., & Abdel-Aty, M. (2014b). An optimal variable speed limits system to ameliorate traffic safety risk. *Transportation Research Part C: Emerging Technologies*, *46*, 235–246.

Yu, R., & Abdel-Aty, M. (2014c). Using hierarchical Bayesian binary probit models to analyze crash injury severity on high speed facilities with real-time traffic data. *Accident Analysis & Prevention*, *62*, 161–167.

Zheng, Z., Ahn, S., & Monsere, C. M. (2010). Impact of traffic oscillations on freeway crash occurrences. *Accident Analysis & Prevention*, *42*(2), 626–636.

VULNERABLE ROAD USERS: CHALLENGES AND SOLUTIONS

CHAPTER 10

PROVIDING FOR PEDESTRIANS

Per Gårder

ABSTRACT

Purpose – *This chapter aims to advise the public as well as municipal, state and national agencies about how pedestrian safety can be improved through changes in our built environment. Higher safety can lead to more walking and thereby a more sustainable society.*

Methodology – *The chapter is based on a review of literature, including a review of published papers and field studies by the author himself.*

Findings – *To reach 'acceptable' safety levels, all arterials and collector roads – at least segments with more than 50 pedestrians a day – should have sidewalks. The sidewalks should be separated from the roadway by a curb if speeds are low and by a barrier or wide separation strip in high-speed areas; that is, where speeds are higher than 50 km/h. Local roads also need sidewalks unless traffic volumes and speeds are very low. The major safety issue for pedestrians is, however, where they cross streets. Elderly pedestrians and pedestrians in a great hurry or under the influence of intoxicants in particular need streets to be narrow and have low speeds for them to be able to cross safely. Marking crosswalks or even signalising them will have only marginal safety effects at best. Posting them for low speed is also not enough unless we have photo speed-enforcement ensuring that everyone drives slowly. Rather, using narrow cross-sections or speed*

Safe Mobility: Challenges, Methodology and Solutions
Transport and Sustainability, Volume 11, 207–228

ISSN: 2044-9941/doi:10.1108/S2044-994120180000011011

cushions at the approaches ensuring that 90-percentile speeds are no more than 30 km/h at crossing points is key to safety. In between crossing points a speed of 50 km/h is acceptable with pedestrians on adjacent sidewalks.

Social implications – *We as a society need to encourage walking as a mode of transportation since walking promotes better health and a cleaner environment; that is, a more sustainable society. However, it has to be safe to walk or people will prefer to drive to their destinations. Also, distances between destination points have to be kept reasonably short and the environment, where people walk needs to be interesting and aesthetically somewhat pleasing to encourage walking.*

Keywords: Pedestrian; safety; crosswalks; speed; elderly

INTRODUCTION

This chapter is the first of two chapters focussed on safety issues related to vulnerable road users. While this chapter focusses on pedestrian safety concerns, it complements Chapter 11, which focusses on the safety concerns of cyclists.

The primary objective of this chapter is to discuss how pedestrian safety can be improved by making enhancements to transport infrastructure design and operation. Obviously, pedestrian safety cannot always be maximised since the safety of bicyclists or motorists may at times conflict with the safety of pedestrians. Efficiency of traffic movements as well as aesthetic and environmental concerns also have to be considered. Naturally, installation and maintenance costs are other important factors when deciding on what to do to get cost-effective solutions. Having said that, using designs which are safe for pedestrians often also leads to high safety for other road-users; safety for everybody combined with short delays for motorists accomplished by the concept 'quick but slow' often go hand in hand, such as when utilising roundabouts along arterials rather than signalised intersections (see Chapters 7 and 8 for safety issues related to intersections and roundabouts, respectively). One of the first studies to quantify the 'quick but slow' concept was performed along South Golden Road in Golden, Colorado in the United States (US) (Isebrands, Hallmark, Fitzsimmons, & Stroda, 2008). This study showed a reduction in travel times by 10 seconds over an 800-m segment (quicker), while the 85-percentile speeds were reduced from 76 to 53 km/h (47–33 mph) and improved safety. Prior to construction of the roundabouts there were 10 injury crashes per year, and in the 4 years after the roundabouts were constructed only one injury crash was reported (Hartman as cited in Isebrands et al., 2008).

This chapter addresses questions such as: Where are sidewalks *not* needed? Where are separate walking paths desired? Where should crosswalks be provided? What should they look like? What design should be used for other potential pedestrian crossing points? What needs to be done with respect to speed, signage, roadway markings and control to make crosswalks reasonably safe? And, how will Intelligent Transportation Systems (ITSs) change these factors?

The chapter has a focus on pedestrian traffic safety. Pedestrian security – that is, how a layout may influence the risk of assault, rape or other violent crimes – is also an important issue. So is environmental safety. If pedestrians are exposed to excessive levels of air pollution from high volumes of traffic their mortality rate may be significantly impacted (Amorim et al., 2013). The attractiveness of facilities is a different – and very important – factor in promoting walking as a mode of transportation, which is essential for a sustainable society. We know that people want direct, aesthetically pleasing and safe facilities. Making the facilities attractive can by itself promote safety. If drivers see many pedestrians along a street, they are more likely to slow down and yield to crossing pedestrians. On the other hand, making facilities attractive may lead not only to more pedestrians in an area but can also lead to more pedestrian crashes unless safe infrastructure is used. Education and enforcement is never a substitute for safe physical layout but can be important complements.

Pedestrian fatalities dominate motor-vehicle fatalities in most developing countries, and worldwide there are over 270,000 pedestrians killed every year (World Health Organization, 2013a). However, pedestrian fatality numbers are often thought of as 'no-big-deal' in industrialised countries such as the US. But the same people who say that traffic fatalities are an acceptable toll of our modern society would typically not say that the number of victims in the September 11, 2001, terrorist attacks in New York City, Washington, DC, and Pennsylvania were not unacceptably high. On September 11, there were 2,977 innocent victims and 19 hijackers killed. That number can be compared to well over 4,000 innocent pedestrians killed *every* year in the US for the last 60 years. We may say that road users are aware of the risks and have themselves to blame when getting killed so they are not victims, but – in my opinion – it should be safe to walk to a destination, just like it should be safe to work in an office tower, but it is not safe to walk along or across many if not most streets in most counties. The main reason for this is that we – the road users – don't want to 'pay' for other people's safety. We may be willing to pay for airbags and other safety features in our own cars, but most of us do not want to pay with money or with time for what makes roads safer for pedestrians.

Benefits of Walking

Walking is maybe the most natural way of getting from one place to another. In some countries walking is still a dominant mode of transportation, but in many industrialised countries – especially the US – walking has been marginalised and many if not most desired activities/destinations are easier to reach by automobile than on foot. If towns and cities were planned accordingly, many more activities could be reached from a person's home by foot and people could incorporate walking into their daily activities, even if they live far away from work and service centres by walking, for example, from shop to shop. Safe, secure, unrestricted and aesthetically attractive facilities are needed to persuade people to walk rather than to drive.

There are many health benefits connected to walking. Walking is probably the form of exercise that has the fewest negative side effects in the form of injuries – as long as injuries caused by collisions with motor vehicles are avoided. Carré (1997) estimates that a person gains one hour of expected life for every hour they are engaged in moderate exercise, for example, brisk walking. If we extrapolate these findings, a person would live forever if they kept on walking. Obviously, that is not true. But the findings do mean that 'no time is wasted' when going by foot to and from activities. Recent empirical research by the US National Cancer Institute and Harvard University, following more than 661,000 adults, found that just over an hour a day of moderate exercise, such as brisk walking, seems to be optimal for longevity; less or more exercise was correlated to a higher mortality rate during the 14-year period studied (Reynolds, 2015).

There are obviously people with health problems who cannot walk, at least not very far. But if we include riding a wheelchair under the general umbrella of 'walking,' then most of us can and do walk, at least sometimes. Then why do we not walk more? First, it takes time; although, as stated above, it does not really take time out of a person's life unless 'excessive' time is spent on walking. However, it is human nature that we do not want to sacrifice minutes today for a gain later in life, and we often feel that we do not have time to spend 20 minutes on walking when we can get there in five minutes by car.

Another reason people do not walk is safety. How people perceive safety for themselves and their families is therefore very important. It is especially important that children are given safe and secure opportunities to walk to and from school. Walking may become a lifelong habit if it becomes the norm at an early age. Parents often worry more about a child being kidnapped or assaulted than about traffic, but as is pointed out by the British author Warwick Cairns, author of *How to Live Dangerously*, 'an average child would have to stand outside alone for 750,000 hours unsupervised to be kidnapped by a stranger. At that point, the child would be 85 years old' (*USA Today*, 2015). So, while we inflate some risks we tend not to focus on others, such as obesity and diabetes, that may be the result of children being driven rather than walking to destinations.

Later in life people also need to have 'inviting' environments to be encouraged to walk. People should not have to cross 'desert-sized' parking lots when walking to shops and restaurants. A building's entrance should be located close to the street with safe, direct access directly from a sidewalk. Parking should be below, above, behind or on the side of buildings – not in front. It is also important for pedestrians to see others walking in that area. To have residences and businesses mixed – with restaurants and shops on the ground floor, offices one flight up and apartments above that – means that there will be pedestrian activity both during daytime and evening hours.

In order to reach a destination by foot people will frequently need to cross streets and roads. The goal should be to provide 'absolute' safety for pedestrians who are crossing streets, but absolute safety may not be a realistic goal. Grade-separated crossing points are not only expensive but may also not be used. So, if absolute safety cannot be provided, what is an acceptable safety level? Vision Zero goals – no fatalities and no serious injuries – should obviously still remain the ultimate long-term goal.

Measurement of Safety

When discussing pedestrian traffic safety in this context, high safety is defined as few pedestrian crashes per million crossing (or passing) pedestrians at an intersection or per million pedestrian-kilometres walked along segments. Obviously, for a given layout, crash rates – such as pedestrian crashes per million [crossing] pedestrians – will be influenced by a number of factors, such as vehicle volume and vehicle speeds as well as by pedestrian volume. Also, it is not just the number of crashes per million pedestrians that is of interest, it is also the severity of those crashes. We can 'live' with pedestrian crashes that are not fatal or resulting in permanent disabilities. So, one measure of low safety at road crossing points is a high number of fatal pedestrian crashes per million crossing pedestrians. For sections of roadway, the measure would be the number of fatal pedestrian crashes per million-kilometres (or million-miles) walked.

User Groups

People should be able to walk outdoors no matter what age they are. Still, we should not expect all walking facilities to be safe for people who are very young, say below age 6, unless they are accompanied by an adult. It is more controversial to state that people who are very old, and functionally not able to cope with fast-moving traffic, may also need assistance when walking – that goes for people of any age with disabilities. Worldwide, over 600 million people live with a disability. Approximately, 70 million individuals are affected by severe hearing disabilities, some 285 million people are estimated to be visually impaired, 100 million people use wheelchairs and potentially another 100 million people need a wheelchair but cannot afford one (World Health Organization, 2013b). There are certainly also non-handicapped people in all age categories who have issues with fast-moving traffic because of temporary disabilities or impairment due to, for example, alcohol. Just being in a hurry to catch a bus on the other side of a street may also make a person inattentive to traffic.

PEDESTRIAN FACILITIES ALONG ROADWAYS

To obtain a sustainable environment, pedestrians should not only be allowed to walk more or less everywhere but be encouraged to do so. Providing sidewalks is an important part in achieving this. The one type of roadway where pedestrians typically are not allowed is the motorway (freeway); however, most other major roads should have sidewalks along them wherever there are substantial numbers of pedestrians, unless there are nearby parallel footpaths. The question becomes what constitutes a 'substantial' number? Is one pedestrian an hour enough to motivate sidewalks for new construction? Can economic reasoning allow us to keep operating an arterial road lacking sidewalks even if there are 300 pedestrians walking along it every day? And, if so, should the speed along it be restricted to 30 km/h (19 mph) no matter how rural it is?

The current standards in Sweden are that rural roads need separate pedestrian and bicycle facilities if the combined pedestrian and bicycle volume is at least 100 per day (during summer months). If the pedestrian and bicycle flow is below 100 per day and vehicle annual average daily traffic (AADT) is above 2,000, paved shoulders are acceptable, whereas mixed traffic is acceptable for such low pedestrian and bicycle flows if AADT is below 2,000 (Trafikverket, 2012). Finnish guidelines are similar but require separation of rural arterials when there are 50 pedestrians or cyclists a day if AADT is above 2,000 (see Chapter 11 for safety issues specifically related to cyclists). Are those standards good enough? Maybe not if we are aiming for Vision Zero.

A sidewalk is more attractive if it is set back from the roadway by a grass or vegetation strip. That strip also makes it clearer for drivers to tell if pedestrians are intending to cross the road or not (see Fig. 1). Making the strip wider, about 2 m or so, and having stem trees and/or utility poles separating vehicles from pedestrians will increase perceived safety and the greater distance will cut back on water and mud splashing from car tires. Fig. 2 shows a 'primitive' example of this. There is a possibility that errant vehicles will go off the travelled roadway and hit pedestrians on a sidewalk, especially if the roadway and sidewalk are not separated by a wide strip. *A Policy on Geometric Design of Highways and Streets* (American Association of State

Fig. 1. Paved Sidewalk along an Exurban Arterial Highway in Readfield, Maine, US.

Fig. 2. Unpaved 'Primitive' Walking Path along an Arterial Highway in Connecticut, US.

Highway and Transportation Officials (AASHTO), 2011a), the so-called 'Green Book', defines a clear zone as the 'unobstructed, traversable area provided beyond the edge of the travelled way for recovery of errant vehicles.' This guideline allows sidewalks to be within the clear zone, but it is obvious that the assumption is that cars may traverse into it with some probability. The Green Book continues to state, for local roads and streets, that 'a clear zone of 2 to 3 m or more from the edge of the travelled way, appropriately graded with relatively flat slopes and rounded cross-sectional design, is desirable' (pp. 5–8).

For higher-speed rural collector roads and arterials, the Green Book refers to the AASHTO *Roadside Design Guide* (AASHTO, 2011b). That publication recommends that clear zones have a width of 9–10 m for flat, level terrain adjacent to a straight section of a 100 km/h [60 mph] highway with an average daily traffic of 6,000 vehicles. For steeper slopes on a 110 km/h [70 mph] roadway the recommended clear zone increases to 12–14 m, whereas on a low-speed, low-volume roadway it drops to 2–3 m. The guide also states that for horizontal curves the clear zone should be increased by up to 50% from these figures. It would be prudent to not have sidewalks within this zone unless there is a barrier separating the roadway from the sidewalk, as shown in Fig. 3. A curb, as shown in Fig. 1, is not a substitute for a barrier. This is what the Federal Highway Administration (FHWA) says on their safety website (FHWA, 2013):

> When the Green Book and the Roadside Design Guide were last updated, the AASHTO committees coordinated to dispel the misunderstanding that 2 ft (actually, 18 in) behind a curb constituted a clear zone. Since curbs are now generally recognised as having no significant containment or redirection capability, clear zone[s] should be based on traffic volumes and speeds, both without and with a curb.

PEDESTRIAN FACILITIES ACROSS ROADWAYS

Stepping off the sidewalk and out into a street is sometimes perceived as being a very dangerous action. Parents with young children may fear the child stepping out into traffic as equally dangerous as them stepping off a cliff, as illustrated in Fig. 4.

In some jurisdictions pedestrians have legal priority wherever they cross a roadway adjacent to an intersection. That may also be the case in between intersections in some living-street environments (Woonerven, Home Zones, etc.) and in shared-space zones. Such areas seem to operate with high safety but more research is needed.

Except for in those rare cases mentioned above, pedestrians typically have formal priority only in marked crosswalks and even there they may not be given priority. It can be noted that there are lots of gap acceptance studies in the literature where researchers have looked at pedestrians accepting

Fig. 3. Walking Facilities on Mount Desert Island, Maine, and in Skåne, Sweden, Separated by Barriers.

Fig. 4. Walking Path along Dangerous Drop-off, on Mount Desert Island, Maine, US.

and rejecting gaps between cars at marked crosswalks, but few if any studies of crosswalks where the study looks at drivers accepting and rejecting gaps between pedestrians. So, most of the time it is obviously up to pedestrians to find gaps even when they have the (formal) right-of-way, and every gap should be of sufficient length. Imagine if we were at a stop sign placed on a minor street and noted that four out of five automobile drivers approaching the stop sign did not even slow down but drove straight across a major highway without looking or even considering yielding. Should we then study how drivers along the main highway accept gaps and look at whether motorists on the main highway continue towards their destination or not without having to stop at that junction? Should drivers along all priority roads have to stop every time they get to a minor intersection, and should drivers who have green lights always stop and look even when they are going straight in case someone may want to turn in front of them? The absurdity of how pedestrians are treated needs to be highlighted. Pedestrians in marked crosswalks should not have to wait for a gap. It should be drivers crossing the crosswalk who look for gaps between pedestrians; in the United Kingdom (UK) such behaviour can be observed at zebra crossings. Until pedestrians are given priority, crosswalks should possibly be eliminated (painted over) since they often instil pedestrians with a false sense of security, as pointed out more than 40 years ago by Herms (1972). Rather, we should provide crossing points at locations where the roadway is not wider than one (narrow) lane between raised refuge islands, at places where cars are travelling at 'low' speed, probably never above 30 km/h (19 mph). Encouraging pedestrians to cross where 85-percentile speeds are above 55 km/h (35 mph) should never be done unless traffic volumes are low, sight distances are good and, if such conditions prevail, marking crosswalks are not needed. Marking crosswalks certainly can lead to more crashes, especially on multi-lane streets (Zegeer, 2005; Zegeer, Stewart, Huang, & Lagerwey, 2002)

Many studies have shown that the lowest level of safety for pedestrians is found when they cross a street (Elvik, Hoye, Vaa, & Sorensen, 2009). Whether the safety of pedestrians is improved by marking a crosswalk or not depends on a range of factors. Paradoxically, in jurisdictions where very few motorists pay attention to pedestrians in crosswalks, marking the crosswalk probably has a negligible effect on safety. In jurisdictions where a majority – but far from all – motorists yield to pedestrians, marking a crosswalk seems to be beneficial to safety. Unexpectedly, in jurisdictions where almost all motorists stop and yield to pedestrians, marking a crosswalk seems to lead to an increase in the number of crashes. The reason for this is that if, say, only one in a thousand drivers does not yield to pedestrians, pedestrians get a feeling of (false) safety. It feels safe to walk out in front of a moving car, since it is (almost) certain to stop. However, a few drivers from 'away', or distracted drivers, may not notice the pedestrian and even if it is only one in a thousand drivers, those few can cause a large number of crashes. This was found to be the case in the previously referenced study by Herms (1972), which has been replicated several times. The Zegeer (2005) nationwide study referenced above found that low safety occurs particularly at marked crosswalks across busy four-lane streets, and especially in states where the yieldrate is high. The effect of giving absolute priority to pedestrians in Sweden (Leden, Wikström, Gårder, & Rosander, 2006) showed similar effects, with safety suffering when yield behaviour improved. A crash analysis based on a macro-study of all of Sweden suggests that the injury risk in 'standard' marked crosswalks increased by 27% for pedestrians as a result of the change in law in May 2000, which gave pedestrians absolute priority in crosswalks, no matter how close or at what speed vehicles are approaching. This increase in crashes occurred in spite of a massive information campaign about the new law and increased enforcement. A conclusion is that 'high' safety cannot easily be achieved through education or enforcement alone. Improving yield behaviour from none to roughly half of all motorists yielding may be beneficial on high-volume, single-lane streets – because pedestrians cannot wait for ever for a safe gap to appear – but improving yield behaviour beyond that may lead to lower safety – especially on multi-lane streets – until we get a 100% yield rate, which would require something beyond a human operator controlling the vehicle (i.e., ITS). Until then, the most effective safety measures will be based on geometric changes to our roadways.

Geometric features that help pedestrian safety include more or less anything leading to lower speeds and shortened crossing distances (Gårder, 2004). A raised refuge island or a flush refuge island protected by pollards (reflectors) is typically enough to break up a roadway into two or more 'crossing' segments. Having raised refuge islands between each lane of traffic may be the most effective safety measure, even if during an introductory phase it could be confusing to pedestrians, making them look the wrong way.

How to Achieve Low Speeds

One reason speed is important for safety is that pedestrians do not operate in a constant time-gap mode. They tend to accept shorter vehicle time gaps when speed increases. Pedestrians, especially the elderly, tend to go for constant distance gaps (Yannis, Papadimitriou, & Theofilatos, 2013). Another issue is related to the overall lower walking speeds of the elderly and people with disabilities. In fact, unlike car drivers, where higher speed is correlated to an increased likelihood and severity of a crash, for pedestrians lower crossing speeds imply greater risk of being involved in a crash (Shinar, 2008).

Low speeds can be achieved simultaneously with narrow crossing distances if the roadway width is so narrow that it by itself slows down traffic. This is difficult to achieve in practice, especially if we want wide, heavy vehicles and busses to be able to pass through. Also, in climates where we need to remove snow through ploughing it is difficult to build roads with, for example, 2.1-m curb-to-curb distance, as has been done at some locations in the UK (see Fig. 5). If curb-to-curb width is much wider than that, for example, 2.8 m (9 ft), many motorists will feel comfortable driving at speeds well over 30 km/h (19 mph).

Another way to reduce speed, besides narrowing the roadway, is to make the geometry induce horizontal or vertical *g*-forces on the vehicle to such an extent that it becomes uncomfortable – or even impossible – to drive fast. Small intersections or mid-block traffic circles can be used for this and, if built correctly, can reduce median speeds to around 25 km/h and 85-percentile

Fig. 5. Example of a Narrow Travel Lane that Slows Drivers Down.

speeds to around 30 km/h. Watt's hump and speed cushions, as seen in Fig. 6, at Regementsgatan and Dragonstigen in Malmö, Sweden, will also achieve speeds in the 30 km/h (19 mph) range, whereas most speed tables with 8–10 cm height have 85-percentile speeds in the 45 – 50 km/h (30 mph) range (Johansson, Gårder, & Leden, 2003; Leden, Gårder, & Johansson, 2006).

A relatively new invention is the 'Actibump', which is a 'trap-door' bump activated only when a vehicle is going a set number of km/h above the posted speed limit. Actibump is a commercially developed product by the Swedish company Edema and can be used at crosswalks as well as toll plazas and other locations, where low speeds need to be ensured for every single motorist. An alternative to this product is a system where drivers get a red light when they speed. The City of Los Angeles, California, has been experimenting with such a system since 2014. Along North Figueroa Street, near Amabel Street, there is a speed feedback (Your Speed) sign which triggers the downstream traffic signal to go red for a brief period if an approaching motorist is exceeding the speed limit by more than 8 km/h (5 mph), forcing the motorist to come to a complete stop. If motorists travel below that speed the light doesn't change unless there is cross-traffic or pedestrians at the small cross-street. A radar unit in the speed feedback sign determines the vehicle speed and sends a message to the controller when a speeding vehicle (see Chapter 3 for additional information about speeding and aggressive driving) is approaching to switch the light to red (S. Skehan, personal communication, March 18, 2015).

Fig. 6. Speed Cushion, Refuge and Marked Pedestrian Crosswalk (Photo by Lars Leden).

A system like this can obviously be coupled with a red-light-running camera system if motorists start ignoring the signal. An advantage with a system like this compared to direct camera enforcement of speeds may be that it is politically easier to implement. However, camera enforcement of speeds at crosswalks is obviously an alternative. Such a system is used at Brunnstorget in Helsinki (see Fig. 7). Here motorists are fined when they exceed the 30 km/h (19 mph) speed limit by just a few km/h when approaching a signalised crosswalk which, prior to this system being installed, had many pedestrian crashes. Fig. 7 also shows a stop line located 10 m before the crosswalk, so that motorists are encouraged to stop early and jaywalking pedestrians can tell whether a car will stop or not.

The current US *Manual on Uniform Traffic Control Devices* (FHWA, 2009) recommends that yield lines at unsignalised mid-block crosswalks be placed 6–15 m (20–50 ft) ahead of the crosswalk, as shown in Fig. 8. A safety issue at the Helsinki location prior to this marking was caused by the left turn and through movements having separate green phases, and jaywalking pedestrians assuming that all cars would be stopped by a red light when only some were.

Fig. 7. Speed Enforcement at a Crosswalk in Helsinki, Finland (Photo by Lars Leden).

A large study focussed on the New York City area found a similar trade-off between pedestrian safety and motorist safety. The study found that two of the strategies studied – split phase timing and signal installation – were effective in reducing multiple vehicle crashes but had the opposite effect on pedestrian safety (Chen, Chen, & Ewing, 2014).

A Swedish study from the 1980s found that scramble – that is, pedestrians having a simultaneous walk phase on all crosswalks while all cars are stopped – is ineffective in promoting pedestrian safety unless red-walking frequencies are very low, and that red-walking frequencies become high when wait times are long as is the case when scramble is being used (Gårder, 1989). The study also found that left-turning vehicles are involved in many pedestrian crashes, so that there is no easy way of using signalisation as a way to improve pedestrian safety (Gårder, 1989). Other studies confirm that left-turning vehicles pose more of a threat to pedestrians than right-turning ones (in countries where people drive on the right). Reasons include higher speeds and opposing traffic blocking view of the crosswalk (Lord, 1996; Lord, Smiley, & Haroun, 1998).

Fig. 8. Yield Line at Crosswalk in Amherst, Massachusetts, US.

Speeds and Safety

There is ample evidence that lower speeds lead to safer environments in general and for pedestrians in particular. It should be kept in mind that it is not the posted speed that will determine the impact of a crash. Nor is it the travelling speed prior to an evasive manoeuvre that will directly be related to the injuries sustained. It is the collision speed that matters. However, there typically is a relationship between posted speed and actual travel speed, and there is a strong relationship between travel speed and impact speed. The faster people drive, the more likely it is that they will not have time to brake before impact, so lowering the average travel speed by 1 km/h will most likely lower the impact speed by more than 1 km/h.

More important than lowering the average speed is to lower the top speeds and, for example, 85-percentile speeds, since the faster drivers are over-involved in crashes. Analysis of crashes in general shows that few multi-vehicle crashes happen without any of the parties taking evasive manoeuvres prior to the impact. That ought to be true for drivers colliding with pedestrians as well. A study by Walz, Hoefliger, and Fehlmann (1983) found that in Zürich, Switzerland, only 18% of the fatal crashes had collision speeds equal to the travelling speed of the striking vehicle. In 62% the collision speed was reduced by at least 20% before striking the pedestrian. Walz et al. also report that a reduction of the speed limit from 60 to 50 km/h (37 to 31 mph) in Zurich was accompanied by a reduction of 20% in pedestrian causalities and a 25% decrease in pedestrian fatalities. These reductions were attributed to the change in travelling speeds caused by the change in the speed limit.

A British Department of Transport study, *Killing speed and saving lives: The government's strategy for tackling the problem of excess speed on our roads,* contains a graph of the rate of pedestrian fatalities and injury severity relating to motor vehicle speed which can be summarised as follows: at a vehicle speed of 32 km/h (20 mph), the chance of a fatality is 5%, at 48 km/h (30 mph) it is 45%, and at 64 km/h (40 mph) it is 85% (Department of Transport, 1992).

Anderson, McLean, Farmer, Lee, and Brooks (1997) conclude from studying 176 fatal pedestrian crashes in Adelaide, Australia, that if all drivers obeyed the speed limit the effect would be a 13% reduction in fatalities among pedestrians; if all drivers were travelling 10 km/h (6 mph) slower the effect would be a 48% reduction.

However, the classic study on these relationships is not British but German. Teichgräber (1983) reported a correlation between the probability of death for a pedestrian hit by a car and the collision speed of the passenger car as 20 km/h (12 mph) ≈ 10%, 30 km/h (19 mph) ≈ 20%, 40 km/h (25 mph) ≈ 30%, 50 km/h (31 mph) ≈ 60%, 60 km/h (37 mph) ≈ 80%, 70 km/h (44 mph) ≈ 95% and 80 km/h (50 mph) ≈ 98%.

A Swedish study, based on the largest in-depth pedestrian accident study ever undertaken (at least at that time), shows that for adult pedestrians hit by the front of passenger cars the fatality risk at 50 km/h (31 mph) is more than twice as high as the risk at 40 km/h (25 mph) and more than five times higher than the risk at 30 km/h (19 mph) (Rosén & Sander, 2009).

A study from Maine (Gårder, 2004) shows that pedestrians face 50 times higher risk when crossing a four-lane street posted at 56 km/h (35 mph) and with frequent speeding, than when crossing a low-speed, two-lane street where drivers are travelling at speeds around 31 km/h (20 mph).

In conclusion, pedestrians are usually not seriously injured when hit by a car moving at a speed of 30 km/h (19 mph) or less at the time of impact. If vehicle speeds are between 30 and 55 km/h (19 and 34 mph), injuries are usually serious, while speeds above 55 km/h (34 mph) at the time of impact often endanger life.

Making Pedestrians Noticed

Ensuring low speeds and short crossing distances means that most pedestrians should be safe. However, a pedestrian who is intoxicated, or just distracted, may walk out in front of a moving car and be seriously injured also at moderate speeds. As previously pointed out, it should not be up to pedestrians to look for gaps between cars but drivers should look for gaps between pedestrians. In order for drivers to yield to pedestrians, drivers need to notice them, or at least to notice that there is a marked crosswalk. There are many ways to make crosswalks more noticeable, one is seen in Fig. 9. The absolute minimum, besides good pavement markings, is to have signs indicating where the crosswalk is located as, especially in locations with snow, the markings may be obstructed or worn off by ploughing efforts. To have additional markings, including speed limits painted in the carriageway, and signs at the approaches to crosswalks may also be beneficial. Speed cushions or humps some 10 m prior to the crosswalk will lead to not only crosswalks being noticed but also to drivers approaching the crosswalks and thereby the pedestrians with lower speeds. Using rumble strips on crosswalk approaches has, however, not been shown to reduce crashes significantly (Gårder, 2004).

Fig. 9. Embedded Pavement Lights and Overhead Flashing Light in Old Town, Maine, US.

Prioritising Pedestrians

In Central Business Districts and residential areas, shared zones can be used to give pedestrians access to streets at all times. At other locations, pedestrians and automobiles may need to be separated in time. It is then important to minimise wait times for pedestrians. Signals that dwell on red mean that the pedestrian gets their walk phase immediately when pushing a button, or being detected automatically. If a conflicting vehicle phase is activated, that phase should be shortened, or at least its duration indicated by a countdown display of the number of seconds until the walk phase. A countdown at the end of the walk phase is also beneficial to both slow-walking pedestrians and younger pedestrians being informed about whether to try to reach a signalised intersection or not before the end of the walk phase (see Chapter 7 for safety issues related to intersections). Poorly designed facilities (or policies) can contribute to a lack of compliance by pedestrians, which is then often interpreted as risk-taking behaviour. Examples include defining crossing roads at mid-blocks as jaywalking and lengthy wait times at pedestrian signals.

INTELLIGENT TRANSPORTATION SYSTEMS

ITS functions can be structured to aid pedestrians in the following five functional requests, or needs, of pedestrians: environmental adaptation (using ITS to change driver behaviour to pedestrian limitations), guidance (leading or navigating pedestrians away from dangerous locations), danger alerts or information (about risks), confidence and security enhancements (to feel safe enough to leave home and walk), contact systems (to get in touch with others and/or being localised) (Leden et al., 2014). In the not-to-far future, pedestrian–vehicle interactions will probably become much less of a threat to pedestrian health and safety. However, in 2018 and for the next several years, ITS will not solve all or most issues and, even in a future with ITS, we will need safe sidewalks where errant vehicles are not injuring pedestrians. Auto-brake and eventually autonomous vehicles will lead to higher safety but we need to design our roads to be safe before autonomous vehicles are mandated.

CONCLUSIONS

We as a society need to encourage walking as a mode of transportation since walking promotes better health and a cleaner environment; that is, a more sustainable society. However, it has to be safe to walk. We need more research to be able to state with certainty how 'acceptable' safety can be reached. An interim hypothesis is that all arterials and collector roads – at least segments with more than 50 pedestrians a day – should have sidewalks. The sidewalks should be separated from the roadway by a curb if speeds are low, and by a barrier or wide separation strip in high-speed areas, that is, where speeds are higher than 50 km/h (31 mph). Local roads also need sidewalks unless traffic volumes and speeds are very low.

The major safety issue for pedestrians is, however, where they cross streets. In particular, elderly pedestrians and pedestrians in a great hurry or under the influence of intoxicants need streets to be narrow and have low speeds for them to be able to cross safely. Marking crosswalks or even signalising them will have only marginal safety effects at best. Posting them for low speed is also not enough unless we have automatic camera speed-enforcement ensuring that everyone drives slowly. Rather, using very narrow cross-sections or speed cushions ensuring that 90-percentile speeds are no more than 30 km/h (19 mph) at crossing points is key to safety.

REFERENCES

American Association of State Highway and Transportation Officials. (2011a). *A policy on geometric design of highways and streets* (6th ed.).

American Association of State Highway and Transportation Officials. (2011b). *Roadside design guide.*

Amorim, J. H., Joana Valente, J., Cascão, P., Rodrigues, V., Pimentel, C., Miranda, A. I., & Borrego, C. (2013). Pedestrian exposure to air pollution in cities: Modeling the effect of roadside trees. *Advances in Meteorology*, *2013*, Article ID 964904.

Anderson, R. W. G., McLean, A. J., Farmer, M. J. B., Lee, B. H., & Brooks, C. G. (1997). Vehicle travel speeds and the incidence of fatal pedestrian crashes. *Accident Analysis and Prevention, 29*, 667–674.

Carré. J.- R. (1997). Mobilité urbaine et déplacements non motorisés: situation actuelle, évolutions, pratiques et choix modal, Volume II-Partie 1: La sécurité des piétons et des cyclistes, INRETS-rapport sur convention Ministères de l'Environnement (DGAD/SRAE) – Ministère de l'Equipement (DRAST).

Chen, L., Chen, C., & Ewing, R. (2014). The relative effectiveness of signal related pedestrian countermeasures at urban intersections—Lessons from a New York City case study. *Transport Policy*, *32*, 69–78.

Department of Transport. (1992). *Killing speed and saving lives: The government's strategy for tackling the problem of excess speed on our roads.* London: Department of Transport.

Elvik, R., Hoye, A., Vaa, T., & Sorensen, M. (2009). *The handbook of road safety measures* (2nd ed.). Emerald Group.

Federal Highway Administration. (2009). *The manual on uniform traffic control devices (MUTCD)*. Washington, DC: Federal Highway Administration.

Federal Highway Administration. (2013). *Clear zones*. Retrieved from: http://safety.fhwa.dot.gov/roadway_dept/clear_zones/cmclearzones/. Accessed on 4 November 2013.

Gårder, P. (1989). Pedestrian safety at traffic signals: A study carried out with the help of a traffic conflicts technique. *Accident Analysis & Prevention*, *21*(5), 435–444.

Gårder, P. (2004). The impact of speed and other variables on pedestrian safety in Maine. *Accident Analysis & Prevention*, *36*(4), 501–696.

Herms, B. (1972). *Pedestrian crosswalk study: Accidents in painted and unpainted crosswalks*. Transportation Research Board Record No. 406, Washington, DC, USA.

Isebrands, H., Hallmark, S., Fitzsimmons, E., & Stroda, J. (2008). *Toolbox to evaluate the impacts of roundabouts on a corridor or roadway network*. Center for Transportation Research and Education, Iowa State University.

Johansson, C., Gårder, P., & Leden, L. (2003). Towards Vision Zero at zebra crossings—A case study in Malmö, Sweden on traffic safety and mobility for children and elderly. *Transportation Research Record*, *1828*, 67–74.

Leden, L., Gårder, P., & Johansson, C. (2006). Safe pedestrian crossings for children and elderly. *Accident Analysis & Prevention*, *38*(2), 289–294.

Leden, L., Gårder, P., Schirokoff, A., Monterde-i-Bort, H., Johansson, C., & Basbas, S. (2014). A sustainable city environment through child safety and mobility – A challenge based on ITS? *Accident Analysis & Prevention*, *62*, 406–414.

Leden, L., Wikström, P.-E., Gårder, P., & Rosander, P. (2006). Safety and accessibility effects of code modifications and traffic calming of an arterial. *Accident Analysis & Prevention*, *38*(3), 455–461.

Lord, D. (1996). Analysis of pedestrian conflicts with left-turning traffic. *Transportation Research Record*, *1538*, 61–67.

Lord D., Smiley, A., & Haroun, A. (1998). Pedestrian accidents with left-turning traffic at signalized intersections: Characteristics, human factors and unconsidered issues. Paper presented at the 77th annual TRB meeting, TRB, National Research Council, Washington, DC.

Reynolds, G. (2015). Exercise in just the right dose. *New York Times*, April, 21, p. D6.

Rosén, E., & Sander, U. (2009). Pedestrian fatality risk as a function of car impact speed. *Accident Analysis and Prevention*, *41*, 536–542.

Shinar, D. (2008). *Traffic safety and human behavior*. Bingley: Emerald Group Publishing Limited.

Teichgräber, W. (1983). Die Bedetung der Geschwindigkeit für die Verkehrssicherheit. *Zeitschrift für Verkehrssicherheit*, *2*, 53–63.

Trafikverket. (2012). *Krav för vägars och gators utformning*. Publikation, 179.

USA Today. (2015). Let 'free range' kids roam to their homes. *USA Today*, April 27, p. 7A.

Walz, F. H., Hoefliger, M., & Fehlmann, W. (1983). Speed limit reduction from 60 to 50 km/h and pedestrian injuries. In *Twenty-seventh stapp car crash conference proceedings with international research council on biokinetics and impacts* (IRCOBI) (pp. 311–318). Warrendale, PA: Society of Automotive Engineers.

World Health Organization (2013a). *WHO calls for action to save lives*. Retrieved from http://www.who.int/mediacentre/news/notes/2013/make_walking_safe_20130502/en/

World Health Organization. (2013b). World report on disability. Technical appendix A: Estimates of disability. Retrieved from http://data.worldbank.org/data-catalog/world-report-on-disability.

Yannis, G., Papadimitriou, E., & Theofilatos, A. (2013). Pedestrian gap acceptance for mid-block street crossing. *Transportation Planning and Technology*, *36*, 450–462.

Zegeer, C (2005). *Safety effects of marked versus unmarked crosswalks at uncontrolled locations*. Final Report and Recommended Guidelines, FHWA Publication Number: HRT-04-100.

Zegeer, C. V., Stewart, J. R., Huang, H. H., & Lagerwey, P. A. (2002). *Safety effects of marked vs. unmarked crosswalks at uncontrolled locations: Executive summary and recommended guidelines*. FHWA-RD-01-075, U.S. Department of Transportation, Federal Highway Administration, McLean, VA.

CHAPTER 11

PROVIDING FOR BICYCLISTS

Narelle Haworth and Jacqueline Fuller

ABSTRACT

Purpose – *Bicycle riding provides a sustainable and affordable solution to many of the significant problems associated with motorised transport and physical inactivity. The provision of infrastructure plays an important role in encouraging people to begin and subsequently continue to ride bicycles and to do so safely.*

Methodology – *This chapter describes different types of on- and off-road infrastructure and reviews studies of their effects on rider numbers and safety. In addition, it looks at the roles that end-of-trip facilities and bike-share programs can play in contributing to bicycle use and general transport sustainability.*

Findings – *Infrastructure characteristics can influence both perceived and objective levels of safety. It is important to identify and avoid treatments that increase perceived safety but are actually less safe. The type of infrastructure needed or desired differs between current and potential riders and according to trip purpose. Well-designed marked bicycle lanes on roads can reduce crash rates. Safety at intersections can be improved by: advanced green lights for cyclists, short cuts for right-hand turns, brightly coloured bicycle paths and advanced waiting positions for cyclists. Off-road facilities are generally safer, but intersections with*

Safe Mobility: Challenges, Methodology and Solutions
Transport and Sustainability, Volume 11, 229–253

ISSN: 2044-9941/doi:10.1108/S2044-994120180000011012

roads must be carefully treated. Shared paths and footpaths are risky for older pedestrians (and older cyclists).

Implications – *In many countries the provision of more infrastructure that increases the perceived safety of riding is needed to encourage cycling, particularly transport cycling and cycling by women.*

Keywords: Bicycle riding; sustainable transport; transport cycling; recreational cycling; cycling infrastructure; literature review

INTRODUCTION

This chapter is the second of two chapters focussed on safety issues related to vulnerable road users. While this chapter focusses on cyclist safety concerns, it complements Chapter 10, which focusses on the safety concerns of pedestrians.

The United States (US), Canada and Australia are some of the most car-reliant countries in the world, with more than 80% of trips made by automobile, compared to 45–70% of trips in Western European nations. Unsurprisingly, countries which have a higher dependency on motorised transport have much higher recorded CO_2 emissions per capita (Buehler & Pucher, 2011). Considering also that motorised transport is one of the fastest growing consumers of non-renewable petroleum products, and as a result of this consumption is the fastest growing contributor of greenhouse gas emissions, our current practices are clearly not sustainable (Maibach, Steg, & Anable, 2009; Tolley, 2003). It is therefore vital that better attention is paid to the development and implementation of more sustainable forms of transport, defined as those which are able to meet the needs of the current population, without compromising the capacity of future generations to also meet their needs (Black, 1998).

In light of the significant problems associated with motorised transport and physical inactivity, cycling has steadily been promoted as a viable, healthier alternative. The advantages of choosing to cycle over driving an automobile have been well documented and range from addressing climate change and pollution to combating obesity (see, e.g., Handy, van Wee, & Kroesen, 2014; Heinen, van Wee, & Maat, 2010; Rojas-Rueda, de Nazelle, Teixidó, & Nieuwenhuijsen, 2012) as well as being inexpensive, accessible to almost everyone and an efficient and reliable mode of transportation (Hudson, 1982; Pucher, Buehler, Merom, & Bauman, 2011). In densely populated areas cycling can be a more efficient mode of transport than driving, allowing cyclists to avoid traffic jams while simultaneously working to reduce congestion (Australian Bicycle Council, 2010; Steinbach, Green, Datta, & Edwards, 2011). Despite these clear benefits however, cycling remains a relatively underused mode of transport, though this does vary considerably from country to country. Indeed, estimates show that cycling accounts for only 1% of all trips made in Australia, the US and United Kingdom (UK), compared with 18% and 27% in Denmark and the Netherlands, respectively (Buehler & Pucher, 2012; de Hartog, Boogaard, Nijland, & Hoek, 2010; Pucher & Buehler, 2008). It has been suggested that one of the reasons that active transportation (which includes cycling) is more prevalent in Europe is because European cities tend to be more compact, meaning that there is less distance to travel (Bassett, Pucher, Buehler, Thompson, & Tolmie, 2008). However, many trips made by car in low-cycling countries are short (less than 7.5 km) and thus are potentially amenable to cycling (de Hartog et al., 2010; Maibach et al., 2009).

The provision of infrastructure plays an important role in encouraging people to begin and subsequently continue to ride bicycles. Indeed, a study by Caulfield, Brick, and McCarthy (2012) reports that, of the persons surveyed, an increased number of off-road bicycle paths (74.1%) and better-connected on-road bicycle paths (56.4%) would encourage them to cycle to work. Similarly, Akar and Clifton (2009) report that both current cyclists and non-cyclists agreed that better bicycle paths would encourage them to continue to cycle or take up cycling. Other studies have also found a positive correlation between the supply of bicycle paths and cycling rates. For example, a US-based study, which considered 42 major cities, found that for every extra linear mile of bicycle lanes per square mile of city area there was an increase of one percentage point in the share of bicycle commuters (Dill & Carr, 2003).

In addition to encouraging bicycling by increasing the perceived safety of riding by current and potential bicycle riders (improving subjective safety), infrastructure provision can potentially improve actual safety (objective safety). Sorensen and Mosslemi (2009) reviewed walking and cycling initiatives in terms of their effects on both kinds of safety. Of the 54 initiatives that they reviewed, 6 had contrasting effects on the different kinds of safety. They identified 16 ‘problem treatments’ that have a positive effect on subjective safety (as reflected in increased use) but a negative effect on objective safety. Among these were: tracks for cycling, improving surface evenness, one-way streets, maintenance of paths during winter and road resurfacing.

The provision of infrastructure for bicycling needs to be guided by an understanding of the heterogeneity of (current and potential) riders and trip purposes. Some research has classified riders according to their degree of confidence and related this to the infrastructure required for these groups.

For instance, in his study of cyclists in Portland, Geller (2006) asserts that cyclists could be placed into one of four categories:

1. *The Strong and the Fearless*: cyclists who bicycle regardless of the road condition.
2. *The Enthused and Confident*: cyclists who are comfortable sharing the road with motorised traffic, but prefer to have their own facilities, such as bicycle lanes.
3. *The Interested but Concerned*: this type of cyclist enjoys cycling and may bicycle within their local vicinity. However, they are worried about sharing the road with motorised vehicles and do not use their bicycle as a mode of transport to work or to major commercial areas.
4. *No Way No How*: individuals falling within this category do not have any intention of cycling for reasons relating to topography, inability or disinterest.

This typology was also used by Dill and McNeil (2013), who explored the infrastructure necessary to increase cycling participation. The authors focussed on the largest group, Interested but Concerned, and noted that bicycle boulevards and cycling paths, in conjunction with signage, have been demonstrated to increase participation (see also Pucher, Dill, & Handy, 2010).

Other research has distinguished between cycling for transport (utilitarian purposes) and recreation (e.g., Heesch & Sahlqvist, 2013) and the types of infrastructure needed to underpin these activities. Transport cyclists, who may also be referred to as 'utility cyclists', are those who choose cycling as their preferred mode of transport. This group can be further divided into two sub-categories: commuter cyclists and non-commuter cyclists; that is, persons who cycle to and from their place of work and those who cycle to destinations other than the workplace. In comparison, as the name would suggest, recreational cyclists are those who choose to cycle for enjoyment or exercise (Heesch & Sahlqvist, 2013). Interestingly, the relative proportion of individuals who cycle for transport and recreation varies across the world (Pucher & Buehler, 2012). In countries where overall bicycle riding rates are low (such as Australia and North America), relatively more trips are made for recreation and less for transport (Australian Bicycle Council, 2017). Many transport cyclists are also recreational cyclists, but the reverse is less common. In comparison, European countries have much higher rates of commuter cycling, which accounts for 28% of all trips made in Germany, 32% in the Netherlands and 35% in Denmark (Danish Ministry of Transport, 2007; German Federal Ministry of Transport, 2003; Statistics Netherlands, 2010).

In countries where overall cycling participation rates are low, fewer women tend to participate in cycling when compared with men (see, e.g., Garrard,

Crawford, & Hakman, 2006; Garrard, Rose, & Lo, 2008; van Bekkum, Williams & Morris, 2011; Winters, Friesen, Koehoorn, & Teschke, 2007). Within European nations, rates of cycling participation for women appear to be more equal (Pucher & Buehler, 2008). Specifically, figures suggest that women constitute 55% of all cyclists in the Netherlands, 45% in Germany and 49% in Denmark, which compares with just 24% in the US (Pucher & Buehler, 2008). Not surprisingly, women are over-represented among less confident cyclists (Dill & McNeil, 2013) and among transport riders, compared to recreational riders (Heesch, Sahlqvist, & Garrard, 2012). Explanations such as activity preferences, time constraints, concerns about appearance and being less risk-averse have been proposed to account for the difference in female participation (Garrard, Handy, & Dill, 2012). It is important, therefore, that cycling policies and infrastructure planning take into account these differences and plan for them accordingly.

There is no doubt that cycling is a viable source of sustainable transport that can be accessible to almost anyone. However, as highlighted in the section above, not everyone is choosing to cycle and differences in participation exist across genders and countries. Thus, in order for more people to be encouraged to participate in cycling, it is important that the various needs of cyclists are identified and met.

HIERARCHY OF OPTIONS IN PROVIDING FOR BICYCLISTS

A five-step hierarchy developed in the UK (IHT (Institution of Highways and Transportation), CTC (Cyclists' Touring Club), Bicycle Association, and Department of Transport, 1996) provides a useful approach to selecting options for providing for bicyclists. These five options are briefly outlined as follows:

1. *Decrease traffic*: this can include traffic management schemes and off-road short cuts for cyclists.
2. *Decrease traffic speed*: reduced speeds can lessen the speed gap between cyclists and motorised vehicles and the risk of injury.
3. *Intersection treatment and traffic management*: issues surrounding the lack of appropriate spaces for cyclists to wait at intersections, narrow shoulders and difficult locations for crossing or turning should be addressed (see Chapter 7 for discussion about intersection design and operational safety).

4. *Reallocation of or additional carriageway space*: often existing roadways have enough space to cater appropriately for cyclists if unnecessary parking areas and median strips are removed.
5. *Specific bicycle infrastructure*: in instances where the above is not appropriate, cyclist-specific infrastructure, such as bicycle lanes and underpasses, may need to be provided.

Parkin and Koorey (2012) note that the hierarchy assumes that the route has already been determined and that it is merely a case of identifying and implementing the correct plan of action. They assert that spatial planning and demand modelling should be considered first.

Government transport authorities in some jurisdictions have adopted decision frameworks regarding the type of cycling infrastructure that should be provided, based on a combination of motorised vehicle speeds and motorised vehicle volumes (e.g., Austroads, 2017). These frameworks generally propose that off-road cycling facilities should be provided in urban areas where the posted speed limit is 80 km/h or greater (regardless of the traffic volume) or where the posted speed limit is 60 km/h or greater if the traffic volume exceeds 5,000 vehicles per day. These documents also recommend that segregated facilities be considered where the rare combination of low speeds and high volumes exist. Application of these frameworks results in a very different mix of cycling spaces than would apply under the Vision Zero road safety philosophy, which espouses that vulnerable road users (cyclists and pedestrians) should be separated from motorised vehicles whose speeds exceed 30 km/h (Johansson, 2009).

As the previous sections have highlighted, bicycle infrastructure plays an important role in encouraging individuals to begin and later continue to cycle. In light of this, the aim of the following sections is to provide a comprehensive overview of two forms of provision of infrastructure for bicyclists: on-road infrastructure and off-road infrastructure. An international approach will be taken, and thus it is beyond the scope of this chapter to discuss the technical details of the different types of infrastructure at any length.

ON-ROAD CYCLING INFRASTRUCTURE

In general, on-road cycling paths encompass a dedicated lane for bicycles on the road, which is often indicated by road markings. They may also include separate cycle paths, whereby bicycles and motorised vehicles are segregated by some form of physical divider (Lieswyn et al., 2012).

Marked Bicycle Lanes

A valuable part of cycling infrastructure, bicycle lanes specify where they are appropriate for motorised vehicles and bicycles to travel (see Fig. 1). They can provide a legal means for cyclists to navigate traffic on the road as well as create awareness among drivers with regard to cyclists as road users (Fowler & Koorey, 2006; Parkin & Meyers, 2010). As touched upon in the preceding

Fig. 1. A Marked Bicycle Lane Leading to an Intersection in Brisbane, Australia; Unfortunately the Marked Lane does not Continue Through the Intersection (*Source*: author).

sections of this chapter, however, there remains considerable anxiety among some groups of cyclists, particularly female and less-experienced cyclists, as to the safety of such on-road facilities (see, e.g., Fishman, Washington, & Haworth, 2012).

Despite this, however, there is some research to suggest that, in terms of well-designed bicycle lanes, such fears may be unfounded. For example, a recent study in New Zealand examined the influence on cycle count and crash numbers of implementing bicycle lanes. The study found that the installation of the lanes was associated with a decreased crash rate of 43% (Parsons & Koorey, 2013). Another study by Hamann and Peek-Asa (2013), which examined crash risk by any on-road bicycle facility and by facility type, including bicycle lanes, found that marked bicycle lanes could reduce the number of collisions between bicycles and motorised vehicles. In contrast, however, Jensen, Rosenkilde, and Jensen (2008), who conducted a before-and-after study of bicycle paths and lanes, found that bicycle lanes resulted in an insignificant increase in the number of accidents (5%) as well as the number of recorded injuries (15%). The results were reflective of a recorded 20% increase in bicycle and moped mileage.

Intersections: Careful Planning

Well-planned intersections are also important to ensure that all cyclists are safe while cycling. A review of the literature regarding infrastructure and risk for cyclists by Harris et al. (2013), for example, found that intersections were associated with higher risks for cyclists. Similarly, a Dutch study by Schepers, Kroeze, Sweers, and Wüst (2011) highlighted that most bicycle–motorised vehicle collisions occur within built-up areas at unsignalised major intersections.

In their review of cycling success stories in the Netherlands, Denmark and Germany, Pucher and Buehler (2008) expressly highlight the importance of the provision of separate cycling facilities along busy roadways and intersections. In particular, the authors suggest that some of the key measures in terms of making intersections safe for cyclists include: advanced green lights for cyclists, short cuts for right-hand turns, brightly coloured bicycle paths and advanced waiting positions for cyclists (Pucher & Buehler, 2008). Considering the risk that intersections pose for cyclists, it is important that more of these measures are adopted throughout the world.

OFF-ROAD CYCLING INFRASTRUCTURE

There are disagreements within the published literature as to what constitutes an off-road bicycle path. For the purposes of this chapter, however, off-road cycling infrastructure is taken to comprise all areas which are outside the roadside kerb. Under this definition, bicycle lanes with painted or physical separators are classified as on-road spaces, while footpaths (sidewalks in North America) are classified as off-road, whether they are specifically marked for bicycle use or not. In terms of the New Zealand lexicon of cycling facilities (Lieswyn et al., 2012), the off-road spaces included in this chapter include: cycle paths (whether Danish cycle tracks or cycle paths at footpath level, called cycle tracks in the European Cycling Lexicon; European Economic and Social Commission, 2011), exclusive or shared paths (beside a road or not) and footpaths. Trails (mountain bike tracks and shared-use trails) belong to the general category of off-road spaces but are not discussed in this chapter.

There is also considerable disagreement regarding the functions of on-road bicycle paths compared to off-road bicycle paths (see Fig. 2). Some cycling advocates argue that all roads should be suitable for cyclists, thereby discounting the need for off-road paths. In comparison, however, some road safety advocates assert that on-road cycling should be permitted only occasionally.

Fig. 2. An Off-Road Bicycle Path in Brisbane, Australia (*Source*: author).

Disagreement also exists within the academic literature regarding the relative safety of off-road and on-road bicycle paths (see, e.g., Aultman-Hall & Hall, 1998). Moreover, as the above sections have highlighted, cyclists are a highly heterogeneous group, with varying skill levels and needs. Australian and US studies have shown that men are more likely to prefer on-road bicycle paths, while female and less-experienced cyclists are more likely to prefer off-road infrastructure (Garrard et al., 2008; Stinson & Bhat, 2003).

An extensive amount of literature has looked at the performance of bicycle tracks. Traditionally, bicycle paths in Denmark are designed parallel and adjacent to the road, with a kerb to the carriageway and an additional kerb to the footpath (Jensen et al., 2008). Research which examined the before-and-after development of the bicycle paths in Denmark found that they were associated with an increase in non-motorised vehicle mileage (20%) and a decrease (10%) in motorised vehicle mileage. Although a non-significant decrease was found in relation to bicycle crashes (10%), a significant (18%) increase in crashes at intersections was found. However, the authors note that the increase was related to an increased number of motor vehicles turning corners to access back streets because of the removal of on-street parking that occurred as a result of the implementation of the bicycle paths (Jensen, 2008).

Similarly, a meta-analysis conducted by Elvik, Vaa, Erke, and Sorensen (2009) examined the safety outcome of bicycle paths in northern Europe. The authors report that the implementation of bicycle paths was associated with an 11% reduction in bicycle crashes on the road and a 24% increase in bicycle crashes at intersections. It was noted that these increases were most likely due to the increased numbers of cyclists from the implementation of bicycle paths.

Lusk, Morency, Miranda-Moreno, Willett, and Dennerlein (2013) completed a more recent study which looked at bicycle facility guidelines and crash rates on bicycle tracks in the US. The bicycle paths were an assortment of two-way, one-way and contra-flow paths. From their analysis of police-reported crash data and bicycle counts on 19 bicycle paths in the US, the authors found a total crash rate of 2.3 per million bicycle kilometres, which they argue is less than other published crash rates on roadways. The authors do, however, point out that injury severity data were unavailable at the time of their study, and thus it is unclear if a difference in severity between the crashes on bicycle paths and roads existed.

Finally, in a Canadian case-crossover study, Teschke et al. (2012) conclude that the risk of injury on a bicycle path is approximately one-ninth of that on a major street route with parked cars and without bicycle infrastructure.

It was further estimated that the risk of injury is 40% less on bicycle paths than on major streets, but this difference is not statistically significant.

In many countries it is common practice for riders on an off-road path to be required to give way to motor vehicles when the path crosses a road (see Fig. 3). While changing this practice would improve cyclist amenity, there is little research on what the outcomes would be in safety terms.

Shared Bicycle Paths

As the opening sections of this chapter have outlined, active transportation, specifically walking and cycling, is vital to sustainable transport, while also playing a key role in our overall health and wellbeing. Thus, in order to maximise the number of people choosing to participate in such activities, governments around the world have committed to implementing off-road facilities. However, because of financial and spatial limitations, some facilities have been designed for the shared use of cyclists and pedestrians (see, e.g., Austroads, 2017), which has caused some concern among cyclists and pedestrians (see Fig. 4). European research highlights that pedestrians, even those who cycle regularly, feel safer when there is a

Fig. 3. Off-Road Bicycle Path Where Riders Need to Give Way to Motor Vehicles to Cross Freeway On- and Off-Ramps (*Source*: author).

Fig. 4. A Shared Off-Road Bike Path in Brisbane, Australia (*Source*: author).

complete separation of pedestrians and cyclists (Bernhoft & Carstensen, 2008; Wennberg, Stahl, & Hyden, 2009).

The findings of research on the safety of shared paths are mixed. For example, the meta-analysis completed by Elvik et al. (2009) found that shared paths were associated with a 2% increase in bicycle injury crashes along the road and a 1% increase in bicycle injury crashes at intersections. In comparison, the analysis documented a 35% decrease in pedestrian injury collisions along the road and a 1% increase in pedestrian crashes at intersections. None of these differences were significant at the 5% level.

An observational study on shared paths in the Australian state of New South Wales observed more than 50,000 pedestrians and 12,000 bicyclists but only five near misses and no actual contact between bicyclists and pedestrians. The study concluded that the perception of danger of shared paths outweighs the actual level of risk (RTA, 2009).

Studies of riders hospitalised in New South Wales (Chong, Poulos, Olivier, Watson, & Grzebieta, 2010) and the UK (Davidson, 2005) have shown that most were males injured in collisions with motor vehicles on roads, and relatively few were injured on shared paths (or other off-road facilities). Chong et al. (2010) note that the severity of injury was greater for people aged 65 and older, regardless of whether they were a pedestrian in a collision with a cyclist or a cyclist in a collision with a pedestrian or a motor vehicle. They conclude that shared bicycle–pedestrian paths may need to be avoided in areas where there are many elderly pedestrians.

Some high-quality shared bicycle paths can not only provide for and encourage bicycling but also become attractions for tourists and locals. One such example is the Hudson River Greenway which opened in 2001. As a completely separate, bi-directional bicycle path, the Greenway provides cyclists and pedestrians with a safe and scenic route along Manhattan's western shore, which is especially important for families and less-experienced cyclists (Pucher, Thorwaldson, Buehler, & Klein, 2010).

Cycling on the Footpath

The use of footpaths by bicyclists has been suggested as another way to separate cyclists from motorised vehicles and to utilise the widespread network provided by footpaths. Unfortunately, however, footpath cycling is illegal or, at the very least, discouraged in many jurisdictions across the world (see Fig. 5). These regulations appear to originate from concerns regarding the risk to cyclists and pedestrians (see, e.g., Haworth & Schramm, 2011; Lusk et al., 2013). Despite this, however, very little research has investigated the issue in any detail.

Fig. 5. Sign Encouraging Cyclists to Use the Street Instead of the Footpath in Washington, D.C. (*Source*: author).

One of the earliest studies on this topic examined police-reported bicycle–motor vehicle collisions at intersections and driveways. The authors found that the increased likelihood of footpath crashes was almost exclusively related to cycling against the direction of traffic (Wachtel & Lewiston, 1994). Interestingly, a more recent review of North American literature by Reynolds, Harris, Teschke, Cripton, and Winters (2009) found that footpaths posed the largest risk of all cycling locations, while appropriate bicycle paths were associated with the lowest risk. Finally, an Australian study, which observed cyclist–pedestrian conflicts across six locations in Brisbane, found that the increased risk of a conflict was associated with several factors, including cycling on the footpath (Haworth, Schramm, & Debnath, 2014).

In an online survey of 2,532 adult riders in Queensland, Australia, where footpath cycling is legal (Haworth & Schramm, 2011), 33.9% of respondents reported riding on footpaths. A larger proportion of the distance ridden by new riders was on footpaths (6.5%) than for continuing (3.9%) or other (4.5%) riders. However, in terms of the mean distance travelled per week, continuing riders actually rode further on footpaths (3.73 km) than new riders (3.22 km) or other riders (3.10 km). Utilitarian riders were the most likely to ride on the footpath, followed by social and then fitness riders. Regardless of trip purpose, about two-thirds of all riders who rode on the footpath reported doing so reluctantly. Thus, it appears that footpaths connected the network for these riders, allowing them to avoid on-road sections that were dangerous or inconvenient. Almost 70% of the self-reported crashes on footpaths did not involve any other bicycle, pedestrian or animal (single-vehicle collision). The study found no increase in crash risk associated with riding on the footpath (about 5% of crashes and 5% of distance travelled), and the severity of injury was lower for footpath crashes. This agrees with slower riding speeds on footpaths reported in other studies (de Rome, Boufous, Senserrick, Richardson, & Ivers, 2011; Kiyota, Vandebona, Katafuchi, & Inoue, 2000), which may underpin some of the rider reluctance to travel on footpaths.

DIRECTIONAL SIGNAGE

At present, directional signage for bicyclists has been implemented across jurisdictions around the world, including New York (Pucher & Buehler, 2008; Pucher et al., 2010). Koh and Wong (2013) note that directional signs are particularly useful for cyclists as they enable those who are unfamiliar with the route to choose the quickest path. Certainly this is a valid point, particularly for transport cyclists, who place greater emphasis on travel time when deciding their route (see, e.g., Sener, Eluru, & Bhat, 2009; Stinson & Bhat, 2003). Arguably, however, such infrastructure could be utilised for much more. In the case of recreational cyclists, for example, directional signs could be used to direct them to a more scenic route, which they often prefer (Downward & Lumsdon, 2001), as opposed to the quickest (see Figs. 6 and 7). Similarly, the signs may indicate an alternative route in the event that the main bicycle path is blocked or otherwise inaccessible. There is also potential to provide guidance to safer routes or routes that are flatter. Recent work in this area has spread to electronic means of assisting cyclists to choose routes, with a proliferation of websites and apps providing suggested routes.

Although very little research has been done in this area (Pucher et al., 2010), there is some evidence to suggest that directional signage and electronic mapping may be attractive additions to bicycle infrastructure for all cyclists (Austroads, 2017).

END-OF-TRIP FACILITIES

One final important consideration in this area relates to end-of-trip facilities for cyclists. In general, end-of-trip facilities refer to dedicated places that

Fig. 6. Directional Signage on a Long-Distance Cycling Route near Cologne, Germany (*Source*: author).

Fig. 7. Location Map on a Popular Shared-Use Path in Brisbane, Australia (*Source*: author).

support people who use non-motorised modes of transport. This can include any number of the following: shower facilities, change rooms, secure store facilities and parking areas (see Fig. 8; Department of Housing and Public Works, 2013; Pucher et al., 2010).

The importance of such facilities in encouraging people to either begin to cycle or continue to cycle has been well recognised (Litman, 2009; Pucher et al., 2010). The Netherlands in particular has a strong track record of providing quality end-of-trip facilities, such as bicycle parking (see Fig. 9). Moreover, improved bicycle parking in the Netherlands has been associated with a significant increase in the number of bicycle trips taken between residences and public transport (Martens, 2007). The importance of end-of-trip facilities in encouraging people to take up cycling has also been highlighted by Akar and Clifton (2009) in their study of the influence of bicycle infrastructure on the decision to cycle. In particular, the authors found that the majority of the non-bicyclists who were surveyed (70%) stated that a suitable place to change clothes and shower would encourage them to cycle. Considering that one of the barriers to women's participation in cycling is concern about appearance (Garrard et al., 2012), the value of end-of-trip

Fig. 8. Bicycle Parking Facility at Amsterdam Railway Station (*Source*: author).

facilities should not be ignored if sustainable transport is to be promoted and accepted within the wider community.

BIKESHARE PROGRAMS

Public bicycle hire (commonly termed 'bikeshare') programs are another form of providing for bicycling. They aim to encourage bicycling, enhance transport choices, improve air quality and ease congestion (see Fig. 10). Moreover, the programs also work to provide people with transport options for the 'first/last mile' – the distance that may be deemed too far to walk between home and public transport, and/or public transport and the place of work (Shaheen, Guzman, & Zhang, 2010). The fundamental principle of such programs is simple – people can make use of the bicycles whenever needed, but without the costs and responsibilities of owning their own bicycle.

There have been several variations of bikeshare over the years but, in general, current bikeshare programs tend to encompass, 'the provision of bicycles to enable short-term rental from one docking station to another' (Fishman, Washington, & Haworth, 2013, p. 148). In comparison to bicycles for private use, shared bicycles tend to be heavier and are designed to be used several times a day and in all weather conditions (Midgley, 2011). Most bicycles are equipped

Fig. 9. Bicycle Parking as Part of End-of-Trip Facilities in a New Office Building (*Source*: author).

with a tracking device to monitor the bicycle's journey from one docking station to another (Fishman et al., 2013; Midgley, 2011). Short-term use is widely encouraged with the provision of a 'free ride period', whereby the bicycle can be ridden at no cost, for a set period of time, before the user is charged. Bicycle hire and payment is usually made via credit card (Fishman et al., 2013).

While bikeshare programs began in the 1960s they have increased dramatically in popularity in the last decade (Shaheen et al., 2010). In 2007, for example, Paris initiated a 20,000-bikeshare program. The Chinese cities of Wuhan and Hangzhou also currently operate the largest programs, with 70,000 and 60,000 bicycles, respectively (DeMaio, 2009; Fishman et al., 2013). More recently, 2010 saw the launch of similar programs in Brisbane and Melbourne (Fishman, Washington, Haworth, & Watson, 2015). At present, there are approximately 700 cities with bikeshare programs, operating in more than 30 countries around the world (Meddin & DeMaio, 2014; Midgley, 2011).

Considering the increasing number of bikeshare programs which have been established on a global scale, as well as the operating and maintenance costs associated with such programs (New York City Department of City Planning as cited in Midgley, 2011), it is important to examine their success in terms of sustainable transport. Many cities have claimed that their bikeshare programs have led to substantial savings in car trips and, consequently, CO_2

Fig. 10. Bikeshare Docking Station in Gothenburg, Sweden (*Source*: author).

emissions (Shaheen et al., 2010). However, these claims have often been based on the assumptions that all bikeshare trips replace car trips and have ignored the distances driven by trucks used to reposition bicycles. A recent study by Fishman et al. (2014) looked at the impact of bikeshare programs on the use of private motorised transport using data from several cities across the world. The study observed a decrease of motorised transport use of approximately 90,000 km/year in Melbourne and Minneapolis/St Paul and 243,291 km for Washington D.C. This is compared with an increase of motorised transport use in London. The increase in London was primarily due to a small car mode replacement rate and considerable truck use for rebalancing of bicycles (Fishman et al., 2015). Thus, while bikeshare programs can contribute to bicycle usage and sustainability, the reduction in car trips that they bring about can be modest or even negligible.

CONCLUSIONS

Bicycle riding provides a sustainable and affordable solution to many of the significant problems associated with motorised transport and physical inactivity. The provision of infrastructure plays an important role in encouraging people to begin and subsequently continue to ride bicycles and to do so safely.

The provision of desired infrastructure is critical for increasing current levels of transport cycling in many countries given that there is less flexibility in destinations and therefore routes for transport cycling. Cycling for transport is arguably more important from the viewpoint of sustainability than cycling for recreation, because it is more likely to replace trips by motorised vehicles and to occur on a more regular basis. While recreational riding has health benefits, these are proportionally less, given the likely lower frequency of riding.

Provision of infrastructure which makes cycling feel safer is also crucial to address the lower rates of bicycle riding by women in many countries. The experience of countries where bicycle riding is a common means of transport suggests that good infrastructure, supplemented by cultural acceptance, can result in similar rates of riding by males and females. The evidence suggests that women are more concerned about the danger of collisions with motor vehicles and therefore prefer infrastructure where they can ride separated from motor vehicles.

Fear of being injured as a result of a collision with a motor vehicle (Fishman et al., 2012; Pucher, Greaves, & Garrard, 2010; Rose, Lo, & Garrard, 2008) and of abusive behaviour by drivers (Daley, Rissel, & Lloyd 2007; O'Connor & Brown, 2008) are major barriers to bicycle riding for many people. Provision of better infrastructure can reduce the fear of being injured in a collision and may go some way towards combating the idea held by some motorists that cyclists are not legitimate road users (see, e.g., Basford, Reid, Lester, Thompson, & Tolmie, 2002; Walker, 2007). However, infrastructure alone will probably be insufficient to deal fully with this issue. O'Connor and Brown (2010) suggest that clear directions to the public about the road being a legitimate place for cyclists and road rules which concern cyclists' behaviour are vital. Daley, Rissel, and Lloyd (2007) further assert that political support and leadership are necessary to support campaigns that legitimise cycling as an important form of sustainable transport that is deserving of quality infrastructure and strong investment.

A new research challenge is how to optimise the hard and soft infrastructure that is being developed for connected and autonomous vehicles in order to maintain and improve the safety and amenity of bicycle riders.

REFERENCES

Akar, G., & Clifton, K. J. (2009). Influence of individual perceptions and bicycle infrastructure on decision to bicycle. *Journal of the Transportation Research Board*, *2140*, 165–172. doi: 10.3141/2140-18

Aultman-Hall, L., & Hall, F. L. (1998). Ottawa-Carleton commuter cyclist on- and off-road incident rates. *Accident Analysis & Prevention, 30*(1), 29–43.

Australian Bicycle Council. (2010). *Gearing up for active and sustainable communities*. National Cycling Strategy 2011-2016 (AP-c85/10). Sydney: Austroads.

Australian Bicycle Council. (2017). *Australian cycling participation: Results of the 2017 national cycling participation survey* (AP-C91-17). Sydney: Austroads.

Austroads. (2017). *Cycling aspects of Austroads guides* (AP-G88-17). Sydney: Austroads.

Basford, L., Reid, S., Lester, T., Thompson, J., & Tolmie, A. (2002). *Drivers' perceptions of cyclists*. Transport Research Laboratory, Wokingham: United Kingdom.

Bassett, D. R., Pucher, J., Buehler, J., Thompson, D. L., & Crouter, S. E. (2008). Walking, cycling, and obesity rates in Europe, North America, and Australia. *Journal of Physical Activity and Health, 5*, 795–814.

Bernhoft, I. M., & Carstensen, G. (2008). Preferences and behaviour of pedestrians and cyclists by age and gender. *Transportation Research Part F: Traffic Psychology and Behaviour, 11*(2), 83–95.

Black, W. (1998). Sustainability of transport. In B. Hoyle & R. Knowles (Eds.), *Modern transport geography* (2nd ed., pp. 337–351). Chichester: Wiley.

Buehler, R., & Pucher, J. (2011). Sustainable transport in Freiburg: Lessons from Germany's environmental capital. *International Journal of Sustainable Transportation, 5*(1), 43–70. doi: 10.1080/15568311003650531

Buehler, R., & Pucher, J. (2012). International overview: Cycling trends in western Europe, North America, and Australia. In J. Pucher & R. Buehler (Eds.), *City cycling* (pp. 287–318). Cambridge: MIT Press.

Caulfield, B., Brick, E., & McCarthy, O. T. (2012). Determining bicycle infrastructure preferences – A case study of Dublin. *Transportation Research Part D, 17*, 413–417. doi: 10.1016/j.trd.2012.04.001

Chong, S., Poulos, R., Olivier, J., Watson, W. L., & Grzebieta, R. (2010). Relative injury severity among vulnerable non-motorised road users: Comparative analysis of injury arising from bicycle–motor vehicle and bicycle–pedestrian collisions. *AAP, 42*, 290–296.

Daley, M., Rissel, C., & Lloyd, B. (2007). All dressed up and nowhere to go? A qualitative research study of the barriers and enablers to cycling in inner Sydney. *Road & Transport Research: A Journal of Australian and New Zealand Research and Practice, 16*(4), 42–52.

Danish Ministry of Transport. (2007). *Danish national travel surveys*. Copenhagen, Denmark: Danish Institute of Transport Research.

Davidson, J. A. (2005). Epidemiology and outcome of bicycle injuries presenting to an emergency department in the United Kingdom. *European Journal of Emergency Medicine, 12*(1), 24–29.

de Hartog, J. J., Boogaard, H., Nijland, H., & Hoek, G. (2010). Do the health benefits of cycling outweigh the risk? *Environmental Health Perspectives, 118*(8), 1109–1116.

De Rome, L., Boufous, S., Senserrick, T., Richardson, D., & Ivers, R. (2011). *The pedal study: Factors associated with bicycle crashes and injury severity in the ACT*. Sydney, Australia: The George Institute for Global Health, University of Sydney Medical School & Australian National University.

DeMaio, P. (2009). Bike-sharing: History, impacts, models of provision, and future. *Journal of Public Transportation, 12*(4), 41–56.

Department of Housing and Public Works. (2013). *End-of-trip facilities*. Retrieved from http://www.hpw.qld.gov.au/SiteCollectionDocuments/EndOfTripFacilities.pdf

Dill, J., & Carr, T. (2003). Bicycle commuting and facilities in major US cities: If you build them, commuters will use them –Another look. *Transportation Research Record, 1828*, 116–123.

Dill, J., & McNeil, N. (2013). Four types of cyclists? Examination of typology for better understanding of bicycling behavior and potential. *Transportation Research Record: Journal of the Transportation Research Board*, *2387*(1), 129–138.

Downward, P., & Lumsdon, L. (2001). The development of recreational cycle routes: An evaluation of user needs. *Managing Leisure*, *6*, 50–60. doi:10.1080/13606710010026368

Elvik, R., Vaa, T., Erke, A., & Sorensen, M. (Eds.). (2009). *The handbook of road safety measures*. Bingley: Emerald Group Publishing.

European Economic and Social Commission. (2011). *European cycling lexicon*. Retrieved from http://www.eesc.europa.eu/resources/docs/eesc-2011-27-en.pdf

Fishman, E., Washington, S., & Haworth, N. (2012). Understanding the fear of bicycle riding in Australia. *Journal of Australasian College of Road Safety*, *23*(3), 19–27.

Fishman, E., Washington, S., & Haworth, N. (2013). Bike share: A synthesis of the literature. *Transport Reviews. A Transnational Transdisciplinary Journal*, *33*(2), 148–165. doi:10.1080/01441647.2013.775612

Fishman, E., Washington, S., & Haworth, N. (2014). Bike share's impact on car use: Evidence from United States, Great Britain and Australia. *Transportation Research Part D: Transport and Environment*, *31*, 13–20.

Fishman, E., Washington, S., Haworth, N., & Watson, A. (2015). Factors influencing bikeshare membership: An analysis of Melbourne and Brisbane. *Transportation Research Part A: Police and Practice*, *71*, 17–30. doi:10.1016/j.tra.2014.10.021

Fowler, M., & Koorey, G. (2006). The effects of the Pages Road cycle lane on cyclist safety and traffic flow operations. In *IPENZ Transportation Group technical conference*, Queenstown, New Zealand, 8–11 October 2006, 13pp.

Garrard, J., Crawford, S., & Hakman, N. (2006). *Revolutions for women: Increasing women's participation in cycling for recreation and transportation.* Victoria: Deakin University.

Garrard, J., Handy, S., & Dill, J. (2012). Women and cycling. In J. Pucher & R. Buehler (Eds.), *City cycling* (pp. 211–234). Cambridge, MA: MIT Press.

Garrard, J., Rose, G., & Lo, S. K. (2008). Promoting transportation cycling for women: The role of bicycle infrastructure. *Preventative Medicine*, *46*, 55–59.

Geller, R. (2006). *Four types of cyclists*. Retrieved from http://www.portlandoregon.gov/transportation/article/264746

German Federal Ministry of Transport. (2003). *German federal travel survey 2002 (MiD)*. Berlin: German Federal Ministry of Transport.

Hamann, C., & Peek-Asa, C. (2013). On-road bicycle facilities and bicycle crashes in Iowa, 2007–2010. *Accident Analysis & Prevention*, *56*, 103–109.

Handy, S., Van Wee, B., & Kroesen, M. (2014). Promoting cycling for transport: Research needs and challenges. *Transport Reviews*, *34*(1), 4–24.

Harris, A. M., Reynolds, C.., Winters, M., Cripton, P. A., Shen, H., Chipman, M. L., …, Teschke, K. (2013). Comparing the effects of infrastructure on bicycling injury at intersections and non-intersections using a case-crossover design. *Injury Prevention*, *19*(5), 303–310. doi:10.1136/injuryprev-2012-040561

Haworth, N., & Schramm, A. (2011). How do level of experience, purpose for riding and preference for facilities affect location of riding? Study of adult bicycle riders in Queensland, Australia. *Transportation Research Record, 2247,* 17–23.

Haworth, N., Schramm, A., & Debnath, A. K. (2014). An observational study of conflicts between cyclists and pedestrians in the city centre. *Journal of the Australasian College of Road Safety*, *25*(4), 31–40.

Heesch, K., & Sahlqvist, S. (2013). Key influences on motivations for utility cycling (cycling for transport to and from places). *Health Promotion Journal of Australia*, *24*(3), 227–233.

Heesch, K., Sahlqvist, S., & Garrard, J. (2012). Gender differences in recreational and transport cycling: A cross-sectional mixed-methods comparison of cycling patterns, motivators, and constraints. *International Journal of Behavioral Nutrition and Physical Activity*, *9*(1), 106–146.

Heinen, E., Van Wee, B., & Maat, K. (2010). Commuting by bicycle: An overview of literature. *Transport Reviews*, *30*(1), 59–96.

Hudson, M. (1982). *Bicycle planning, policy and practice*. London: Architectural Press.

IHT (Institution of Highways and Transportation), CTC (Cyclists' Touring Club), Bicycle Association, and Department of Transport. (1996). *Cycle friendly infrastructure: Guidelines for planning and design*. Goldalming: Cyclists' Touring Club.

Jensen, S. U. (2008). Bicycle tracks and lanes: A before-after study. Paper presented at TRB annual meeting. Retrieved from http://www.ibiketo.ca/sites/default/files/bicycle%20tracks%20and%20lanes.pdf. Accessed on 28 June 2014.

Jensen, S., Rosenkilde, C., & Jensen, N. (2008). Bicycle tracks and lanes: A before–after study. In *87th annual meeting of the Transportation Research Board*. Transportation Research Board, Washington, D.C.

Johansson, R. (2009). Vision zero – Implementing a policy for traffic safety. *Safety Science*, *47*, 826–831.

Kiyota, M., Vandebona, U., Katafuchi, N., & Inoue, S. (2000). Bicycle and pedestrian traffic conflicts on shared pavements. In *fourteenth velo-city international conference proceedings*, Munich.

Koh, P. P., & Wong, Y. D. (2013). Influence of infrastructural compatibility factors on walking and cycling route choices. *Journal of Environmental Psychology*, *36*, 202–213.

Lieswyn, J., Macbeth, A., Wilke, A., Ward, J., Ashford, J., Houghton, R., & Lloyd, W. (2012). An illustrated lexicon of cycle facilities. In *2 Walk & cycle conference*, Hastings, New Zealand, 22–24 February 2012, p. 13.

Litman, T. (2009). *Bicycle parking, storage, and changing facilities*. Retrieved from: http://www.vtpi.org/tdm/tdm85.htm.

Lusk, A. C., Morency, P., Miranda-Moreno, L. F., Willett, W. C. & Dennerlein, J. T. (2013). Bicycle guidelines and crash rates on cycle tracks in the United States. *American Journal of Public Health*, *103*(7), 1240–1248.

Maibach, E., Steg, L., & Anable, J. (2009). Promoting physical activity and reducing climate change: Opportunities to replace short car trips with active transportation. *Preventative Medicine*, *49*, 326–327.

Martens, K. (2007). Promoting bike-and-ride: The Dutch experience. *Transportation Research Part A*, *41*, 326–338.

Meddin, R., & DeMaio, P. (2014). *The bike-sharing world map*. Retrieved from http://www.bikesharingworld.com

Midgley, P. (2011). *Bicycle-sharing schemes: Enhancing sustainable mobility in urban areas* (pp. 1–12). New York: United Nations, Department of Economic and Social Affairs.

O'Connor, J. P., & Brown, T. D. (2010). Riding with the sharks: Serious leisure cyclists' perceptions of sharing the road with motorists. *Journal of Science and Medicine in Sport*, *13*, 53–58.

Parkin, J., & Koorey, G. (2012). Network planning and infrastructure design. In J. Parkin (Ed.), *Cycling and sustainability* (pp. 131–160). Bingley: Emerald Publishing.

Parkin, J., & Meyers, C. (2010). The effect of cycle lanes on the proximity between motor traffic and cycle traffic. *Accident Analysis & Prevention*, *42*(1), 159–165.

Parsons, J., & Koorey, G. (2013). The effect of cycle lanes on cycle numbers and safety. In *IPENZ Transportation Group conference*, Dunedin, New Zealand, 14–16 April 2013, 15pp.

Pucher, J., & Buehler, R. (2008). Cycling for everyone. Lessons from Europe. *Journal of the Transportation Research Board*, *2074*, 58–65.

Pucher, J., & Buehler, R. (2012). International overview: Cycling trends in western Europe, North America, and Australia. In J. Pucher & R. Buehler (Eds.), *City cycling* (pp. 9–30). Cambridge, MA: MIT Press.

Pucher, J., Buehler, R., Merom, D., & Bauman, A. (2011). Walking and cycling in the United States, 2001–2009: Evidence from the national household travel surveys. *American Journal of Public Health*, *101*(1), 310–317.

Pucher, J., Dill, J., & Handy, S. (2010). Infrastructure, programs, and policies to increase bicycling: An international review. *Preventive Medicine*, *50*, S106–S125.

Pucher, J., Greaves, S., & Garrard, J. (2010). Cycling down under: A comparative analysis of bicycling trends and policies in Sydney and Melbourne. *Journal of Transport Geography*, *19*(2), 332–345.

Pucher, J., Thorwaldson, L., Buehler, R., & Klein, N. (2010). Cycling in New York: Innovative policies at the urban frontier. *World Transport Policy and Practice*, *16*, 7–50.

Reynolds, C. C., Harris, M. A., Teschke, K., Cripton, P. A., & Winters, M. (2009). The impact of transportation infrastructure on bicycling injuries and crashes: A review of the literature. *Environmental Health*, *8*(1), 47.

Rojas-Rueda, D., de Nazelle, A., Teixidó, O., & Nieuwenhuijsen, M. J. (2012). Replacing car trips by increasing bike and public transport in the greater Barcelona metropolitan area: A health impact assessment study. *Environment International*, *49*, 100–109.

Rose, G., Lo, S. K., & Garrard, J. (2008). Promoting transportation cycling for women: The role of bicycle infrastructure. *Preventive Medicine*, *46*(1), 55–59.

RTA. (2009). *Study of bicyclist and pedestrian safety on shared paths.* Report by Taverner Research for NSW Roads and Traffic Authority.

Schepers, J. P., Kroeze, P. A., Sweers, W., & Wüst, J. C. (2011). Road factors and bicycle motor vehicle crashes at unsignalized priority intersections. *Accident Analysis & Prevention*, *43*(3), 853–861.

Sener, I. N., Eluru, N., & Bhat, C. R. (2009). An analysis of bicycle route choice preferences in Texas, US. *Transportation*, *36*(5), 511–539.

Shaheen, S. A., Guzman, S., & Zhang, H. (2010). Bikesharing in Europe, the Americas, and Asia: Past, present, and future. *Journal of the Transportation Research Board*, *2143*, 159–167. doi:10.3141/2143-20.

Sorensen, M., & Mosslemi, M. (2009). *Subjective and objective safety: The effect of road safety measures on subjective safety among vulnerable road users.* Oslo: The Institute of Transport Economics.

Statistics Netherlands. (2010). *Transportation statistics*. Amsterdam: Statistics Netherlands.

Steinbach, R., Green, J., Datta, J., & Edwards, P. (2011). Cycling and the city: A case study of how gendered, ethnic and class identities can shape healthy transport choices. *Social Science & Medicine*, *72*, 1123–1130. doi:10.1016/j.socscimed.2011.01.033

Stinson, M. A., & Bhat, C. R. (2003). Commuter bicyclist route choice: Analysis using a stated preference survey. *Transportation Research Record: Journal of the Transportation Research Board*, *1828*(1), 107–115.

Teschke, K., Harris, M. A., Reynolds, C. C., Winters, M., Babul, S., Chipman, M., … Cripton, P. A. (2012). Route infrastructure and the risk of injuries to bicyclists: A case crossover study. *American Journal of Public Health, 102*(12), 2336–2343.

Tolley, R. (2003). Ubiquitous, everyday walking and cycling: The acid test of a sustainable transport policy. In I. Docherty & J. Shaw (Eds.). *A new deal for transport? The UK's struggle with the sustainable transport agenda* (pp. 178–197). Oxford: Blackwell Publishing Ltd.

van Bekkum, J. E., Williams, J. M., & Morris, P. G. (2011). Cycle commuting and perceptions of barriers: Stages of change, gender and occupation. *Health Education, 111*(6), 476–497. doi: 10.1108/09654281111180472

Wachtel, A., & Lewiston, D. (1994). Risk factors for bicycle–motor vehicle collisions at intersections. *ITE Journal (Institute of Transportation Engineers), 64*(9), 30–35.

Walker, I. (2007). Drivers overtaking bicyclists: Objective data on the effects of riding position, helmet use, vehicle type and apparent gender. *Accident Analysis & Prevention, 39*(2), 417–425.

Wennberg, H., Stahl, A., & Hyden, C. (2009). Older pedestrians' perceptions of the outdoor environment in a year-round perspective. *European Journal of Ageing, 6*, 277–290.

Winters, M., Friesen, M. C., Koehoorn, M., & Teschke, K. (2007). Utilitarian bicycling: A multilevel analysis of climate and personal influences. *Journal of Preventative Medicine, 32*(1), 52–58. doi:10.1016/j.amepre.2006.08.027

METHODS FOR UNDERSTANDING AND PREDICTING SAFETY PERFORMANCE

CHAPTER 12

CROSS-SECTIONAL MODELLING

Fred Mannering

ABSTRACT

Purpose – *Information collected from police crash reports has long been the primary source of data for the analysis of factors that determine the likelihood of a crash (crash frequency) and its resulting severity (measured in terms of the extent of injuries to vehicle occupants). Proper cross-sectional analyses techniques, covered in this chapter, are important for guiding safety policy and countermeasures.*

Methodology – *This chapter provides an overview of some of the more commonly used cross-sectional statistical and econometric methods, and discusses the nuances and their limitations with regard to how they are applied to typical crash-report data.*

Findings – *The wide variety of analytic methods available to safety researchers makes the selection of appropriate methods critical. This chapter provides important guidance for safety researchers in their choice of methodological approach.*

Implications – *Understanding the importance of proper model specification, unobserved heterogeneity, endogeneity and other factors covered in this chapter is extremely important in analysing safety data and must be given full consideration before any results are finalised.*

Safe Mobility: Challenges, Methodology and Solutions
Transport and Sustainability, Volume 11, 257–277

ISSN: 2044-9941/doi:10.1108/S2044-994120180000011013

Keywords: Cross-sectional data; unobserved heterogeneity; specification error; count-data models; discrete outcome models

INTRODUCTION

This chapter is the first of six chapters that collectively represent state of the practice methodologies for both understanding and predicting the safety performance of transport networks. This first chapter describes how to best examine and understand cross-sectional crash data, and is a complement to the appropriate management of time-series crash data (Chapter 13), crash data from a sample of sites with many sites recording zero crashes (Chapter 14), crash data where examining severity is of prime importance or interest (Chapter 15), methods for identifying high-risk sites or blackspots (Chapter 16), and finally methods for using surrogate measures of crashes to examine risk (Chapter 17).

Although cross-sectional methods are commonly applied in vehicle-crash research, one must realise that the data are not cross-sectional in the traditional sense; that is, in true cross-sectional data, a sample of data would be collected at one point in time (such as a survey of households collected at a specific point in time) so that the time-varying effects need not be addressed. While a survey of driver opinions about the effectiveness of airbags, for example, could be collected over a very short period of time, and could thus be considered cross-sectional, vehicle crashes are relatively rare events that accumulate over time and require longer periods of data collection. So, while traditional cross-sectional methods are typically applied to the analysis of crash data there is a temporal component (the fact that the data are based on an accumulation of crashes over an extended time period) that can sometimes complicate the estimation of statistical models.

In fact, in addition to observed crashes accumulating over time, crash data also accumulate over space. For example, at the end of a one-month period an analyst would have the number of accumulated crashes that have occurred and the spatial distribution of these crashes. Spatially, crashes are typically assigned to a roadway entity (such as a specific section of roadway or a specific intersection) and the frequency of crashes that occur on specific roadway entities is what is used to assess the likelihood of a crash and the factors that may affect this likelihood.

In addition to the likelihood of a crash (measured by the number of crashes observed to accumulate over some time period on a roadway entity), once a crash has occurred an analyst will have access to detailed information on the crash, including resulting injury severities and other information that can be gathered from crash reports filed by police, or possibly even hospital data. This is an important distinction, because data that may be used to predict the likelihood of a crash cannot include the data that are available only after a crash has occurred. For example, the number of severely injured vehicle occupants observed over the time period cannot be used to predict the number of crashes over the time period because injury severity is only observed conditional on a crash having occurred. To study the likelihood of a crash the number of crashes over some pre-determined time period is most often used as the dependent variable in a statistical analysis. To study the severity of a crash once a crash has occurred the probability of various injury-severity outcomes (killed, severely injured, injured, no injury) becomes the focus of interest.

The intent of this chapter is to introduce some of the basic cross-sectional modelling methodologies for assessing the frequency and severity of crashes, as well as discuss a variety of modelling issues that may be encountered in the process of analysing traditional crash data. Please keep in mind that the intent is not to provide a comprehensive review of the literature. There are relatively recent review papers that cover crash-frequency modelling methods (Lord & Mannering, 2010; Mannering & Bhat, 2014) and crash-severity modelling methods (Savolainen et al., 2011) and the reader is referred to those sources for an exhaustive review of the literature.

ANALYSIS OF CRASH-FREQUENCY DATA

One of the most common forms of data (usually from police crash reports) is the frequency of crashes that occur on a roadway entity (such as a specific section of roadway or a specific intersection) over some pre-determined time period (such as a week, month or year). As mentioned above, while these data stretch over a period of time, and thus could technically be treated as time-series data, they are most often treated as cross-sectional data. However, when treating these data as cross-sectional the issue of time can create some interesting statistical issues. For example, consider a data sample of 1,500 intersections in a city, and the analyst has the number of crashes that occurred in the month of January for each of these intersections. More common is that the analyst will have the number of crashes per month for the 1,500 intersections over a period of say 2 years. In this case, each intersection generates 24 frequency observations (one for each month). The fact that there are 24 observations for each intersection (that vary over time) suggests that some time-series approaches may be appropriate, but having just 24 observations from each intersection is not sufficient to conduct a true time-series analysis. As a result, data of this type (repeat observations from the same highway entity) are generally treated as cross-sectional data, but the fact that the data are repeated from specific roadway entities (for example 24 from each intersection) needs to be addressed in the analysis, because the repeat observations from each entity will likely be correlated (more on this in the correlation-of-observations sub-section later in this chapter).

Consider a case where the number of crashes that occurred on the 1,500 highway intersections only includes data from the month of January (so we do not have more than one monthly observation from each intersection). The number of crashes at each intersection in January is count data and will consist of only non-negative integer values. As such, linear regression analyses (such as ordinary least squares) would not be appropriate, because such methods are used to deal with continuous data, which could give predicted values that are in fractions and possibly negative (if some sort of logarithmic transformation or censoring technique is not used). As a result, the Poisson and negative binomial regression models are the most common statistical approaches for analysing such crash-count data. In a Poisson regression model the probability of intersection i having n_i crashes in January, per the example above, (where n_i is a non-negative integer) is given by:

$$P(n_i) = \text{EXP}(-\lambda_i)\lambda_i^{n_i}/n_i! \tag{1}$$

where $P(n_i)$ is the probability of intersection i having n_i crashes in January and λ_i is the Poisson parameter for intersection i, which is equal to intersection i's expected number of crashes in January, $E[n_i]$. To introduce explanatory variables that will allow the analyst to statistically determine what intersection geometric, traffic and other conditions make crashes more or less likely to occur the Poisson parameter is made to be a function of the intersection, traffic and environmental characteristics. In looking at Eq. (1) it is clear that, in determining the relationship between intersection characteristics and the Poisson parameter (λ_i), a functional form must be chosen to ensure that λ_i is not negative. The exponential form is an obvious choice so that:

$$\lambda_i = \text{EXP}(\beta \mathbf{X}_i) \tag{2}$$

where $\mathbf{X}_i$ is a vector of explanatory variables associated with intersection i and β is a vector of the estimated parameters. This model is estimated using standard maximum likelihood methods, with the likelihood function given as (Washington et al., 2011):

$$L(\beta) = \prod_i \frac{\mathrm{EXP}\left[-\mathrm{EXP}(\beta \mathbf{X}_i)\right]\left[\mathrm{EXP}(\beta \mathbf{X}_i)\right]^{n_i}}{n_i!} \tag{3}$$

While Poisson regression can serve as a basis for the analysis of crash-frequency data, it has inherent limitations that make its direct application to crash data almost always inappropriate. The primary limitation is the property of the Poisson distribution that restricts the mean, $E[n_i]$, to be equal to the variance of the distribution, $\mathrm{var}[n_i]$. This restriction rarely if ever holds in crash data, which tend to be almost always over-dispersed ($\mathrm{var}[n_i] >> E[n_i]$). In such cases, the Poisson model is extended to the negative binomial (or the Poisson-Gamma, as this model is sometimes referred to). This is done by adding an error term, ζ, to Eq. (2) such that:

$$\log \lambda_i = \beta \mathbf{X}_i + \zeta_i \tag{4}$$

where $\mathrm{EXP}(\zeta_i)$ is gamma-distributed with a mean of one and a variance of γ. The addition of this term produces a negative binomial model that allows the mean to differ from the variance, such that $\mathrm{var}[n_i] = E[n_i][1 + \gamma E[n_i]]$, where γ is an additional estimated parameter. If, during model estimation, γ is not significantly different from zero then the data are not over-dispersed and the simple Poisson model is appropriate.

As research has progressed in the analysis of crash frequencies many other models and variations have been applied. Going back to the example of the number of crashes on 1,500 intersections in January, one would expect that many of these intersections would have zero-observed crashes in January. Count models allow for the zero possibility, but there may be a very large number of zeros in some databases. This has led to the consideration of several types of models for analysing such data. Chapter 14 in this book covers this topic and describes how to handle datasets with a very large number of zeros.

A variety of other count-data models and variations have also been considered over the years to potentially improve the statistical defensibility of the estimations and account for abnormalities in the data (such as under-dispersed data, $\mathrm{var}[n_i] << E[n_i]$, as opposed to over-dispersed data). Some of these efforts have included the Conway–Maxwell–Poisson model, and the negative binomial-Lindley model, while other work has the count process as a variation-ordered probability model. A complete review of count-data methods is provided in Lord and Mannering (2010) and this review is updated in Mannering and Bhat (2014).

Continuous and Duration Approaches to Crash Frequency Data

Crash frequency data can also be analysed by transforming data counts into a continuous variable for analysis using traditional non-count statistical methods. For example, if a segment of highway is of a known length (in kilometrès for example) and the daily traffic volume is known, the number of crashes observed in a particular time period (say one month) can be used to arrive at a crash rate (this would be calculated by dividing the crash frequency by the vehicle-kilometres travelled, which would be the total traffic volume over the month multiplied by the length of the highway segment). However, because crash rates on specific highway segments are assessed over some finite time period, some highway segments may have no crashes reported during the analysis period. Thus, modelling crash rates by standard ordinary least squares (as one would do with a continuous variable such as crash rate) would result in biased and inconsistent parameter estimates (Washington et al., 2011). The solution to this is to consider crash rates as a censored dependent variable (censored at zero) and apply a tobit model (Tobin, 1958). The tobit model is expressed (for roadway segment *i*) using a left limit of zero (which will be the case for crash rates) as:

$$\begin{aligned} Y_i^* &= \beta \mathbf{X}_i + \varepsilon_i, \quad && i = 1, 2, \ldots, N, \\ Y_i &= Y_i^* && \text{if } Y_i^* > 0, \\ &= 0 && \text{if } Y_i^* \leq 0 \end{aligned} \tag{5}$$

where N is the number of observations, Y_i is the dependent variable (for example, crashes per 100-million vehicle-kilometres travelled), $\mathbf{X}_i$ is a vector of independent variables (traffic and roadway segment characteristics), β is a vector of the estimated parameters and ε_i is a normally and independently distributed error term with zero mean and constant variance σ^2. It is assumed that there is an implicit, stochastic index (latent variable) equal to Y_i^* which is observed only when positive. The left limit of zero allows the tobit model to capture excess zeros and may avert potential conceptual issues when dealing with two-state models such as the zero-inflated Poisson and negative binomial models. Chapter 14 provides further discussions on this topic.

Yet another way of framing the crash-frequency problem is to consider the time between crashes, as opposed to the frequency of crashes over some time period. The frequency of crashes and the time between crashes are obviously interrelated. In fact, count-data models (such as the standard Poisson model) imply an underlying distribution of time between crashes (for the standard Poisson the underlying time distribution between crashes is exponentially distributed), and a model of the duration of time between crashes can be aggregated to produce an expected frequency in any given period of time.

The most common duration-model approach is a hazard-based model that considers the conditional probability of a crash happening at some time, $t + dt$, given that it has been time t since the last crash occurred. The hazard function is defined as:

$$h(t) = f(t)/[1 - F(t)] \tag{6}$$

where $F(t)$ is the cumulative distribution function of crashes and is equal to $P(T < t)$, with P being the probability of a crash before some transpired time t and T being a random time variable and $f(t)$ is the density function corresponding to $F(t)$ (the first derivative of the cumulative distribution with respect to time, $dF(t)/dt$). The hazard function ($h(t)$ in Eq. (6) gives the rate at which crash-free time periods are ending at time t (implying a crash will occur).

Hazard-based models can be estimated under a wide variety of distributional assumptions and non-parametric forms, and allow important inferences to be made on how the probabilities of having a crash change over time, with explanatory variables influencing the hazard function by either increasing or decreasing the probability of a crash in time $t + dt$ given that it has been time t since the last crash occurred (see Washington et al., 2011).

Hazard-based duration models can be quite sophisticated in terms of their ability to handle data and common problems associated with crash data (unobserved heterogeneity, etc.) and can offer insights relating to duration effects. As an example, instead of considering the number of crashes individual drivers have had over their lifetime, Mannering (1993) used a hazard-based duration model to study the factors affecting the time between their crashes. The results of this study found interesting duration effects, in that the longer males went without having a crash, the less likely they were to have a crash soon, but that the length of time females went without a crash was found to have no statistically significant effect on their crash probabilities.

There is considerable potential for the future application of duration models to crash frequency analysis, but the level of data required in terms of the timing of crashes and the values of explanatory variables and how they change over time can be prohibitive in many instances.

ANALYSIS OF CRASH-INJURY SEVERITY DATA

Crash data (most commonly sourced from police records) generally provide levels of vehicle occupant injury severity in discrete categories, such as no injury, possible injury, evident injury, disabling injury and fatality. Again these data are modelled as cross-sectional data but since all crashes do not occur at a single point in time they are really aggregated temporally over some time period, such as a month or year. This type of discrete data is well suited to standard discrete-outcome models, such as ordered probability models and unordered multinomial models.

Ordered probability models define an unobserved variable z, which is used as a basis for modelling the ordinal ranking of the data. This unobserved variable is specified for each crash observation i as:

$$z_i = \boldsymbol{\beta}\mathbf{X}_i + \varepsilon_i \tag{7}$$

where $\mathbf{X}_i$ is a vector of characteristics determining crash severity, $\boldsymbol{\beta}$ is a vector of the estimated parameters and ε_i is a random disturbance. Using this equation, observed injury severities, y_i, are defined as:

$$\begin{aligned} y_i &= 0 \quad \text{if } z_i \le \mu_0 \\ &= 1 \quad \text{if } \mu_0 < z_i \le \mu_1 \\ &= 2 \quad \text{if } \mu_1 < z_i \le \mu_2 \\ &= \ldots \\ &= j \quad \text{if } z \ge \mu_{j-1} \end{aligned} \tag{8}$$

where μ's are the estimated parameters that define y, which corresponds to integer rankings. If the disturbance term in Eq. (7) is assumed to be standard normal (with mean = 0 and variance = 1) then an ordered probit model results, and if the disturbance is assumed to be standard logistic then an ordered logit model results.

An alternative approach to modelling injury severity is not to consider the ordering of the data (i.e., that 'no injury' is less severe than 'possible injury') and instead consider a richer model specification where each injury-severity level has its own function that determines injury-outcome probabilities (as opposed to the single function used in ordered probability models). This unordered approach defines an injury-severity function as:

$$S_{ij} = \boldsymbol{\beta}_{ij}\mathbf{X}_{ij} + \varepsilon_{ij} \tag{9}$$

where S_{ij} is a severity function for crash observation i determining the occupant-injury-severity category j, $\mathbf{X}_{ij}$ is a vector of explanatory variables that affect the probability of injury-severity category j for crash observation i, β_{ij} is a vector of the estimated parameters for injury-severity i for crash observation i and ε_{ij} is an error term. A common approach to arrive at an estimated model is to assume that ε_i are generalised extreme value distributed. If this is done, the multinomial logit model results (see McFadden, 1981) with:

$$P_i(j) = \frac{\text{EXP}\left[\beta_{ij}\mathbf{X}_{ij}\right]}{\sum_{\forall J} \text{EXP}\left[\beta_{ij}\mathbf{X}_{ij}\right]} \tag{10}$$

where $P_i(j)$ is the probability that crash i will be of injury-severity level j and other terms are as previously discussed.

The choice of methodological approach (ordered vs. unordered) is difficult and likely data-dependent. Unordered discrete-modelling frameworks (such as a multinomial logit) provide great flexibility in how variables affect injury-severity probabilities (and can even be immune to under-reporting of less severe crashes if multinomial logit assumptions are strictly met. However, they do not account for the ordering of the data. In contrast, ordered probability models account for the ordering of the data, but place potentially unrealistic restrictions on how variables can affect severity-outcome probabilities (there is additional discussion of this later in the paper; for further discussions of these points see also Mannering & Bhat, 2014; Savolainen et al., 2011).

UNOBSERVED HETEROGENEITY

Available crash databases are often quite limited in terms of the number of explanatory variables they have available to statistically study the frequency and severity of crashes. Sometimes missing explanatory variables are just the result of poor or inadequate data collection. Other times missing explanatory variables evolve from factors that are not readily observable or virtually impossible to collect. If these potentially important missing explanatory variables are correlated with observed explanatory variables (as will almost certainly be the case) biased parameters will be estimated during statistical analysis and incorrect inferences could result. Missing explanatory variables may not be known to the analyst, but they can be broadly classified as unobserved heterogeneity.

As an example of such unobserved heterogeneity and how it may interact with a measurable explanatory variable, consider a statistical model of crash frequency that has snowfall accumulation as a measurable explanatory variable. The effect that snowfall accumulation can have on a roadway segment or intersection can vary widely from one location to the next, because snowfall can interact in complex ways with subtle elements of roadway geometrics that may not be readily measurable (angle of the sun at certain times of the day, etc.) and, in addition, drivers' responses to snowfall in terms of modifying their driving behavior can also vary greatly from one location to the next. By assuming that snowfall accumulation has the same effect on crash frequencies at all locations a potentially significant restriction is being imposed on the model (assuming a constant effect across all observations instead of having the snowfall effect vary from one location to the next as a result of unobserved factors) and erroneous parameter estimates will result across the model because not accounting for the variability in snowfall effects will result in biased parameter estimates across the model.

The unobserved heterogeneity effect can be resolved in a number of ways. One is to develop a model structure that allows the parameters to vary across observations to account for possible unobserved heterogeneity (unobserved factors that may vary across observations, such as the variability in the impact of snowfall accumulation). Allowing such parameter variation is often referred to as a random parameters approach because the distribution of parameters across observations is assumed to follow some probabilistic distribution. To allow for such random parameters in count-data models, for each observation *i* estimated parameters are written as (see Anastasopoulos & Mannering, 2009; Greene, 2007):

$$\beta_{\mathrm{i}} = \beta + \omega_i \tag{11}$$

where ω_i is a randomly distributed term (for example a normally distributed term with mean zero and variance σ^2). This formulation can be used as a basis to alter the Poisson parameter to $\lambda_i|\omega_i = \mathrm{EXP}(\beta_i\mathbf{X}_i)$ and $\lambda_i|\omega_i = \mathrm{EXP}(\beta_i\mathbf{X}_i + \varepsilon_i)$ in the negative binomial, with the corresponding probabilities for Poisson or negative binomial now $P(y_i|\omega_i)$ (see Eqs. (1), (2) and (4)). With this random parameters version of a count-data model the log-likelihood is written as (Anastasopoulos & Mannering, 2009):

$$\mathrm{LL} = \sum_{\forall I} \ln \int_{\omega_i} g(\omega_i) P(y_i \mid \omega_i)\, d\omega_i, \tag{12}$$

where $g(.)$ is the probability density function of the ω_i. Functional forms of the parameter density functions typically include normal, lognormal, triangular, uniform or Weibull distributions. Because probability estimations are computationally cumbersome, a simulation-based maximum likelihood method is used (more on this below).

For injury-severity models unobserved heterogeneity can be handled in similar ways. Consider the mixed-logit model (random-parameter multinomial logit model). To arrive at the mixed-logit model, random parameters are introduced with $f(\beta_i|\varphi)$, where φ is a vector of parameters of the chosen density function (mean and variance). The resulting mixed-logit injury-severity probabilities are (see McFadden & Train, 2000; Train, 2003):

$$P_i(j\,|\,\varphi) = \int \frac{e^{\beta_{ij}\mathbf{X}_{ij}}}{\sum_{\forall J} e^{\beta_{ij}\mathbf{X}_{ij}}} f(\beta_{ij}\,|\,\varphi)\,d\beta_{ij}, \tag{13}$$

where $P_i(j\,|\,\varphi)$ is the probability of injury severity j conditional on $f(\beta_{ij}|\varphi)$. If the variance in φ is determined to be significantly different from zero there will be crash-specific variations of the effect of $\mathbf{X}$ on injury severity across crash observation n, with the density function $f(\beta_{ij}|\varphi)$ used to determine the values of β_i across crashes (see Train, 2003).

As with the random-parameter count-data models, a simulated maximum likelihood is used to estimate mixed logit models with logit probabilities approximated by drawing values of βi from $f(\beta_i|\varphi)$ for given values of φ. Research by Bhat (2003) has shown that an efficient way of drawing values of β_i from $f(\beta_i|\varphi)$ to compute logit probabilities is to use a Halton sequence approach (for more on this technique see Bhat, 2003; Halton, 1960; Train, 2003).

Another way of capturing unobserved heterogeneity is to use latent classes (finite mixtures). In contrast to random-parameter models, which account for heterogeneity from one observation to the next, finite mixture models adopt a semi-parametric way to allow the possibility that distinct classes (subgroups of data) with homogeneous features may exist. To this end, the latent class approach replaces the continuous distribution assumed by a random-parameter model with a discrete distribution that is represented by a finite and specified number of classes. This accounts for common heterogeneities among groups of observations and eliminates the need to make a distributional assumption on β_i. In crash-severity analysis this means that crash-involved drivers can be divided into C distinct classes and that each of these classes will have their own parameter-estimate values.

As an example, the resulting outcome probabilities for a latent class multinomial logit model of injury severities are given by (see Greene & Hensher, 2003):

$$P_i(j \mid c) = \frac{\text{EXP}(\beta_{jc}\mathbf{X}_{ij})}{\sum_{\forall J} \text{EXP}(\beta_{jc}\mathbf{X}_{ij})}, \tag{14}$$

where $P_i(j|c)$ is the probability of injury-severity level j, for crash i, which is a member of unobserved class c. The unconditional class probabilities $P_i(c)$ are also determined by the multinomial logit form as:

$$P_i(c) = \frac{\text{EXP}(\alpha_c \mathbf{Z}_i)}{\sum_{\forall C} \text{EXP}(\alpha_c \mathbf{Z}_i)}, \tag{15}$$

where $\mathbf{Z}_n$ is a vector of characteristics that determine class c probabilities for crash i and α_c is a corresponding vector of estimated parameters. With Eqs. (14) and (15), the unconditional probability of crash i resulting in injury-severity category j is simply:

$$P_i(j) = \sum_{\forall C} P_i(c) \times P_i(j \mid c). \tag{16}$$

Latent class models can be a powerful alternative for capturing unobserved heterogeneity and are readily estimated with maximum likelihood procedures (see Greene & Hensher, 2003). There is also the possibility of combining latent classes and random parameters where, instead of having the parameters being fixed across observations within each class, they can vary across the observations within each class (Xiong & Mannering, 2013). This gives a very powerful tool to capture unobserved heterogeneity in multiple ways (for a complete review of models and issues related to unobserved heterogeneity in safety data see Mannering, Shankar, & Bhat, 2016).

An alternative approach to a latent class model applied in the literature is to introduce a random mixing distribution based on the Diritchlet process. With this approach the number of classes and their characteristics are directly governed by the data (Heydari, Fu, Lord, & Mallick, 2016; Shirazi, Lord, Dhavala, & Geedipally, 2016).

SPATIAL AND TEMPORAL CONSIDERATIONS IN CROSS-SECTIONAL DATA

A concern with all models is whether or not their estimated parameters are transferable spatially (across roadway segments or intersections) or temporally (over time). For traditional police-reported crash data the spatial concern is obvious because roadway entities (roadway segments and intersections) are distributed over areas, such as a city or state. As discussed at the beginning of this chapter, one would not expect temporal considerations to play a role in the application of cross-sectional methods, but since crash data require some time period to accumulate a sufficient number of crashes temporal considerations may be a factor.

From a spatial perspective, transferability is desirable because it means that parameters of models estimated in other places can be used, thus saving the cost of additional data collection and estimation. Temporal transferability ensures that forecasts made with the model have some validity in that the estimated parameters are stable over time. When testing spatial and temporal transferability, likelihood ratio tests can be readily applied (most models in crash-data analysis are estimated using maximum likelihood methods; see Washington et al., 2011). Suppose the transferability of parameters between two geographic areas (or alternatively two time periods), a and b, is being tested. The likelihood ratio test can be applied to this as:

$$X^2 = -2[\mathrm{LL}(\beta_T) - \mathrm{LL}(\beta_a) - \mathrm{LL}(\beta_b)] \tag{17}$$

where $\mathrm{LL}(\beta_T)$ is the log-likelihood at convergence of the model estimated with the data from both geographic areas (or both time periods), $\mathrm{LL}(\beta_a)$ is the log-likelihood at convergence of the model using geographic-area/time-period a data, and $\mathrm{LL}(\beta_b)$ is the log-likelihood at convergence of the model using geographic-area/time-period b data. In this test the same variables are used in all three models (total model, geographic-area/time-period a model and geographic-area/time-period b model). This X^2 statistic is χ^2 distributed with degrees of freedom equal to the summation of the number of estimated parameters in all geographic or time-period models (a and b in this case but additional geographic areas can be added to this test) minus the number of estimated parameters in the overall model. The resulting X^2 statistic provides the confidence level that the null hypothesis (that the parameters are the same) can be rejected.

Alternatively, one could conduct the following test:

$$X^2 = -2[\mathrm{LL}(\beta_{ba}) - \mathrm{LL}(\beta_a)], \tag{18}$$

where $LL(\beta_{ba})$ is the log-likelihood at convergence of a model using the converged parameters from geographic/time-period b (using only geographic/time-period b's data) on geographic/time-period a's data (restricting the parameters to be geographic/time-period b's estimated parameters), and $LL(\beta_a)$ is the log-likelihood at convergence of the model using geographic/time-period a data. This test can also be reversed using $LL(\beta_{ab})$ and $LL(\beta_b)$. The statistic is χ^2 distributed, with the degrees of freedom equal to the number of estimated parameters in β_{ba} and the resulting X^2 statistic provides the probability that the models have different parameters. The combination of these two tests yields a good assessment of the model's transferability over time and space.

EXPLICITLY ADDRESSING TEMPORAL CONSIDERATIONS IN CROSS-SECTIONAL CRASH DATA

Because crash data are accumulated over time, there may be an unobserved temporal heterogeneity which is similar in concept to the unobserved heterogeneity discussed earlier, but with a temporal element (for a detailed discussion see Mannering, 2018). Unobserved temporal considerations could include factors such as weather, which would vary over the data collection period. As an example, consider the sample of 1,500 intersections collected over a two-year period, discussed earlier in this chapter. If we wish to model the number of crashes per month this will generate 24 observations, but these observations will be distributed over time (the two-year period). Temporal unobserved heterogeneity could thus be present, because a number of unobserved factors could vary over time and not accounting for this time-varying heterogeneity will lead to biased parameter estimates and erroneous inferences when variability over time is present (Lord & Persaud, 2000; Xiong, Tobias, & Mannering, 2014).

In the crash literature, temporal heterogeneity has been addressed by considering hidden-state, Markov-switching models, which account for unobserved heterogeneity across time periods by assuming the frequency and severity of crash transitions between two or more states over time (such as the 24 monthly time periods in the two-year data example presented earlier in this chapter). Theoretically, there are a number of reasons why multiple hidden states could exist, including variations in drivers' responses to changing weather conditions over time, which would manifest as unobserved heterogeneity in traditional statistical models. For example, Malyshkina and Mannering (2009) considered the existence of two highway-safety states resulting from distinct combinations of driving-environment conditions, driver reactions and other factors that interact and change over time (resulting in roadway segments changing from one state to another over time). Using a Markov-switching multinomial logit model of crash severity they showed statistically the presence of two states with the less safe state (the state likely to result in more severe crash injuries) being highly correlated with adverse climatic conditions relating to precipitation, low temperatures, snow and visibility. It is likely that the presence (or combined presence) of these less-than-ideal weather-related conditions (many of which are typically not available for empirical analyses) results in drivers reacting in different ways as they struggle to estimate the change in crash probabilities, and develop strategies to modify their speeds and car-following distances accordingly.

A typical Markov-switching model assumes that this temporal heterogeneity follows a stationary, multiple-state Markov chain process over time. For example, if two hidden states are assumed to exist ($s_t = 0$ and $s_t = 1$) the time-dependent transition probabilities can be written as:

$$P\left(s_{t+1} = 1 \mid s_t = 0\right) = p_{0\to 1}\text{ , and } P\left(s_{t+1} = 0 \mid s_t = 1\right) = p_{1\to 0} \tag{19}$$

where $P\left(s_{t+1} = 1 \mid s_t = 0\right)$ is the conditional probability of $s_{t+1} = 1$ at time $t + 1$ given that the observations are in state $s_t = 0$ at time t, $P\left(s_{t+1} = 0 \mid s_t = 1\right)$ is the conditional probability of $s_{t+1} = 0$ at time $t + 1$ given that the observations are in state $s_t = 1$ at time t, and the transition probabilities $p_{0\to 1}$ and $p_{1\to 0}$ are transition probabilities that can be estimated from the crash data.

Although estimation of Markov-switching models is complex and typically requires Bayesian Markov Chain Monte Carlo (MCMC) methods to sample the hidden states, the potential to track temporal, unobserved heterogeneity in data that are typically viewed as cross-sectional (by crash-data analysts) is a very powerful tool that can yield important new insights into the determinants of crashes and their resulting injury severities.

OTHER MODEL ESTIMATION CONCERNS

In the estimation of crash frequency and severity data, it is important to keep in mind the basic econometrics. Specifically, during model estimation close attention must be paid to the following model specification errors.

Correlation of Observations

As mentioned at the beginning of the chapter, the accumulation of crashes over time can create cause correlation issues that must be addressed. Consider the example at the beginning of this chapter where the analyst has access to crash data on 1,500 intersections over a two-year period. If the analyst seeks to estimate a per-month intersection crash-frequency model this implies each intersection will generate 24 monthly observations. Almost all statistical and econometric estimation methods are predicated on the assumption that the unobserved characteristics (the ε's in many of the above equations) are independent from one observation to the next. This will not be the case with 24 observations from each intersection, because the unobserved characteristics of a specific intersection in January (for example the angle of the sun at sunset causing visibility problems) is likely to be correlated with the unobserved intersection characteristics for this intersection in other months. It is also worth noting that such correlation could also occur in injury-severity data, because a crash that results in multiple injuries (multiple observations) would share unobserved effects that affect all injuries in that specific crash (for example exact amount of energy dissipated based on actual speed). This sort of correlation can be readily handled using standard panel-data techniques (see Greene, 2007), but it is important that the analyst recognises this as a problem (for a review of work that has addressed this issue see Mannering & Bhat, 2014).

Omitted Variables

If a statistically significant explanatory variable is omitted during estimation from the **X** vector, the model will be miss-specified, resulting in a biased estimate of the parameter vector β. This is an important problem for databases that may not have all of the explanatory variables that will produce significant parameters, or for analysts who want to make simplistic models (with few variables). For example, a model of crash frequency using only travel volume as an explanatory variable will give a biased estimate of the travel-volume parameter that can easily lead to erroneous conclusions and inferences (see Washington et al., 2011).

Irrelevant Variables

If a statistically insignificant parameter is left in the model, higher standard errors will result on all parameter estimates (which in turn will result in lower *t*-statistics). This could cause the analyst to ignore potentially significant relationships (variables producing parameters with understated *t*-statistics).

Non-Linearities

The assumption often made is that $\beta\mathbf{X}$ is linear (e.g., $\beta_1 X_1 + \beta_2 X_2 + \ldots$). If this is not the case, and the model is estimated as if it were, the estimate of the parameter vector β will be biased. However, untangling non-linearities from unobserved heterogeneity can present a formidable methodological challenge. The effect of a variable may appear to be non-linear when really the non-linearities appear because the methodological approach did not consider the effects of unobserved heterogeneity (thus giving spurious non-linearities; for a discussion of this point see Mannering et al., 2016).

Multicollinearity

Multicollinearity is often not the severe specification problem that many researchers believe. If the variables comprising the explanatory variable vector $\mathbf{X}$ are highly correlated, standard errors of β estimates will become large. However, estimates of β will still be unbiased. It is mathematically possible, depending on the variance in the data, that two very highly correlated variables can both produce statistically significant parameter estimates. While multicollinearity can be used to explain why some variables are not statistically significant in the model, it should never be used as a basis for not considering a variable in model estimation (a variable should only be excluded after it has been found to produce a statistically insignificant parameter).

Selectivity-bias/Endogeneity

Selectivity-bias/endogeneity is frequently overlooked in the analysis of crash data, and it can take on very subtle forms. For example, consider the effect of speed limits on crash frequency and severity. One may expect that a roadway with a higher speed limit would have a higher frequency and severity of crashes. However, roadways with higher speed limits tend to have more favourable geometric and traffic conditions that make the roads inherently safer, and these conditions warrant the use of a higher speed limit (higher speed limits are placed on select roads with favourable crash-reducing characteristics).

Thus, including speed limit in the model would result in a downward bias in the effect of speed limits that may make the speed-limit parameter estimate insignificant or in some cases negative (higher speed limits reduce the frequency in severity of crashes). In the negative-parameter case, it is not that higher speed limits are safer (which would not make sense from a physics perspective), it just means that the speed limit is capturing unobserved heterogeneity relating to missing data on traffic conditions, geometrics and speed-limit determination. Clearly, great caution must be exercised in interpreting this and other variables that may have this selectivity-bias/endogeneity issue (for additional discussion of this point, see Mannering & Bhat, 2014).

Under-reporting of Crashes with Less Severe Injuries

Police records and data based on reported crashes are the most common data sources in the study of crash frequency and severity. With such data, crashes resulting in no injuries or less severe injuries are much less likely to be reported. This means that the frequency of crashes will be under-reported (perhaps to varying degrees from one highway section/intersection to the next depending on the severity distribution) and that the distribution of crashes in databases among the injury-severity categories will differ from the actual distribution of crashes among the severity categories (which would include non-police-reported crashes). The absence of minor crashes from crash databases has clear consequences for both frequency and severity. If anything happens to fundamentally shift the distribution of injury severities, both frequency (because a greater or lesser number of crashes will appear in the database) and severity estimates will be affected. The potential bias from this source can be addressed in a number of ways and should be considered when interpreting results (see Washington et al., 2014; Mannering & Bhat, 2014).

Choice of Methodological Approach

As shown in this chapter, there are a number of methodological alternatives that can be applied to both crash frequency and crash severity models. For many of these methodological alternatives, one can statistically demonstrate the superiority of one over the other. For example, a random-parameter negative binomial, a random-parameter multinomial logit model and a random-parameter-ordered probit model can be readily statistically compared to their fixed-parameter counterparts. Thus, there is no statistical justification for estimating a conventional fixed-parameter unless it can be demonstrated that random parameters are not statistically justified. This is much the same as it is for a simple Poisson versus a negative binomial regression, because the negative binomial model is a generalisation of the Poisson model; if the dispersion parameter on the negative binomial model is statistically significant the Poisson model is statistically incorrect. Again, the same is true for random-parameter models (and latent-class models), which are generalisations of their fixed-parameter counterparts.

However, for other models it is not possible to directly determine if one is statistically superior to another. For example, in severity analysis one cannot say that a random-parameter ordered probit model is statistically superior to a random-parameter multinomial logit model because these models are not nested (derived from the same core econometric model). One could do a comparison of predicted values, but often analysts are less interested in prediction and more interested in the fundamental insights gained from the interpretation of individual parameters. It is extremely important to note that any claim of superiority of one non-nested approach over another can often be data specific and must be viewed with extreme caution.

SUMMARY

This chapter has provided a fundamental understanding of the basic issues encountered in the analysis of the frequency and severity of crash data, as well as some of the subtler issues associated with the application of cross-sectional methods in the crash-analysis literature. The chapter has also presented a sample of some statistical and econometric modelling techniques that have historically been applied to analyse such data.

As with many subject areas that use statistical and econometric methods to gain fundamental insights into complex phenomena, cross-sectional modelling methods will undoubtedly remain the workhorse methods for analysing the frequency and severity of vehicle crashes for many years to come. However, emerging data sources promise to present new cross-sectional method opportunities and challenges. For example, highly instrumented vehicles that can provide in-vehicle data collected in real time (with video-recording technologies and on-board vehicle sensors that record a wide array of data), data from vehicles' event-data recorders (which are currently installed on all late-model cars and record a significant amount of data prior to and during the crash), and other emerging sources have the potential to greatly expand our understanding of factors affecting the frequency and severity of crashes in a cross-sectional setting and enable us to better design safe and sustainable transportation systems. However, many of the fundamental cross-sectional methodological issues associated with the analysis of existing data (as outlined in this chapter) will still need to be addressed with the exciting new data sources that are on the horizon.

REFERENCES

Anastasopoulos, P., & Mannering, F. (2009). A note on modeling vehicle-accident frequencies with random-parameters count models. *Accident Analysis and Prevention*, *41*(1), 153–159.

Bhat, C. (2003). Simulation estimation of mixed discrete choice models using randomized and scrambled Halton sequences. *Transportation Research Part B*, *37*(1), 837–855.

Greene, W. (2007). *Limdep, Version 9.0*. Plainview, NY: Econometric Software Inc.

Greene, W., & Hensher, D. (2003). A latent class model for discrete choice analysis: Contrasts with mixed logit. *Transportation Research Part B*, *37*(5), 681–698.

Halton, J. (1960). On the efficiency of evaluating certain quasi-random sequences of points in evaluating multi-dimensional integrals. *Numerische Mathematik*, *2*(1), 84–90.

Heydari, S., Fu, L., Lord, D., & Mallick, B. (2016). Multilevel Dirichlet process mixture analysis of railway grade crossing crash data. *Analytic Methods in Accident Research*, *9*, 27–43.

Lord, D., & Mannering, F. (2010). The statistical analysis of crash-frequency data: A review and assessment of methodological alternatives. *Transportation Research Part A*, *44*(5), 291–305.

Lord, D., & Persaud, B. (2000). Accident prediction models with and without trend: Application of the generalized estimating equations procedure. *Transportation Research Record*, *1717*, 102–108.

Malyshkina, N. V., & Mannering, F. L. (2009). Markov switching multinomial logit model: An application to accident-injury severities. *Accident Analysis and Prevention*, *41*(4), 829–838.

Mannering, F. (1993). Male/female driver characteristics and accident risk: Some new evidence. *Accident Analysis and Prevention, 25*(1), 77–84.

Mannering, F. (2018). Temporal instability and the analysis of highway accident data. *Analytic Methods in Accident Research*, *17*, 1–13.

Mannering, F., & Bhat, C. (2014). Analytic methods in accident research: Methodological frontier and future directions. *Analytic Methods in Accident Research*, *1*, 1–22.

Mannering, F., Shankar, V., & Bhat, C. (2016). Unobserved heterogeneity and the statistical analysis of highway accident data. *Analytic Methods in Accident Research*, *11*, 1–16.

McFadden, D. (1981). Econometric models of probabilistic choice. In C. F. Manski & D. McFadden (Eds.), *A structural analysis of discrete data with econometric applications* (pp. 198–272). Cambridge, MA: The MIT Press.

McFadden, D., & Train K. (2000). Mixed MNL models for discrete response. *Journal of Applied Econometrics*, *15*(5), 447–470.

Savolainen, P., Mannering, F., Lord, D., & Quddus, M. (2011). The statistical analysis of crash-injury severities: A review and assessment of methodological alternatives. *Accident Analysis and Prevention*, *43*(5), 1666–1676.

Shirazi, M., Lord, D., Dhavala, S. S., & Geedipally, S. R. (2016). A semiparametric negative binomial generalized linear model for modeling over dispersed count data with a heavy tail: Characteristics and applications to crash data. *Accident Analysis and Prevention, 91*, 10–18.

Tobin, J. (1958). Estimation of relationships for limited dependent variables. *Econometrica, 26*(1), 24–36.

Train, K. (2003). *Discrete choice methods with simulation*. Cambridge: Cambridge University Press.

Washington, S., Karlaftis, M., & Mannering, F. (2011). *Statistical and econometric methods for transportation data analysis* (2nd ed.). Boca Raton, FL: Chapman & Hall/CRC.

Washington, S., Karlaftis, M., & Mannering, F. (2014). Applying quantile regression for modeling equivalent property damage only crashes to identify accident blackspots. *Accident Analysis and Prevention, 66*, 136–146.

Xiong, Y., & Mannering, F. (2013). The heterogeneous effects of guardian supervision on adolescent driver-injury severities: A finite-mixture random-parameters approach. *Transportation Research Part B, 49*, 39–54.

Xiong, Y., Tobias, J., & Mannering, F. (2014). The analysis of vehicle crash injury-severity data: A Markov switching approach with road-segment heterogeneity. *Transportation Research Part B, 67*, 109–128.

CHAPTER 13

TIME-SERIES REGRESSION MODELS FOR ANALYSING TRANSPORT SAFETY DATA

Mohammed Quddus

ABSTRACT

Purpose – *Time-series regression models are applied to analyse transport safety data for three purposes: (1) to develop a relationship between transport accidents (or incidents) and various time-varying factors, with the aim of identifying the most important factors; (2) to develop a time-series accident model in forecasting future accidents for the given values of future time-varying factors and (3) to evaluate the impact of a system-wide policy, education or engineering intervention on accident counts. Regression models for analysing transport safety data are well established, especially in analysing cross-sectional and panel datasets. There is, however, a dearth of research relating to time-series regression models in the transport safety literature. The purpose of this chapter is to examine existing literature with the aim of identifying time-series regression models that have been employed in safety analysis in relation to wider applications. The aim is to identify time-series regression models that are applicable in analysing disaggregated accident counts.*

Safe Mobility: Challenges, Methodology and Solutions
Transport and Sustainability, Volume 11, 279–296

ISSN: 2044-9941/doi:10.1108/S2044-994120180000011015

Methodology/Approach – *There are two main issues in modelling time-series accident counts: (1) a flexible approach in addressing serial auto-correlation inherent in time-series processes of accident counts and (2) the fact that the conditional distribution (conditioned on past observations and covariates) of accident counts follow a Poisson-type distribution. Various time-series regression models are explored to identify the models most suitable for analysing disaggregated time-series accident datasets. A recently developed time-series regression model – the generalised linear autoregressive and moving average (GLARMA) – has been identified as the best model to analyse safety data.*

Findings – *The GLARMA model was applied to a time-series dataset of airproxes (aircraft proximity) that indicate airspace safety in the United Kingdom. The aim was to evaluate the impact of an airspace intervention (i.e., the introduction of reduced vertical separation minima, RVSM) on airspace safety while controlling for other factors, such as air transport movements (ATMs) and seasonality. The results indicate that the GLARMA model is more appropriate than a generalised linear model (e.g., Poisson or Poisson-Gamma), and it has been found that the introduction of RVSM has reduced the airprox events by 15%. In addition, it was found that a 1% increase in ATMs within UK airspace would lead to a 1.83% increase in monthly airproxes in UK airspace.*

Practical applications – *The methodology developed in this chapter is applicable to many time-series processes of accident counts. The models recommended in this chapter could be used to identify different time-varying factors and to evaluate the effectiveness of various policy and engineering interventions on transport safety or similar data (e.g., crimes).*

Originality/value of paper – *The GLARMA model has not been properly explored in modelling time-series safety data. This new class of model has been applied to a dataset in evaluating the effectiveness of an intervention. The model recommended in this chapter would greatly benefit researchers and analysts working with time-series data.*

Keywords: Time-series; serial correlation; Poisson; Poisson-Gamma; GLARMA; airprox

INTRODUCTION

This chapter is the second of six chapters that collectively represent state of the practice methodologies for both understanding and predicting the safety performance of transport networks. This chapter describes how to best examine time-series crash data and is a complement to the appropriate management of cross-sectional crash data (Chapter 12), crash data from a sample of sites with many sites recording zero crashes (Chapter 14), crash data where examining severity is of prime importance or interest (Chapter 15), methods for identifying high-risk sites or blackspots (Chapter 16) and finally methods for using surrogate measures of crashes to examine risk (Chapter 17).

Time-series regression models are employed to develop an association between accident (or incident) counts for an entity (e.g., country, region) and their time-varying influencing factors, such as macroeconomic conditions, socio-demographic conditions and transport conditions. Discrete valued time-series safety data, such as the counts of transport accidents and incidents over an entity, are also utilised to evaluate the effectiveness of various engineering and non-engineering-based transport interventions. Time-series safety data are, however, considered to be 'interrupted', as the data of an outcome of interest are collected at multiple time points at equal intervals with the aim of evaluating whether the intervention has had a significantly greater effect than any underlying trend. For instance, Houston and Richardson (2002) quantified the impact of the use of seat-belts by law, on accidents, deaths and injuries. An interrupted time-series is not only affected by the intervention but also other underlying factors (e.g., traffic levels and weather conditions with respect to safety) that influence the outcome. Controlling for the underlying factors and relevant exposure is imperative so as to ensure that a change in safety relates to the intervention rather than other confounding factors. A time-series statistical model is therefore needed in order to accurately isolate the effectiveness of an intervention on safety.

While time-series methods for analysing real-valued continuous data are well developed (e.g., Box & Jenkins, 1970), this is not true for analysing non-negative, integer-valued accident or incident data (Quddus, 2008). There are some key issues in modelling time-series accident data, namely: (a) the conditional distribution of accident counts given past observations and covariates normally follow a Poisson distribution, (b) a common feature of accident count data is over-dispersion (i.e., the variance is larger than the mean) and (c) accident counts of a geographic entity over time may be serially correlated. Therefore, the application of real-valued time-series regression methods – autoregressive integrated moving average (ARIMA) models – may be questionable, especially if the application involves relatively rare events when the average value of the accident counts is low.

The purpose of this chapter is to investigate existing time-series statistical models utilised in analysing transport accidents. This chapter reviews wider literature so as to explore recent developments in time-series regression models to examine whether such models could be applied in analysing transport safety data.

TIME-SERIES REGRESSION MODELS AND THEIR APPLICATIONS IN SAFE MOBILITY

As discussed, there are two purposes of employing a time-series regression model in analysing transport safety data: (1) developing relationships between accident counts and time-varying regressors for the purpose of forecasting accidents (i.e., accident forecasting models), with the assumption that previous observations specifically influence these future observations; and (2) evaluating the effectiveness of an intervention in improving safety.

ARIMA models were commonly applied to analyse time-series safety data in early studies (e.g., Votery & Shapiro, 1983; Houston & Richardson, 2002). The popularity of ARIMA models in accident analysis relates to their flexibility and capability in handling serial autocorrelation through autoregressive (AR) and moving average (MA) terms, along with seasonal and non-seasonal differences. For instance, Rock (1996) quantified the impact of the Illinois Child Passenger Protection Act on fatalities; Houston and Richardson (2002) developed an accident forecast model by using an ARIMA time-series model to examine whether the change of an existing seat-belt law from secondary to primary enforcement enhanced traffic safety. As discussed, the fundamental issues of applying such models in accident analysis relate to the assumption of Gaussian distribution and discreteness of accident counts.

There are a few fundamental issues with the application of ARIMA models to integer-valued time-series processes (e.g., monthly aviation accidents). A pure ARIMA model with only an AR(1) term can be written as

$$y_t = \rho y_{t-1} + u_t$$

where y_t is the monthly (or other timeframe) accident or incident count at discrete time t (e.g., week, month, quarter or year), ρ is the first-order AR term and u_t is the white noise. Since y_t is a non-negative integer then u_t has to be an integer and ρ has to be either 1, 0 or -1 (or an integer). Therefore, imposing such constraints may limit the application of ARIMA models in analysing count variables (Quddus, 2008). The second problem relates to the commonly made assumption of Gaussian distribution. ARIMA models may be appropriate for highly aggregated time-series processes with a large mean (e.g., annual road traffic fatalities in Great Britain) in which the distribution is usually found to be an approximate normal. Consequently, the Gaussian assumption is less questionable. For highly disaggregated time-series of accident counts (e.g., monthly fatal traffic accidents in a small geographic entity) forecasting errors from an ARIMA are, however, large, as the underlying distribution is skewed to the right (Quddus, 2008).

In order to justify the application of ARIMA models, many studies changed their dependent variable to accident rates (i.e., a continuous variable), namely accidents per capita (Votery & Shapiro, 1983) and accidents per km of road length. Others have employed the Prais-Winston AR model on accident rates (Chi, Cosby, Quddus, Gilbert, & Levinson, 2010) or state-space models (Antoniou & Yannis, 2013) or a Poisson-Gamma model with a trend variable (Noland & Quddus, 2004). There is, however, no guarantee that this kind of model can explicitly account for serial correlation in the case of a long time-series count data (Quddus, 2008).

In 2008, Brijs, Karlis and Wets (2008) and Quddus (2008) introduced integer-valued autoregressive (INAR) Poisson models to handle both issues in modelling time-series analysis of accident counts. An INAR(1) Poisson model is expressed as follows:

$$y_t = \alpha \circ y_{t-1} + u_t$$

in which $\alpha \circ y_{t-1}$ is a binomial thinning variable and α is the probability of successes in each of y_{t-1} independent trials. It is assumed that the thinning operation of α on y_{t-1} is independent of u_t, which consists of the elements entering the system as innovations during the interval $[t-1, t]$. The innovations u_t have an independent and identical Poisson distribution, denoted as

$$u_t \sim \text{Poisson}(\mu_t)$$

$$\mu_t = \exp(\beta x_t)$$

in which μ_t is the Poisson parameter at time t, x_t is the time-varying exogenous variables and β are their corresponding coefficients. Although the Poisson INAR(1) model is better in analysing inter-valued count data, the assumption is that the underlying time-series process is stationary. In addition, the modelling structure of the INAR(1) Poisson model is not flexible enough to properly account for serial autocorrelation, as only the first-order autoregressive AR(1) is allowed and there is no MA term. Moreover, this model cannot handle over-dispersed accident or incident data.

TIME-SERIES REGRESSION MODELS: RECENT DEVELOPMENTS

After conducting a review of wider literature on time-series regression models for count data, there are primarily three new classes of models that have been explored in safety analysis in recent years: (1) the Poisson autoregressive (PAR) model; (2) the NB Integer-valued generalised autoregressive conditional heteroscedastic (NBINGARCH) model; and (3) an observation-driven model, termed the *generalised linear autoregressive and moving average* (GLARMA) model. These models are briefly discussed below.

Poisson Autoregressive Model

Brandt and Williams (2001) developed a linear PAR model of order p – that is, AR(p) – that can account for autocorrelation in the time-series counts and handle the problem of over-dispersion in accident counts. A PAR model assumes a linear function of lagged values of the dependent variable and an exponential function of the explanatory variables so as to ensure that the Poisson parameter takes positive values only. The model is written as

Observation equation:

$$y_t \sim \text{Poisson}\left(\mu_t\right) \tag{1}$$

State equation:

$$u_t = \sum_{i=1}^{p} \rho_i y_{i-1} + \left(1 - \sum_{i=1}^{p} \rho_i\right) \exp\left(\beta x_t\right) \tag{2}$$

It is assumed that the density of the state variable has a gamma-distributed conjugate prior, as follows:

$$u_t \mid y_{t-1} \sim \Gamma\left(\tau_{t-1} u_{t-1}, \tau_{t-1}\right), \quad u_{t-1} > 0, \sigma_{t-1} > 0 \tag{3}$$

where,

$$u_{t-1} = E\left(y_t \mid y_{t-1}\right) \text{and } \tau_{t-1} = \text{Var}\left(y_t \mid y_{t-1}\right)$$

in which p is the number of AR lags, ρ_i is the AR parameter for lag i and τ_{t-1} is a parameter that allows for over-dispersion. Readers are referred to Brandt and Williams (2001) for the details of model specification and estimation.

NBINGARCH Models

Both INAR(1) Poisson and PAR models fail to control the effect of over-dispersion. To address this issue, Davis and Wu (2009) employed an NB logit regression model that can control both the over-dispersion and serial autocorrelation inherent in time-series count data. An extension of this model, termed the NBINGARCH model, was proposed by Zhu (2011) and concluded that an NBINGARCH model provides a better goodness-of-fit performance relative to the NB logit regression model. Ye, Garcia, Pourahmadi and Lord (2012) modified Zhu's model by re-parameterising the mean function (i.e., μ_t) and relaxing the condition on the dispersion parameter. They then applied it to a road safety analysis by incorporating covariates so as to properly model the relationship between the time-series counts and external factors. Their re-parameterised NBINGARCH model is briefly discussed below:

$$P(Y_t = y_t) = \frac{\Gamma(y_t + k)}{y_t!\,\Gamma(k)} \left\{ \frac{k}{\exp(\mu_t) + k} \right\}^k \left\{ \frac{\exp(\mu_t)}{\exp(\mu_t) + k} \right\}^{y_t} \tag{4}$$

in which

$$\mu_t = \beta_0 + \sum_{i=1}^{p} \beta_i Y_{t-1} + \sum_{m=1}^{r} \gamma_t X_{t,m} + \sum_{j=1}^{q} \delta_j \mu_{t-j} \tag{5}$$

where $\beta_0, \ldots, \beta_p, \delta_1, \ldots, \delta_q \geq 0$ for $p \geq 1, q > 0$ and $t > p$; $k = \exp(r)$, in which r is any real value.

Variance of this NBINGARCH model is $\exp(\mu_t) + \exp^2\left(\frac{\mu_t}{k}\right)$ and the mean is $\exp(\mu_t)$, indicating that the model can account for over-dispersion. The log-likelihood function of the model does not have a close-form, requiring a non-linear optimisation technique (e.g., Newton–Raphson algorithm) to be employed (Ye et al., 2012).

GLARMA Models

A state-space model for a time-series $\{y_t : t = 1, 2, \ldots, n\}$ comprises of two equations, referred to as the observation or measurement and state equations. The observation equation specifies a distribution of y_t given the state vectors and the state equation specifies how the state vectors evolve in the observation process. In this context of time-series modelling, Cox (1981) specifies generalised state-space models into two classes: observation-driven and parameter-driven (i.e., latent process) specifications. The observation equation is the same for both specifications but the state equations are different.

In a simple form, an observation-driven model for the case of the first-order Autoregressive Conditional Heterskedasticity (ARCH) can be written as follows (Engle, 1982):

$$y_t = \sigma_t \varepsilon_t, \qquad \varepsilon_t \sim N(0,1) \tag{6}$$

$$\sigma_t^2 = \sigma_0 + \sigma_1 y_{t-1}^2$$

$$y_t \mid y_{t-1} \sim N\left(0, \sigma_0 + \sigma_1 y_{t-1}^2\right)$$

A parameter-driven model for the case of the first-order volatility model can be written as follows (Shephard, 2005):

$$y_t = \sigma_t \tau_t, \qquad \tau_t \sim N(0,1) \tag{7}$$

$$\text{In}\,\sigma_t^2 = \delta_0 + \delta_1 \sigma_{t-1}^2 + \eta_t, \qquad \eta_t \sim N(0,1)$$

$$y_t \mid \sigma_{t-1} \sim N\left(0, \delta_0 + \delta_1 \sigma_{t-1}^2 + \eta_t\right)$$

As can be seen, the parameter values in the observation-driven model at the current epoch are obtained as deterministic functions of lagged dependent variables (and contemporaneous and lagged independent variables for other models). Although the parameters are stochastic (i.e., the parameters evolve randomly over time), they are perfectly predictable one-step ahead given past information. The likelihood function is therefore available in closed form through the use of the prediction error decomposition. This simplifies the likelihood evaluation of the observation-driven model, making it popular in the applied econometric and statistics fields (Creal, Koopman, Lucas, 2008). On the other hand, it is very difficult to construct the likelihood in the parameter-driven model, because σ_t is unobserved and is subject to its own source of error, and the parameters are not perfectly predictable (Creal et al., 2008). Readers are referred to Creal et al. (2008) for a detailed comparison between these two specifications. Since an observation-driven model is relatively easy to estimate, this specification is further explored for modelling time-series accident or incident data.

Since transport accidents are random, rare and non-negative, count data models (e.g., Poisson, Poisson-lognormal and Poisson-Gamma) are more appropriate (Lord & Mannering, 2010), as documented in Chapter 14. For each time-series of accident count, a Poisson-based distribution should be used to address over-dispersion arising from unobserved or unmeasured heterogeneity. It is assumed that y_t are independently distributed as follows:

$$y_t \mid \mathbb{F}_t \sim \text{Poisson}(\mu_t) \tag{8}$$

in which, $\mathbb{F}_t\left(\mathbb{F}_t = \{y_s : s = 1,2,\cdots,t-1; x_s : s = 1,2,\cdots,t\}\right)$ represents the information set consisting of all observations up to time epoch t. In the observation-driven specification, the state process $\log(\mu_t)$ is assumed to follow a linear model in the independent variables, with residuals that have an MA structure as shown below (Davis, Dunsmuir, & Streett, 2013; Quddus, 2016):

$$\log(\mu_t) = W_t = \beta_0 + \sum_{i=1}^{p} \beta_i X_{ti} + \vartheta_t + Z_t + f(I_t) \tag{9}$$

in which W_t is the state vector, μ_t is the expected number of accident counts; β_0 is the intercept to be estimated; X_t is a p-dimensional vector of regressors at time t; $\vartheta_t \sim \text{Gamma}\left(\frac{1}{k}, k\right)$ in which k is the over-dispersion parameter (the larger the value of k, the larger the dispersion; the Poisson model corresponds to $k = 0$); and Z_t is the state variable, which is considered to be a driving mechanism, as if Z_t were not present in W_t the above model would reduce to the standard generalised linear model (GLM) (i.e. Poisson-Gamma model in this case) in which observations y_t would be independent given the regressors. Therefore, Z_t accounts for serial dependence in the accident counts by using the linear combinations of past predictive residuals (e_t) (Davis et al., 2013) as follows:

$$Z_t = \sum_{j=1}^{\infty} \gamma_j e_{t-j} \tag{10}$$

where the predictive residuals are defined as: $e_t = \frac{Y_t - \mu_t}{\mu_t^{\lambda}}$ (with $\lambda \geq 0$) known as the Martingale difference sequence, hence zero mean and uncorrelated; γ_i are the parameter vectors representing the weights for the infinite MA.

$\{Z_t\}$ is derived using the autoregressive moving average (ARMA)-like recursions; that is, the ARMA(p,q) process with noise given by the Martingale difference sequence $\{e_t\}$ as shown below:

$$Z_t = \sum_{i=1}^{p} \phi_i \left(Z_{t-i} + e_{t-i}\right) + \sum_{i=1}^{q} \theta_i \, e_{t-i} \tag{11}$$

in which p and q are the orders of AR polymonial ϕ and MA polymonial θ, respectively. $\{Z_t\}$ can be thought of as the best linear predictor of a stationary invertible ARMA process. Standardisation of past observed counts is necessary to avoid instability, and this can be achieved through the selection of appropriate type of residuals.

$f(I_t)$ is the intervention function (also known as a transfer function) included to examine the impact of any intervention on the response variable. In some specifications, this intervention variable is included in vector x. Policy- or engineering-related safety interventions may produce both the onset (i.e. abrupt or gradual) and duration (permanent or temporary) effects, implying that any of the four combination effects is possible. The connection between an intervention and its likely effects is normally termed a transfer function. For instance, an impulse transfer function is likely to occur once with an abrupt onset and temporary duration. On the other hand, a step transfer function is likely to produce an effect with the abrupt onset and permanent or long duration (i.e., an immediate impact continuing over the long term). A step function can be written as follows:

$$f\left(I_t\right) = \varpi_0 \, I_t \tag{12}$$

where ϖ_0 is a constant, and I_t is the intervention variable which takes a value of 0 for every time period before the implementation date and a value of 1 for every time period thereafter.

The log-likelihood function for the model represented by Eqs. (1) and (2) can be written as follows (Dunsmuir & Scott, 2015):

$$l(\vartheta) = \sum_{t=1}^{n} \log f\left(y_t \mid W_t(\vartheta)\right) = \sum_{t=1}^{n} \left\{ y_t W_t(\vartheta) - a_t b\left(W_t(\vartheta)\right) + c_t \right\} \tag{13}$$

in which a, b, and c are additional parameters. For the negative binomial response distribution, the log-likelihood is based on the probability density function of the negative binomial distribution. Therefore, the over-dispersion parameter is estimated along with β, ϕ and θ (i.e. $\vartheta = (\beta, \phi, \theta, k)$. The likelihood can be maximised from a suitable starting value of the parameter ϑ using a version of the Fisher scoring iteration or by the Newton-Raphson iteration (Dunsmuir & Scott, 2015).

The parameters of the model presented in Eqs. (1) and (2) can be obtained through the GLARMA R-package that supports three underlying distributions of counts, such as Poisson, binomial and negative binomial (Dunsmuir & Scott, 2015). Three options, namely 'Pearson', 'score type' and 'unscaled', are available for the residuals. Diagnostic and graphical methods as well as two statistical tests (log-likelihood ratio and Wald) are available to check whether counts are serially correlated in generalised model settings.

In summary, it is noticeable that there is a degree of similarity between the NBINGARCH and GLARMA models, and both models can simultaneously control for serial autocorrelation as well as over-dispersion in modelling time-series count data by employing similar modelling mechanisms. It can therefore be concluded that both models are equally applicable in analysing and modelling time-series transport safety data.

AN APPLICATION OF THE GLARMA MODEL TO AIRSPACE SAFETY

Since NBINGARCH models have been previously employed by Ye et al. (2012), the GLARMA model presented in Eqs. (8)–(13) has been chosen as the method to analyse UK airspace safety.

With the unprecedented increase in air travel demand over the last few decades, the aviation industry and governments have continuously explored ways to increase airspace capacity without compromising safety. This is because aviation stakeholders place great emphasis on aviation safety as this is a major selling point, and a poor safety record in the aviation industry is a natural threat to corporate survival. Increasing airspace capacity in order to facilitate the rising demand for air travel has been brought about by the introduction of Reduced Vertical Separation Minima (RVSM) in 2002 within European airspace. On 24 January 2002, at 01:00 UTC, 6,000 air traffic controllers in 41 States (including the UK, Europe and Western Africa) allocated aircraft to six new flight levels between FL290 to FL410, where aircrafts were safely separated by 1,000 feet instead of 2,000 feet. This allows aircraft to fly at their optimum flight level with respect to fuel consumption. According to EUROCONTROL, that was one of the most important air traffic management projects Europe had ever seen. With the implementation of RVSM, Europe's upper airspace capacity grew by 25% and the average yearly benefits for aircraft operators were estimated at around €3.9 billion due to the reduction in delays and fuel consumption.

There were concerns that the introduction of RVSM might affect safety levels. There has been limited scientific research conducted to evaluate its impact since its implementation, especially in the context of UK airspace. The aim therefore is to analyse whether aviation safety has been affected by the introduction of RVSM while controlling for air traffic levels (Air Transport Movements or ATMs) and seasonality. Airproxes[1] are employed as the key safety performance indicator. Historical airprox (January 1999–January 2011) data were obtained from the UK Airprox Board, and air travel demand data for the same period were gathered from the UK Civil Aviation Authority. An intervention model based on the GLARMA model discussed above was applied to understand the relationship between airprox and air traffic levels, seasonality and the introduction of RVSM as an intervention.

Fig. 1 shows time-series plots of monthly airproxes and the corresponding ATMs; that is, take-offs and landings (ATMs in 10,000s). The vertical reference line represents the date of intervention (i.e., January 2015). The average monthly airproxes before the intervention is 17 and this reduced to 14 after the implementation of RVSM.

It is also visually noticeable that the monthly airproxes exhibit a downward trend despite the increase in ATMs from 2002 to 2007 (year 2008 marked the beginning of the economic recession). Seasonal effects are also visible and these are controlled by the use of a categorical variable with four categories, in which: 'Winter' represents December, January and February; 'Spring' represents March, April and May; 'Summer' represents June, July and August; and 'Autumn' represents September, October and December. The intervention variable – the introduction of RVSM – is employed as a dummy variable in which this variable takes 0 for observations before February 2002 and 1 for observations after February 2002.

Poisson- or Poisson-Gamma-based GLARMA models (Eqs. (1)–(6)) were developed to establish a relationship between airproxes and the introduction of RVSM while controlling for air traffic levels and seasonality. Different model specifications in terms of the lags of AR and MA terms, methods for the model fit (Newton-Raphson iteration (i.e., NR) vs. Fisher Scoring iteration (i.e., FS)) and predictive residuals ('Pearson' in which $\lambda = 0.5$ and 'Score' in which $\lambda = 1.0$) were tested and the best fit models were identified based on the AIC (Akaike Information Criterion) value.

In order to see whether time interdependence in airprox counts needs to be taken into account, the LRT and Wald tests were employed. Both tests fit a GLM that matches the GLARMA model regression structure with the same conditional distribution. Both tests rejected the null hypothesis that there is no serial autocorrelation in the monthly airproxes, indicating that the use of

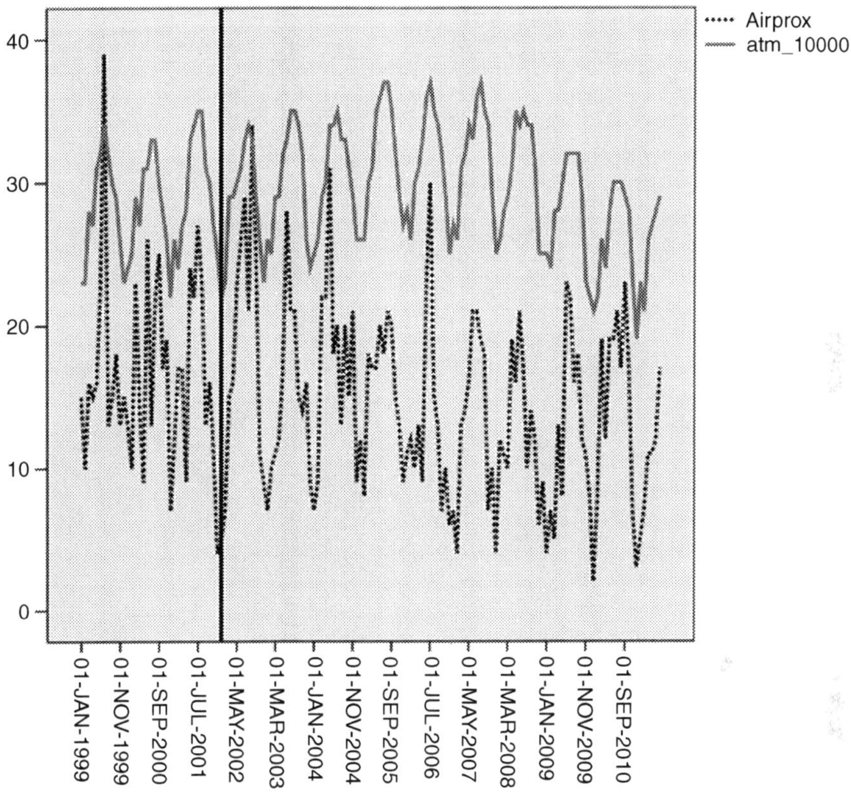

Fig. 1. Time Sequence Plots for Monthly Airproxes and Air Transport Movements (1999–2011).

the GLM, such as a simple Poisson-Gamma or simple Poisson, would have been inappropriate. The inherent serial correlation was controlled by using the AR term at lag 1 and the MA terms at lags 2 and 5.

As can be seen in Table 1, the over-dispersion (or heterogeneity) parameter in the GLARMA with the Poisson-Gamma conditional distribution was estimated to be 0.02, which is quite small. This indicates that monthly airproxes in UK airspace follow a Poisson distribution. This is not surprising, as the primary sources of over-dispersion relate to inherent heterogeneity as observations are sampled independently (i.e., heterogeneity across observations). This is not the case for our airprox data as they are observed in a single entity (i.e., UK airspace) over time so as to investigate the system-wide effect.

Table 1. Poisson- and Poisson-Gamma-based GLARMA Model Estimation.

Dependent Variable: Monthly Airproxes	Weekly: GLARMA NB		Weekly: GLARMA Poisson	
	Coefficient	*t*-Statistic	Coefficient	*t*-Statistic
Ln (Air Transport Movements)	1.8175	4.44	1.8306	5.79
Winter	−0.2615	−2.00	−0.2657	−2.73
Spring	−0.0324	−0.44	−0.0348	−0.51
Summer	0.0787	1.01	0.0727	1.00
Intervention: Introduction of RVSM	−0.1721	−1.90	−0.1669	−1.91
Intercept	−20.0368	−3.88	−20.2042	−5.07
Over-dispersion parameter	0.0200	1.77	na	
Autoregressive (AR) term at lag 1	0.0536	2.92	0.0620	2.29
Moving average (MA) term at lag 2	0.0749	4.26	0.0826	3.44
MA term at lag 5	0.0764	3.68	0.0831	3.33
Estimation methods				
Residuals	Pearson		Pearson	
Model fit	NR		FS	
Log-likelihood at convergence	−424.46		−427.04	
Akaike Information Criterion (AIC)	866.91		872.07	
Number of observations	150		150	
Tests for serial correlation:				
Likelihood ratio test (LRT)	22.82	*p*-value = 0.00	7.40	*p*-value = 0.00
Wald test	24.85	*p*-value = 0.00	7.66	*p*-value = 0.00

1. *n/a*: not applicable;
2. For the LRT and Wald tests, Null hypothesis is '*model is a GLM with the same regression structure*'; alternative hypothesis is '*model is a GLARMA process*'.

Consequently, the over-dispersion is not too obvious. Therefore, controlling time interdependencies among the airprox counts may be more important than accounting for over-dispersion while dealing with time-series count data related to an entity. However, no significant difference is observed in terms of the set of statistically significant variables and the values of the corresponding coefficients. The Poisson-based GLARMA model has been chosen as the most parsimonious model.

According to Table 1, it can be concluded that the introduction of RVSM has a significant positive impact on UK airspace safety. More specifically, the implementation of RVSM in UK airspace has brought about, *ceteris paribus*, a 15% reduction in monthly airproxes. This finding is in line with the view of EUROCONTROL, as according to Joe Sultana, Chief Operating Officer of EUROCONTROL's Network Management Directorate, 'By reducing the number of aircraft sharing a given flight level, this change not only enhanced safety, by allowing more aircraft to operate at their preferred flight levels, but it also underpinned better en route performance, reduced delays, lowered fuel consumption and boosted capacity' (www.eurocontrol.int).

The elasticity of airproxes with respect to air traffic levels is 1.83, indicating that a 1% increase in ATMs within UK airspace would lead to a 1.83% increase in monthly airproxes. During the winter months, there is, on average, a 23% reduction in airprox occurrences in the UK.

CONCLUSIONS

This chapter has examined the current literature on the application of time-series regression models in analysing safe mobility data. A range of different time-series models have been explored in the literature, namely: (1) ARIMA, (2) Prais-Winton AR model, (3) Poisson or Poisson-Gamma models with time trends, (4) state-space models and (5) Poisson INAR(1) models. The relative merits and limitations of these models have been explored in analysing time-series accident count data. This chapter has applied a recently developed GLARMA model to analyse UK airspace safety.

Based on the review of the current literature and the analysis carried out in this chapter, the following recommendations are drawn in terms of using an appropriate time-series regression model in analysing integer-valued accident or incident counts:

1. *Highly aggregated time-series datasets*: when both the spatial entity and the temporal resolution are large (e.g., national-level monthly fatal accidents) with a high value of the mean count (say greater than 50) then the data may be termed as highly disaggregated data. In analysing such datasets, ARIMA models and their extensions, such as seasonal ARIMA and ARIMA with exogenous variables, should be explored and used in forecasting or developing a time-series regression model.

2. *Disaggregated time-series datasets*: when the spatial entity is large (e.g., country) and the temporal resolution is low (e.g., daily) or vice versa, along with a moderate value of the mean count (say between 10 and 20), then Poisson INAR(1) or NBINGARCH or GLARMA models should be examined and the best-fit model should be identified by using different

'measures of accuracy' statistics. These are the mean absolute percentage error, the mean absolute deviation, the mean-squared deviation, the root mean-squared error and relative forecast error.

3. *Highly disaggregated time-series datasets*: when the spatial entity is small (e.g., local area) and the temporal resolution is low (e.g., daily), with a low value of the mean count (say less than 10), then NBINGARCH or GLARMA models should be examined.

The above recommendations shall be used as an 'approximate' indicator, as the final selection of the model should be based on sound statistical rationales, inherent assumptions and relevant modelling mechanisms, along with the consideration of different model goodness-of-fits.

Finally, the GLARMA model employed in this chapter has some weaknesses in dealing with time-series processes. They are: (1) there is no way to control for inherent seasonal variation (e.g., seasonal MA and seasonal AR terms) in the time-series process, except through the use of seasonal dummies; (2) the model assumes that the underlying process is stationary. In order to address these problems a new class of models – generalised seasonal autoregressive integrated moving average models – have recently been developed in R (Briet, Amerasinghe, & Vounatsou, 2013). Other sets of models that are becoming increasingly popular in time-series analysis are the non-linear and non-Gaussian state-space models of non-stationary time-series (Kitagawa, 1987). There are practically no applications of such models in the current safety literature. Future research should focus on exploring these models.

NOTE

1. According to the UK Airprox Board an airprox is defined as, 'An AIRPROX is a situation in which, in the opinion of a pilot or air traffic services personnel, the distance between aircraft as well as their relative positions and speed have been such that the safety of the aircraft involved may have been compromised.'

REFERENCES

Antoniou, C., & Yannis, G. (2013). State-space based analysis and forecasting of macroscopic road safety trends in Greece. *Accident Analysis & Prevention*, *60*, 268–276.

Box, G., & Jenkins, G. (1970). *Time series analysis: Forecasting and control.* San Francisco, CA: Holden-Day.

Brandt, P. T., & Williams, J. T. (2001). A linear Poisson autoregressive model: The Poisson AR(p) model. *Political Analysis*, *9*(2), 164–184.

Briet, O. J. T., Amerasinghe, P. H., & Vounatsou, P. (2013). Generalized seasonal autoregressive integrated moving average models for count data with application to malaria time series with low case numbers. *PLoS ONE*, *8*(6), e65761.

Brijs, T., Karlis, D., & Wets, G. (2008). Studying the effect of weather conditions on daily crash counts using a discrete time series model. *Accident Analysis and Prevention*, *40*(3), 1180–1190.

Chi, G, Cosby, A. G., Quddus, M. A., Gilbert, P. A., & Levinson, D. (2010) Gasoline prices and traffic safety in Mississippi. *Journal of Safety Research*, *41*(6), 493–500.

Cox, D. R. (1981). Statistical analysis of time series: Some recent developments. *Scandinavian Journal of Statistics*, *8*, 93–115.

Creal, D., Koopman, S. J., & Lucas, A. (2008). *A general framework for observation driven time-varying parameter models*. Technical report, Global COE Hi-Stat Discussion Paper Series 038. Retrieved from http://hdl.handle.net/10086/17131. Accessed on 26 July 2017.

Davis, R. A., Dunsmuir, W. T., & Streett, S. B. (2013). Observation-driven models for Poisson counts. *Biometrika*, *90(4)*, 777–790.

Davis, R. A., & Wu, R. (2009). A negative binomial model for time series of counts. *Biometrika*, *96*(3), 735–749.

Dunsmuir, W. T. M., & Scott, D. J. (2015). Package GLARMA: A reference manual for R-package. Retrieved from https://cran.r-project.org/web/packages/glarma/glarma.pdf. Accessed on 30 July 2015.

Engle, R. F. (1982). Autoregressive conditional heteroskedasticity with estimates of the variance of United Kingdom inflation. *Econometrica*, *50*, 987–1007.

Houston, D. J., & Richardson, L. E. (2002). Traffic safety and the switch to a primary seat belt law: The California experience. *Accident Analysis and Prevention, 34*(6), 743–751.

Kitagawa, G. (1987). Non-Gaussian state-space modeling of nonstationary time series. *Journal of the American Statistical Association*, *82*, 1032–1063.

Lord, D., & Mannering, F. L. (2010). The statistical analysis of crash-frequency data: A review and assessment of methodological alternatives. *Transportation Research – Part A, 44*(5), 291–305.

Noland, R. B., & Quddus, M. A. (2004). A spatially disaggregate analyses of road casualties in England. *Accident Analysis and Prevention, 36*(6), 973–984.

Quddus, M. A. (2008). Time series count data models: An empirical application to traffic accidents. *Accident Analysis and Prevention*, *40*(5), 1732–1741.

Quddus, M. (2016). Non-Gaussian interrupted time series regression analysis for evaluating the effect of smart motorways on road traffic accidents. Paper presented at the *95th Annual Meeting of the Transportation Research Board*, Paper # 16-0157.

Rock, S. (1996). Impact of the Illinois child passenger protection act: A retrospective look. *Accident Analysis and Prevention, 28*(4), 487–492.

Shephard, N. (1995). *Generalised linear autoregressions*. Technical report. Nuffield College, University of Oxford, Oxford.

Shephard, N. (2005). *Stochastic volatility: Selected readings* (*Advanced Texts in Econometrics*). Osford: Oxford University Press.

Votery, H. L., & Shapiro, P. (1983). Highway accidents in Sweden: Modelling the process of drunken driving behaviour and control. *Accident Analysis & Prevention*, *15*(6), 523–533.

Ye, F., Garcia, T. P., Pourahmadi, M., & Lord, D. (2012). Extension of a negative binomial GARCH model: Analyzing the effects of gasoline price on fatal crashes in Texas. *Transportation Research Record*, *2279*, 31–39.

Zhu, F. (2011). A negative binomial integer-valued GARCH model. *Journal of Time Series Analysis*, *32*(1), 54–67.

CHAPTER 14

SAFETY PREDICTION WITH DATASETS CHARACTERISED WITH EXCESS ZERO RESPONSES AND LONG TAILS

Dominique Lord and Srinivas Reddy Geedipally

ABSTRACT

Purpose – *This chapter provides an overview of issues related to analysing crash data characterised by excess zero responses and/or long tails and how to overcome these problems. Factors affecting excess zeros and/or long tails are discussed, as well as how they can bias the results when traditional distributions or models are used. Recently introduced multi-parameter distributions and models developed specifically for such datasets are described. The chapter is intended to guide readers on how to properly analyse crash datasets with excess zeros and long or heavy tails.*

Methodology – *Key references from the literature are summarised and discussed, and two examples detailing how multi-parameter distributions and models compare with the negative binomial distribution and model are presented.*

Safe Mobility: Challenges, Methodology and Solutions
Transport and Sustainability, Volume 11, 297–323

ISSN: 2044-9941/doi:10.1108/S2044-994120180000011016

Findings – *In the event that the characteristics of the crash dataset cannot be changed or modified, recently introduced multi-parameter distributions and models can be used efficiently to analyse datasets characterised by excess zero responses and/or long tails. They offer a simpler way to interpret the relationship between crashes and explanatory variables, while providing better statistical performance in terms of goodness-of-fit and predictive capabilities.*

Research implications – *Multi-parameter models are expected to become the next series of traditional distributions and models. The research on these models is still ongoing.*

Practical implications – *With the advancement of computing power and Bayesian simulation methods, multi-parameter models can now be easily coded and applied to analyse crash datasets characterised by excess zero responses and/or long tails.*

Keywords: Count distributions; highly dispersed data; excess zero responses; multi-parameter models

INTRODUCTION

This chapter is the third of six chapters that collectively represent state of the practice methodologies for both understanding and predicting the safety performance of transport networks. This chapter describes how to best examine crash data that contain a relatively large number of zeros (numerous sites recording no crashes in a particular time period), and is a complement to the appropriate management of cross-sectional crash data (Chapter 12), the management of crash data observed at the same sites over time (Chapter 13), crash data where examining severity is of prime importance or interest (Chapter 15), methods for identifying high-risk sites or blackspots (Chapter 16) and finally methods for using surrogate measures of crashes to examine risk (Chapter 17).

There has been a substantial amount of work devoted to the development and application of innovative count models for analysing crash data. These models can be used for different purposes. They include establishing relationships between crashes and different variables, such as traffic, geometric, occupant- or vehicle-oriented attributes, predicting crashes or their severity on existing or new facilities, and identifying risk factors that influence the number and severity of crashes. Mannering, in Chapter 12 of this book, provides the latest information about cross-sectional models and analysis tools. This chapter supplements recent review and discussion of manuscripts that have been documented by Lord and Mannering (2010), Mannering and Bhat (2014) and Mannering, Shankar and Bhat (2016). These authors have provided a detailed review of the key issues associated with crash data, as well as the strengths and weaknesses of the various methodological approaches that researchers have used to address these issues.

One important characteristic that has been recently documented in the literature is related to datasets that contain a large amount of zeros or zero responses (Lord & Geedipally, 2011; Miaou, 1994). For such datasets, the number of sites where no crash is observed is so considerable that classical distributions and regression models, such as the Poisson and negative binomial (NB) models, cannot be utilised efficiently (Geedipally, Lord, & Dhavala, 2012; Lambert, 1992). In some cases, the number of sites with zero crashes may not be the actual issue, but rather the fact that the dataset is characterised by a long tail, which significantly increases the dispersion (and skewness) found in the data. Those datasets are often referred to as 'highly dispersed datasets'. Traditional models also cannot handle this kind of dataset efficiently (Zou, Wu, & Lord, 2015). In fact, Guo and Trivedi (2002) note that NB regression models have problems modelling highly dispersed data, because a negligible probability is usually assigned to high counts. It should be pointed out that, in many cases, the presence of excess zero responses leads to datasets characterised by a long tail, since the high number of zero responses lowers the sample mean (ceteris paribus), which increases the spread of the observations.

Over the last few years, many new innovative methods and tools have been proposed for overcoming this problem. Many of these tools have fallen under the newly introduced three or more parameter models, most of which have been specifically developed for handling datasets with excess zero responses; we will refer to them as multi-parameter models (i.e., models with many shape parameters) in the rest of the chapter. These models offer a better way to interpret the relationship between crashes and risk factors, in addition to providing superior statistical performance. The objectives of this chapter are therefore to discuss challenges related to analysing datasets with excess zero responses and summarise the latest research on how to handle such datasets. The chapter ends with future avenues for research.

CRASH DATA CHARACTERISTICS

In order to reliably conduct safety studies, relevant data need to be collected, compiled and subsequently analysed. Various sources of data have been used to conduct safety studies. They include crash data produced from police reports (AASHTO, 2010), hospital injury data (NHTSA, 2011), traffic conflicts resulting from the interactions between roadway users (St-Aubin, Saunier, Miranda-Moreno, & Ismail, 2013) and based on microsimulation output (Young, Sobhani, Lenné, & Sarvi, 2014) among others. Tarko presents an overview of surrogate measures of safety in Chapter 17 of this book.

Although a wide variety of sources have been used in the past, motor vehicle crashes collected from police reports, or in some cases reported by the drivers themselves, still remain the most helpful and extensive source of data that can be used for conducting safety studies (Hauer, 1997, 2015). Several chapters in this book document how crash data can be used for different safety analyses. With crash data, there is clear evidence that a failure occurred in some manner on the investigated highway network, although the information is collected after people are injured or killed and properties damaged, both in terms of the vehicle(s) involved in the collision and structures located at the site of the event. As opposed to other data, crash records provide evidence about the circumstances of the crash (e.g., time of day, weather, collision pattern, object struck, etc.), roadway characteristics (e.g., location, highway class, number of lanes, roadside design, etc.), the vehicle occupants (e.g., age and gender of driver, number of occupants, severity of injuries, involvement of alcohol, etc.) and the features of the vehicle (e.g., size and age of the vehicle(s), equipment failure, etc.). These explain why crash data are still the most widely used kind of data that are utilised for conducting safety studies (Hauer, 1997; Lord & Mannering, 2010; Mannering & Bhat, 2014; Savolainen, Mannering, Lord, & Quddus, 2011).

Although crash data provide relevant information, the amount of data that can be assembled is usually not as plentiful as other research areas. In these areas, it is not uncommon to assemble datasets in the order of millions of observations, such as the number of power outages after a hurricane landfall (Han, Guikema, & Quiring, 2009) or burglaries within a large city (Levine & Lord, 2013). Collecting safety data is an expensive endeavour, since it usually involves extracting information from electronic files or hard copies, conducting site visits and using tools such as Google Earth from the comfort of one's office space. It is therefore very common to see about 50–70% of research funds being allocated to the collection and reduction of data. Because of these high costs, the amount of data assembled is therefore often limited (Lord & Bonneson, 2005). Furthermore, it is not uncommon to collect a large amount of data but, in the end, only a small proportion of these data are being used to develop statistical models.

Furthermore, it has been well-established that the severity of the collision can significantly affect its reportability (Elvik & Mysen, 1999; Hauer & Hakkert, 1988). More specifically, the amount of data that can be collected is influenced by the severity of the crash, in which more severe crashes are more likely to be reported than less severe crashes, especially those that do not result in an injury. The under-reporting rates for different severity levels are very difficult to estimate, even with improved computing tools (see Watson, Watson, & Vallmuur, 2015), but the general consensus shows that as many as 50–75% of property-damage-only (PDO) crashes may not be reported to relevant authorities (Elvik & Mysen, 1999). The under-reporting problem can affect the development of count- and severity models commonly used in safety (Ye & Lord, 2011) but, most importantly, influences the amount of data that can be collected.

As described above, using crash data provides unique challenges that are not commonly observed in other areas of research. Some of the challenges include statistical models negatively influenced by the small sample size and low sample bias (Lord, 2006; Ye & Lord, 2014), under-reporting (Kumara & Chin, 2005; Ma, 2009), endogeneity (Kim & Washington, 2006), omitted variable biases (Lord & Mannering, 2010; Mitra & Washington, 2012) and aggregation bias (Davis, 2004) among others. Those are not covered here, but the curious reader can find additional information about many of these issues in the following studies: Lord and Mannering (2010), Savolainen et al. (2011), Mannering and Bhat (2014) and Mannering et al. (2016).

An important concern that has not yet been covered extensively is the one related to excess zero responses. Although the issues described above can have a direct impact on the number of zero responses that are part of a dataset, other factors can also influence the large amount of zero responses. Such datasets can be common in other areas (e.g., industrial processing plants, as described in Lambert, 1992), but the large amount of zero responses are often caused by a distinct, systemic data-generating process, which is not the case for crash data. Factors affecting over-dispersion and the amount of zero responses are discussed in the next section.

OVER-DISPERSION AND EXCESS ZERO RESPONSES

Crashes are classified as non-negative, discrete events or integers. Consequently, these data cannot be analysed efficiently using continuous distributions, since the unequal variances associated with the observations can be difficult to manipulate and the estimated values can technically be negative (Hilbe, 2011). Normally, as the mean of the observations increases, the variance increases either at the same, smaller or larger rate. When it increases at the same rate, the counts are referred to as equi-dispersed, at a lower rate as under-dispersed and a higher rate as over-dispersed. It has been well established that crash data are usually over-dispersed (Hauer, Ng & Lovell, 1988; Jovanis & Chang, 1986). This characteristic means that the sample variance is greater than the sample mean. On occasion, crash data may be characterised by under-dispersion (i.e., the variance is smaller than the mean), but this occurs in unique circumstances, such as when the sample mean is very low (Francis et al., 2011; Khazraee, Saez-Castillo, Geedipally, & Lord, 2015). Under-dispersion can also be observed after the model is fitted, and this usually occurs when the sample mean and variance are very close to each other (Lord, Geedipally, & Guikema, 2010). Hence, when the regression model is fitted, under-dispersion can be noted since the fitted values are conditioned upon the mean.

A lot of work has been conducted to understand the nature of the dispersion in count data. Researchers have hypothesised or reported that over-dispersion could be explained by the clustering of data (neighbourhood, regions, wiring boards, etc.) (Poormeta, 1999), unobserved factors and unaccounted temporal correlation (Gourieroux & Visser, 1986), and model misspecification (Cameron & Trivedi, 1998). However, although these factors can contribute to over-dispersion, more fundamentally, it can be attributed to the nature of the crash process. Lord, Washington and Ivan (2005) describe that crashes can be defined as the product of Bernoulli trials with unequal probability of events. They state that as the number of trials increases and becomes very large, the distribution may be approximated by a Poisson process (hence the use of Poisson-based or mixed-Poisson models), where the magnitude of the over-dispersion is dependent upon the characteristics of the Poisson trials (Neldelman & Wallenius, 1986). Barbour, Holst and Janson (1992) propose several approaches for determining if the unequal events of independent probabilities can be approximated by a Poisson process and to quantify the amount of dispersion in the dataset as a function of the traits of the Poisson trials.

Count data should always include observations with the value zero, since this defines the lowest possible boundary of the data. Not including zero responses would bias the results unless a truncated model is used (Cameron & Trivedi, 1998; Hickman, 1976). In safety analyses, zeros mean that the site or entity did not experience a crash for the time period under study. However, this does not mean that the site cannot experience one or more crashes in the future. The problem with zeros becomes important when the number of observations with zero crashes is so large that traditional models, such as the Poisson or NB model, cannot handle the number of zero responses adequately. Usually, this means that the Poisson or NB model estimates or predicts fewer zeros than what is observed in the dataset, given the mean and variance found in the data. How such datasets in safety can have too many zero responses is described next.

Lord et al. (2005) proposed four factors that can influence the presence of excess zero responses in the data: (1) spatial or time scales that are too small, (2) under-reporting or misreporting of crashes, (3) sites characterised by low exposure and high risk and (4) important omitted variables describing the crash process. Let us discuss these four in greater detail.

Spatial and Time Scales

Given the nature of the crash process, described above as an approximation of the Poisson process, crash data are always expressed as a count over space and time. For example, the most frequent approach is to define crashes as the number of crashes per year that occurred on a mile for segments or crashes per year that happened at intersections (where the boundary to assign the crash at the intersection can vary, e.g., crashes located within 250 or 50 ft from the centre of the intersection). The same kind of definition is applicable to observations related to drivers or vehicle occupants. By changing the spatial and/or time scales, the number of zero responses when the count of crashes remains the same will also be adjusted. For the same crash events on a transportation network, it is expected that if the segmentation changes from a 1-mile basis to a 10th of a mile, the number of segments where no crash occurred will increase accordingly. Similarly, if the scale changes from a number of crashes per year to a number of crashes per minute, the proportion of zero responses will increase drastically. Fig. 1 illustrates the general idea of how the change in time scale affects the number of zero responses.

Before using sophisticated tools to overcome the bias caused by the number of zero responses, the safety analyst should examine if the number of zero responses can be changed by modifying the space and time scales, while still maintaining the homogeneity of the observations (e.g., lane width, shoulder width, etc.). This approach was, for example, successfully used by da Costa, Freitas and Pereira (2016), where the authors experimented with the space scale to reduce the proportion of zero responses before developing regression models.

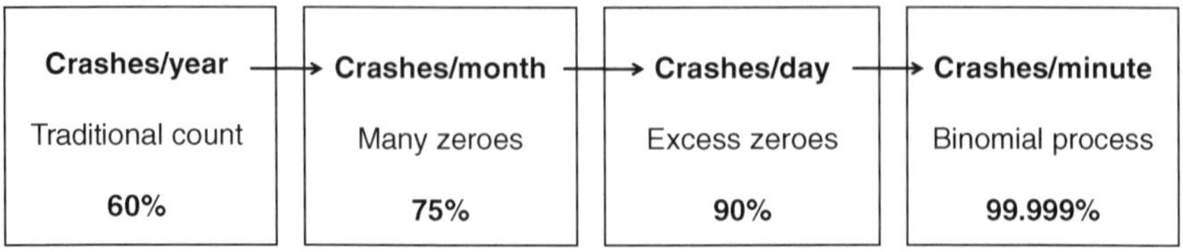

Fig. 1. Percentage of Zero Responses When Changing the Time Scale.

Under-Reporting of Crashes

As discussed above, not all crashes are reportable nor are they all reported. There is usually a minimum threshold where crashes are included in government databases (Hauer & Hakkert, 1988). For example, it is not uncommon to see specific sets of criteria, such as a minimum damage of $1,200 or one of the vehicles involved in the crash has to be towed away, in order for the crash to be included in official statistics. Even if the inclusion criteria are met, many drivers may 'elect' not to report a crash for fear of an increase in automobile insurance premiums or criminal prosecutions. Although a fair amount of work has been done on under-reporting (Watson et al., 2015), the magnitude of the problem is still very difficult to quantify. From a modelling standpoint, some work has been done to overcome the potential biases associated with under-reporting. However, not including crashes that actually occurred in the system artificially create sites with zero collision. Unless the safety analyst has a very good knowledge about the magnitude of the under-reporting rates for different severity levels based on different sources of data (Watson et al., 2015), it may be impossible to identify and adjust the number of observations with zero responses.

Sites Characterised by Low Exposure and High Risk

Datasets with excess zero responses are defined as being excessive as a function of the characteristics of the data. In other words, the number of zero responses is much higher than the expected value of zero responses produced by the distribution given the variance and skewness of the data. Examples are provided in the next section. Using simulation, Lord et al. (2005) have shown that the excess zero responses can be created when sites characterised by higher risk that have large crash counts (i.e., hence high risk) are mixed with sites that are classified with lower risk. These datasets are usually characterised by very long tails, which significantly increases the skewness of the data.

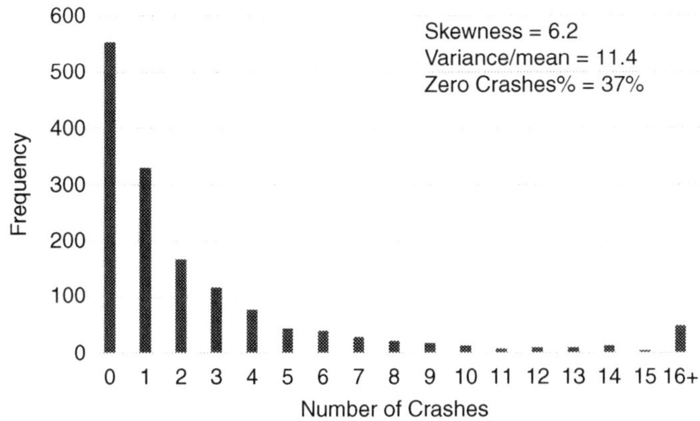

Fig. 2. Texas Dataset (Rural Multilane Highways) for 1997–2001 (*Source*: Adapted from Zha, Lord & Zou, 2015).

The Texas dataset that was used by Lord et al. (2008) and Zou, Zhang and Lord (2012) is characterised as such, as seen in Fig. 2.

Important Omitted Variables

As discussed above, the excess zero responses can be defined as the function of the characteristics of the data and the model used for analysing these data. It has been well documented that models can suffer from the 'omitted-bias variables' when they do not include variables that are known to influence safety (Lord & Mannering, 2010; Mitra & Washington, 2012). This could lead to poorly fitted models and can under-predict values located at the boundary conditions (extreme ends of the distribution). Miaou and Song (2005) and Mitra and Washington (2007) have shown that properly specified models could significantly reduce the amount of dispersion (conditioned on the mean) of the analysed dataset. However, the specification of the model is also a function of the functional form used to link crash data with the covariates (Geedipally, Lord, & Park, 2009).

EFFECTS OF EXCESS ZERO RESPONSES AND LONG TAILS

Several authors have examined the performance of the traditional distributions, such as the Poisson and NB, when excess zero responses exist in the data

(Aryuyuen & Bodhisuwan, 2013; Saengthong & Bodhisuwan, 2013; Zamani & Ismail, 2010; Zamani, Ismail, & Faroughi, 2014). Lord and Geedipally (2011) used single-vehicle, run-off-the-road crash data to examine the performance of the Poisson and NB distributions. This dataset contained about 90% zero responses. The results are shown in Table 1, which indicate that both distributions have difficulties predicting values, especially at the tail end of the distribution. Overall, the Poisson distribution tends to significantly under-predict zero responses and counts at the tail end, whereas the NB distribution tends to over-predict zero responses and counts at the tail end. Similar results were found with simulated data.

Zou et al. (2015) examined the performance of the dispersion parameter of the NB model when analysing over-dispersed crash data with a long tail. Although they specifically did not examine excess zero responses, they were able to quantify the biases for highly dispersed datasets. Using simulation, they compared the NB and the Sichel (SI) models for different levels of dispersions. The SI model was recently introduced for analysing crash data and has been shown to be more flexible than the NB model (Zou, Lord, Zhang, & Peng, 2013, Zou et al., 2015), but not as efficient as the recently introduced multi-parameter models. In one experiment, they simulated data using the SI model and applied both the NB and SI models to the dataset. The results

Table 1. Frequency and Predicted Values for the Poisson and NB Distributions.

Crashes	Observed Frequency	Poisson	NB
0	29,087	28,471.6	29,204.8
1	2,952	3,918.0	2,706.0
2	464	269.6	567.4
3	108	12.4	141.1
4	40	0.4	37.8
5	9	0.0	10.6
6	5	0.0	3.0
7	2	0.0	0.9
8	3	0.0	0.3
9	1	0.0	0.1
10+	1	0.0	0.0
Parameters		$\mu = 0.138$	$\mu = 0.138$ $\phi = 0.284$
Chi-square		2297.31	57.47
Log-likelihood		−14,208.1	−13,557.7

Source: Adapted from Lord and Geedipally (2011).

showed the dispersion parameter, α, of the NB model (i.e., $Y = \mu + \alpha\mu^2$) became biased very quickly, as shown in Fig. 3 and Table 2. Fig. 3 illustrates the level of bias for different sample mean values, while Table 2 quantifies the bias. Even the dispersion parameter of the SI model became biased for extreme conditions.

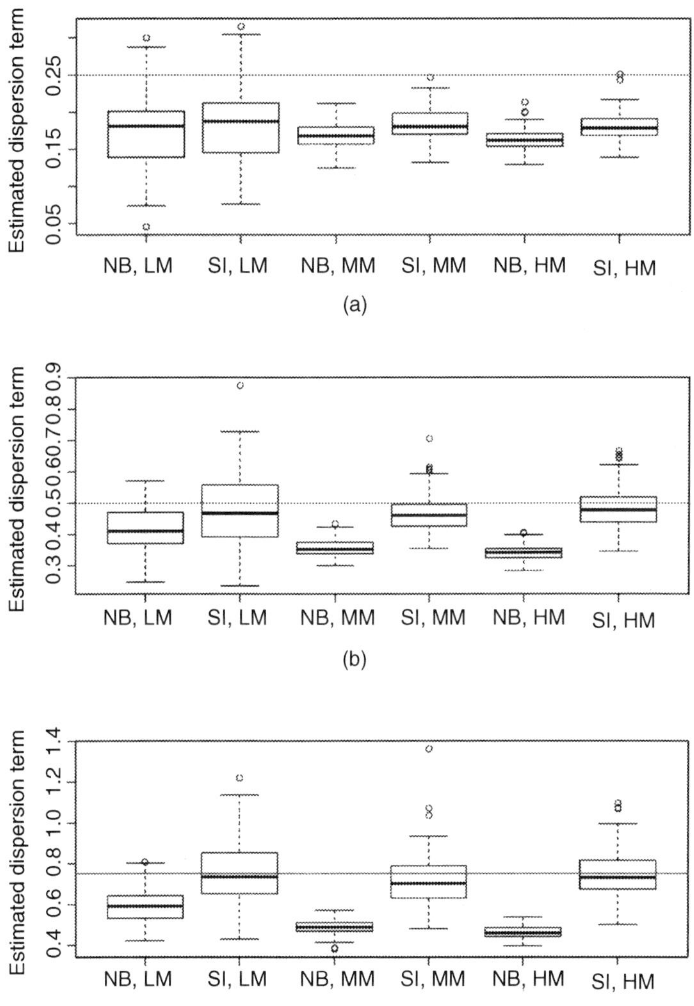

Fig. 3. (*Continued*)

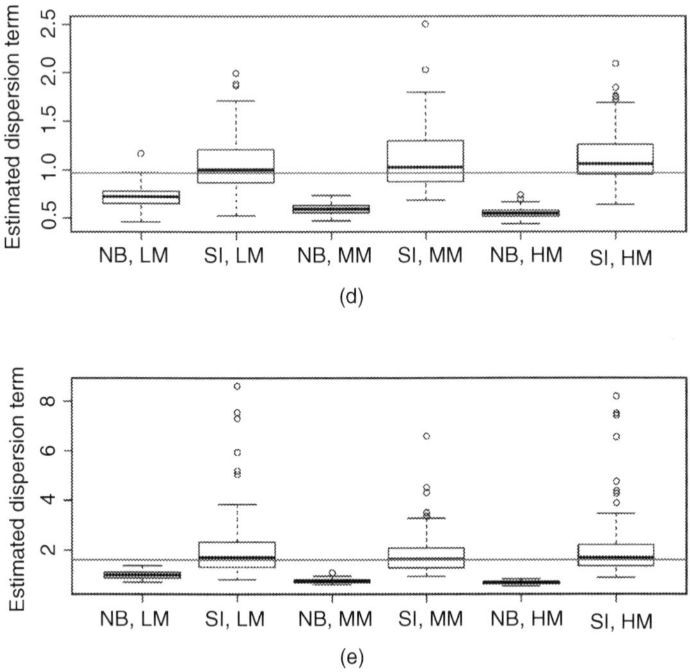

Fig. 3. Boxplots of Estimated Values for Dispersion Term under Different Scenarios Based on the Sichel Simulation. (a) Dispersion Term = 0.25. (b) Dispersion Term = 0.5. (c) Dispersion Term = 0.75. (d) Dispersion Term = 0.95. (e) Dispersion Term = 1.55. NB, negative binomial; SI: Sichel; LM, low sample mean; MM, medium sample mean; HM, high sample mean (*Source*: Zou et al., 2015).

MULTI-PARAMETER MODELS

The previous sections have discussed sources of excess zero responses and highly dispersed data and how traditional regression models cannot be used efficiently for analysing crash datasets. Lambert (1992) was the first to propose a method for handling datasets with excess zero responses. The basic principles of her method are that the zero responses arise from two distinct distributions that are coming from perfect and imperfect states, with probability p and $1–p$, respectively. The method proposes an approach based on a logistic regression model for estimating the probability that the observation with the value zero comes from one of the two states. It should be pointed out

Table 2. Estimation Bias for the Dispersion Term Based on the Sichel Simulation.

Dispersion Parameter	Scenario	Estimation Bias	
		NB	SI
0.25	HM	−0.09	−0.07
	MM	−0.08	−0.07
	LM	−0.08	−0.07
0.5	HM	−0.16	−0.02
	MM	−0.14	−0.03
	LM	−0.09	−0.02
0.75	HM	−0.29	0.00
	MM	−0.26	−0.03
	LM	−0.16	0.01
0.95	HM	−0.40	0.17
	MM	−0.36	0.16
	LM	−0.23	0.11
1.55	HM	−0.86	0.52
	MM	−0.80	0.31
	LM	−0.55	0.53

Notes: NB, negative binomial; SI: Sichel; LM, low sample mean; MM, medium sample mean; HM, high sample mean.
Source: Adapted from Zou et al. (2015).

that the zero-inflated model is a special case of mixture or latent class models (Park & Lord., 2009; Park, Lord, & Hart, 2010), in which one state or class has a long-term mean equal to zero. Her work has led to numerous improvements and extensions, such as truncated (Couturier & Victoria-Feser, 2010) and multivariate zero-inflated models (Fox, 2013), and zero-inflated models with repeated measurements (Gupta, Szczesniak, & Macaluso, 2015). This type of model has been used in a wide variety of areas (more than 1,200 citations as of late 2017). In highway safety, zero-inflated models were first introduced by Miaou (1994) (but who has since not published any more papers on that model). Since then, several other safety researchers have applied zero-inflated models (Dong, Richards, Clarke, Zhou, & Ma, 2014; Shankar, Milton, & Mannering, 1997).

Lambert (1992), however, discusses an important limitation of her proposed model, which is related to the difficulty interpreting the results. More specifically, because the data are supposed to be generated from two distinct data-generating processes, it can be difficult to determine with certainty which observation comes from which data-generating process or state.

Only probabilities can be provided or estimated. Furthermore, in the perfect state, the long-term mean is equal to zero, which entails that the generating process can only produce zero responses (Lord et al., 2005). The only time this can happen is when no road users are using the facility. More specifically, as discussed by Hauer (1999), highway entities cannot be perfectly safe (see also PEO, 1997).

Given the limitations noted above, a few researchers have started, over the last five or six years, to examine or try to develop new distributions that could handle datasets with excess zero responses and/or having long tails (Geedipally et al., 2012; Saengthong & Bodhisuwan, 2013; Zamani & Ismail, 2010). This work has focused on generalising the NB distribution by mixing it with other known distributions, so that the new combined distribution contains higher degrees of freedom (hence the aptly named multi-parameter model), which in turn leads to a more flexible distribution. As expected, the mathematical manipulation can become very complex and even more so if the generalised linear model (GLM) is developed based on the new distribution. The distribution and model, as discussed above, can then be used to analyse relationships between crashes and explanatory variables or for prediction purposes. Below are the latest distributions and models that have been proposed for analysing datasets characterised by excess zero responses and/or long tails.

The Negative Binomial–Lindley Model

The Negative Binomial–Lindley (NB-L) distribution is a mixture of NB and Lindley distributions. This mixed distribution has a heavy tail and performs well when the data encompass a large number of zero responses and/or are highly dispersed (often characterised by long tails). In other cases (e.g., less dispersed data), it functions similar to that of the NB distribution (see Shirazi, Dhavala, Lord & Geedipally, 2017, for a discussion about when should a model be used over the other). This is the first distribution that was specifically developed for analysing count data with excess zero responses (Zamani & Ismail, 2010).

The probability mass function (pmf) of the NB-L distribution is given as (Zamani & Ismail, 2010):

$$P(Z=z;r,\theta)=\frac{\Gamma(r+z)}{\Gamma(r)\times z!}\frac{\theta^2}{\theta+1}\sum_{j=0}^{z}\frac{\Gamma(z+1)}{\Gamma(j+1)\times\Gamma(z+j+1)}(-1)^j\frac{\theta+r+j-1}{(\theta+r+j)^2} \quad (1)$$

The parameter 'r' is the shape parameter of the NB-L distribution, similar to the inverse dispersion parameter 'ϕ' of the NB distribution. The parameter 'θ', in combination with the shape parameter '*r*', dictates the mean and variance of the NB-L distribution.

The NB-L model can be defined as a mixture of NB and Lindley distributions such that:

$$P(Y = y; \mu, \phi, \theta) = \int \mathrm{NB}(y; \phi, \varepsilon\mu)\mathrm{Lindley}(\varepsilon; \theta)\ d\varepsilon \tag{2}$$

Here, $f(\mu; a,\ b)$ means that f is the distribution of the variable μ, with parameters a and b. In the context of crash modelling, the parameter μ is the mean response for the number of crashes and is assumed to have a log-linear relationship with the covariates and is structured as (Geedipally et al., 2012):

$$\ln(\mu) = \beta_0 + \sum_{i=1}^{q} \beta_i X \tag{3}$$

where:

X = traffic and geometric variables of a particular site;
βs = regression coefficients to be estimated; and
q = total number of covariates in the model.

The parameter ε follows the Lindley distribution. The Lindley distribution is a mixture of exponential and gamma distributions (Ghitany, Atieh, & Nadarajah, 2008; Lindley, 1958; Lord & Geedipally, 2011; Zamani & Ismail, 2010). The probability density function (pdf) of the Lindley distribution can be defined as follows (Ghitany et al., 2008):

$$f(X = x; \theta) = \frac{\theta^2}{\theta + 1}(1 + x)e^{-\theta x};\ \theta > 0,\ x > 0 \tag{4}$$

The first moment (i.e., the mean) of the Lindley distribution is given as:

$$E(X) = \frac{\theta + 2}{\theta(\theta + 1)} \tag{5}$$

The variance of the Lindley distribution is given as:

$$\mathrm{Var}(X) = \frac{\theta^2 + 4\theta + 2}{\theta^2(\theta + 1)^2} \tag{6}$$

If the number of crashes Y is assumed to follow an NB-L distribution, then the mean function can be given as:

$$E(Y) = \mu \times E(X) = e^{\beta_0 + \sum_{i=1}^{p} \beta_i X_i} \frac{\theta + 2}{\theta(\theta + 1)} \tag{7}$$

The crash variance is given as:

$$\mathrm{Var}(Y) = \mu \times \frac{\theta + 2}{\theta(\theta + 1)} + \mu^2 \times \frac{2(\theta + 3)}{\theta^2(\theta + 1)} \times \frac{(1 + \varphi)}{\varphi} - \left(\mu \times \frac{\theta + 2}{\theta(\theta + 1)} \right)^2 \tag{8}$$

The Negative Binomial–Generalised Exponential Model

As the name implies, the Negative Binomial–Generalised Exponential (NB-GE) distribution is a mixture of the NB and generalised exponential distributions. The NB-GE distribution was initially introduced by Aryuyuen and Bodhisuwan (2013). They presented the basic characteristics of the NB-GE distribution, such as the mean, variance, skewness and kurtosis. The authors also described an approach to estimate the parameters of the mixed distribution based on the maximum likelihood estimate.

Similar to the NB-L distribution (Lord & Geedipally, 2011), this mixed distribution has also a heavy tail and performs well when the data encompass a large number of zero responses or are highly dispersed. It should be pointed out that the classical NB and NB-exponential distributions are considered a special case of the NB-GE distribution.

The pmf of NB-GE distribution is given as (Aryuyuen & Bodhisuwan, 2013):

$$f(Z = z; r, \alpha, \lambda) = \binom{r + z - 1}{z} \sum_{j=0}^{z} \binom{z}{j} (-1)^j \left(\frac{\Gamma(\alpha + 1)\Gamma\left(1 + \frac{r + j}{\lambda}\right)}{\Gamma\left(\alpha + \frac{r + j}{\lambda} + 1\right)} \right); \tag{9}$$

$$r, \alpha, \lambda > 0$$

The shape parameter 'r' represents the mean and variance of the NB-GE distribution.

The NB-GE distribution is defined as a mixture of NB and GE distributions such that:

$$P(Y = y; \mu, \phi, \alpha, \lambda) = \int \mathrm{NB}(y; \phi, \varepsilon\mu) \mathrm{GE}(\varepsilon; \alpha, \lambda) \; d\varepsilon \tag{10}$$

Similar to the NB-L model, the parameter μ is the mean response for the number of crashes and is assumed to have a log-linear relationship with the covariates. The parameter ε follows the GE distribution. The pdf of the GE distribution can be defined as follows (Gupta & Kundu, 1999):

$$f(X=x;\alpha,\lambda)=\alpha\lambda(1-e^{-\lambda x})^{\alpha-1}e^{-\lambda x};\ x>0 \tag{11}$$

The mean of the GE distribution is given as:

$$E(X)=\frac{1}{\lambda}\left[\psi(\alpha+1)-\psi(1)\right] \tag{12}$$

The variance of the GE distribution is given as:

$$\mathrm{Var}(X)=\frac{1}{\lambda^2}\left[\psi'(1)-\psi'(\alpha+1)\right] \tag{13}$$

where $\psi(.)$ and its derivatives are the digamma and polygamma functions.

If the number of crashes Y is presumed to follow an NB-GE distribution, the mean response can be estimated this way:

$$E(Y)=\mu\times E(X)=e^{\beta_0+\sum_{i=1}^{p}\beta_i X_i}\frac{1}{\lambda}\left(\psi(\alpha+1)-\psi(1)\right) \tag{14}$$

The crash variance is given by Eq. (15):

$$\mathrm{Var}(Y)=E(Y)+\left[\mathrm{Var}(X)+\left(E(X)\right)^2\right]\frac{(1+\varphi)}{\varphi}-\left(E(Y)\right)^2 \tag{15}$$

where $E(X)$ and $\mathrm{Var}(X)$ are described in Eqs. (12) and (13), respectively.

The Negative Binomial–Crack Model

Saengthong and Bodhisuwan (2013) proposed the Negative Binomial-Crack (NB-CR) distribution, which is achieved by combining the NB with a CR distribution. The authors provided the properties of the NB-CR distribution, which included the factorial moments, mean, variance, skewness and kurtosis. They also presented an approach to estimate the coefficients of the mixed distribution using the maximum likelihood method. This distribution has not yet been applied to a crash dataset nor has a GLM been formally proposed and tested.

The pmf of the NB-CR distribution is given as (Saengthong & Bodhisuwan, 2013):

$$f(Z=z;r,\lambda,\theta,\gamma)=\binom{r+z-1}{z}\sum_{j=0}^{z}\binom{z}{j}(-1)^{j}\frac{\exp\left[\lambda\left(1-\sqrt{1+2\theta(r+j)}\right)\right]}{\sqrt{1+2\theta(r+j)}}$$
$$\times\left[1-\gamma\left(1-\sqrt{1+2\theta(r+j)}\right)\right];\ r,\alpha,\lambda,\gamma>0 \tag{16}$$

The mean and variance of the NB-CR distribution are given in Eqs. (17) and (18), respectively:

$$E(Z)=r\left(\frac{\left(1-\gamma\left(1-\sqrt{1-2\theta}\right)\exp\left(\lambda\left(1-\sqrt{1-2\theta}\right)\right)\right)}{\sqrt{1-2\theta}}\right)-r \tag{17}$$

$$\mathrm{Var}(Z)=\exp(\lambda)\left(\frac{\left(r^{2}+r\right)\left(1-\gamma\left(1-\sqrt{1-4\theta}\right)\right)\exp\left(-\lambda\sqrt{1-4\theta}\right)}{\sqrt{1-4\theta}}\right.$$
$$-\frac{r\left(1-\gamma\left(1-\sqrt{1-2\theta}\right)\right)\exp\left(-\lambda\sqrt{1-2\theta}\right)}{\sqrt{1-2\theta}} \tag{18}$$
$$\left.-r^{2}\left(\frac{\left(1-\gamma\left(1-\sqrt{1-2\theta}\right)\right)\exp\left(-\lambda\sqrt{1-2\theta}\right)}{\sqrt{1-2\theta}}\right)^{2}\exp(\lambda)\right)$$

The NB-CR distribution can be defined as a mixture of NB and CR distributions such that:

$$P(Y=y;\mu,\phi,\alpha)=\int \mathrm{NB}(y;\phi,\varepsilon\mu)\mathrm{CR}(\varepsilon;\lambda,\theta,\gamma)\ d\varepsilon \tag{19}$$

The parameter μ is the mean response for the number of crashes and is assumed to have a log-linear relationship with the covariates. The parameter ε follows the CR distribution. The pdf of the CR distribution can be defined as follows (Bowonrattenaset & Budsaba, 2011):

$$f(X=x;\lambda,\theta,\gamma)=\frac{1}{\theta\sqrt{2\pi}}\left[\gamma\lambda\left(\frac{\theta}{x}\right)^{3/2}+(1-\gamma)\left(\frac{\theta}{x}\right)^{1/2}\right]\times\exp\left[-\frac{1}{2}\left(\sqrt{\frac{x}{\theta}}-\lambda\sqrt{\frac{\theta}{x}}\right)^{2}\right];\ x>0 \tag{20}$$

The moment-generating function of the CR distribution is given as:

$$M_X(t) = \frac{\exp\left(\lambda\left(1-\sqrt{1-2\theta t}\right)\right)}{\sqrt{1-2\theta t}}\left[1-\gamma\left(1-\sqrt{1-2\theta t}\right)\right];\ t < \frac{1}{2\theta} \tag{21}$$

APPLICATION OF MULTI-PARAMETER MODELS

In this section, we present the application of the NB-L and NB-GE distributions and models for analysing crash data with excess zero responses and/or with long tails.

Comparing Distributions: An Example

The dataset includes single-vehicle fatal crashes that occurred on divided, multilane and rural highways between 1997 and 2001. The data were assembled as a part of the National Cooperative Highway Research Program (NCHRP) 17–29 research project titled '*Methodology for estimating the safety performance of multilane rural highways*' (Lord et al., 2008). The data contained 1,721 segments, which fluctuated between 0.10 and 11.21 miles. The average segment length was equal to 1.01 miles, while the sample mean was equal to 0.13. About 89% of the segments experienced no fatal crash. The statistical performance of the models, based on the log-likelihood and chi-square, is summarised in Table 3.

Table 3. Single-Vehicle Fatal Crashes on Divided Multilane Rural Highways between 1997 and 2001.

Crashes	Observed Frequency	Poisson	NB	NB-L	NB-GE
0	1,532	1,509.2	1,534.4	1,532.9	1,532.6
1	162	198	154.7	158.3	158.9
2	19	13.0	25.8	23.7	23.6
3	6	0.6	4.9	4.6	4.5
4	2	0	1.2	1.4	1.8
Parameters		$\mu = 0.131$	$\mu = 0.131$	$\Theta = 1{,}532.9$	$r = 1.28$
			$\phi = 0.434$	$R = 1.851$	$\alpha = 1.5$
					$\lambda = 13.569$
Chi-square		102.99	2.73	1.68	**1.39**
Log-likelihood		−715.1	−696.1	−695.6	**−694.5**

Note: Bold character shows a better fit.
Source: Adapted from Lord and Geedipally (2011).

Comparing Models: An Example

A highly dispersed dataset that contained crashes for a 5-year period (1995–1999) at 338 rural interstate road sections in the state of Indiana is used to compare the models. In this dataset, about 36% of highway segments did not have any reported crashes over the 5-year period. For more details, the reader is referred to Geedipally et al. (2012). Table 4 summarises the comparison of various models. The models were assessed using the Deviance Information Criterion (DIC) (Spiegelhalter, Best, Carlin, & Van Der Linde, 2002), Mean Absolute Deviance (MAD) (Oh, Lyon, Washington, Persaud, & Bared, 2003) and the Mean Predicted Square Error (MPSE) (Oh Lyon et al., 2003). In Table 4, the intercepts for the NB-L and NB-GE models can be adjusted with the mean of the Lindley and GE distributions respectively to directly compare them with the intercept of the NB model. For more details about

Table 4. Modelling Results for the Indiana Data.

Variable	NB		NB-L		NB-GE	
	Value	Std. err	Value	Std. err	Value	Std. err
INTERCEPT	−4.779	0.979	−3.739	1.115	−3.233	0.423
Ln(ADT)	0.7219	0.091	0.630	0.106	0.570	0.067
FRICTION	−0.02774	0.008	−0.02746	0.011	−0.0284	0.010
PAVEMENT	0.4613	0.135	0.4327	0.217	0.481	0.158
MW	−0.00497	0.001	−0.00616	0.002	−0.00651	0.002
BARRIER	−3.195	0.234	−3.238	0.326	−3.240	0.324
RUMBLE	−0.4047	0.131	−0.3976	0.213	−0.349	0.154
α	–	–	–	–	2.339	0.42
λ	–	–	–	–	1.526	0.284
ϕ	–	–	–	–	2.459	0.6855
$\varpi = 1/\phi$	0.934	0.118	0.238	0.083	–	–
DIC	1900		**1701**		1784	
MAD[a]	6.91		**6.89**		7.04	
MSPE[b]	206.76		**195.54**		202.93	

Stand err: posterior credible standard error.
Bold character shows a better fit.

[a] $\text{MAD} = \frac{1}{n}\sum_{i=1}^{n}\left|\hat{y}_i - y_i\right|.$

[b] $\text{MSPE} = \frac{1}{n}\sum_{i=1}^{n}(\hat{y}_i - y_i)^2.$

Source: Adapted from Vangala et al. (2015).

the adjustment for the NB-L model, the reader is referred to Geedipally et al. (2012), and for the NB-GE model, the reader is referred to Vangala, Lord and Geedipally (2015). Other than the intercept, all the models' coefficients are almost the same. From Table 4, it can be noted that the NB-L and NB-GE perform better than the NB distribution, and the NB-L seems to perform slightly better than the NB-GE distribution (Vangala et al., 2015).

The cumulative residual plot for the Indiana data is given in Fig. 4. As described in Vangala et al. (2015), the cumulative residual values have been adjusted so that the final value ends at zero. Fig. 4 shows that the NB-GE model performs slightly better than the NB-L model. In addition, both the NB-L and NB-GE models performed better than the NB model.

SUMMARY AND FURTHER RESEARCH NEEDS

This chapter has described the unique characteristics associated with crash data that are not commonly observed in many other areas of research or analysis. Under particular characteristics, crash data can be characterised with excess zero responses and/or long tails, which can negatively affect the

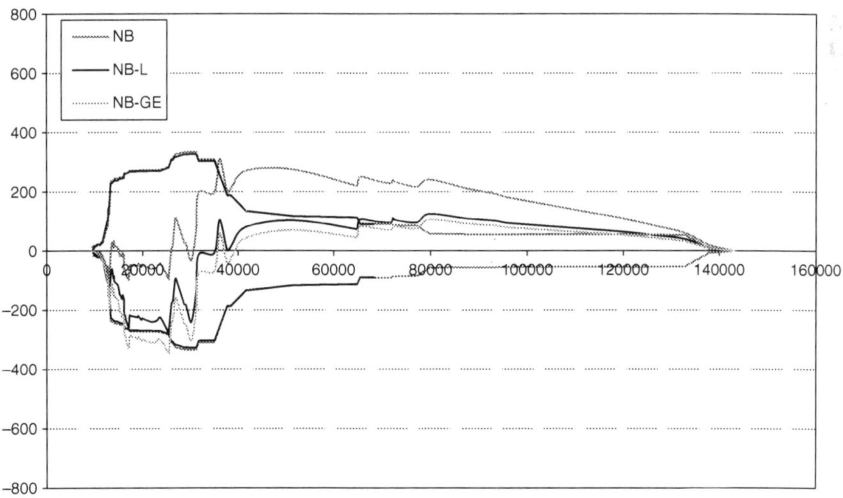

Fig. 4. Cumulative Residual Plot for Indiana Data
Note: Dotted/Dashed lines represent ± 2 Std. Dev.
(*Source:* adapted from Geedipally et al. (2012)).

use of traditional distributions and models. Sources of over-dispersion and excess zero responses were discussed, as well as how they can bias the results when traditional distributions or models are used. These sources can be grouped under four categories: (1) spatial and time scales, (2) under-reporting of crashes, (3) sites characterised by low exposure and high risk and (4) omitted variables in the models. Suggestions have been proposed to reduce the proportion of zero responses and long tails, so that traditional distributions and models could be used.

In the event that the characteristics of the crash dataset cannot be changed or modified, recently introduced multi-parameter distributions and models have been described. These multi-parameter analysis tools have been developed to specifically handle datasets characterised by excess zero responses and/or long tails. They offer a simpler way to interpret the relationship between crashes and explanatory variables while providing excellent statistical performance in terms of goodness-of-fit and predictive capabilities. This superior performance is expected since all these new distributions or models have higher degrees of freedom and are in fact a generalisation of the traditional NB distribution. This means that the NB distribution is actually a special case when the additional shape parameters are either equal to zero, to unity or to infinity depending on the pmf of the model.

This chapter has presented the characteristics of the NB-L, the NB-GE and the NB-CR distributions or models. With the exception of the latter, these models can effectively be used to analyse crash data. Although they are more complex to implement than the NB model, the codes and procedures for applying the distribution and models are readily available in published documents. For example, Hallmark, Qiu, Pawlovitch and McDonald (2013), Xu and Sun (2015), Shaon and Qin (2016) and Shaon, Qin, Shirazi, Lord and Geedipally (2017), who have introduced a random parameters N-BL, have independently used the codes described in Geedipally et al. (2012) to analyse crash data. In all four papers, the NB-L performed better than all the existing models tested in the research, including the zero-inflated model. Finally, two examples describing the performance and comparison analysis for some of these distributions and models were presented.

Since multi-parameter models are relatively new, there is still much more work that can be performed on this topic. On the horizon, we can see four- and five-parameter models, such as the NB–beta Weibull (Ortega, Cordeiro, & Kattan, 2012), the McDonald Weibull model (Cordeiro, Hashimoto, & Ortega, 2014) and the transmuted exponential–Weibull distribution (Saboor, Kamal, & Ahmad, 2015), or semi-parametric model, such as the NB-Dirichlet Process (NB-DP) (Shirazi, Lord, Dhavala, & Geedipally, 2016), that have started to appear in the literature. These distributions and models could offer even more flexibility while still maintaining the essential interpretability with the modelling results. Although several researchers have proposed new and innovative distributions and models, there is still a crucial need to evaluate these models for other issues associated with crash data, such as the low sample mean and sample size biases (Lord, 2006). Furthermore, it would still be important to determine the conditions when these models may not be necessary compared to the NB model or could produce biased results (Shirazi, Dhavala, Lord & Geedipally, 2017; Vangala, 2015). Nonetheless, it is expected that multi-parameter models may become the next set of traditional models in the years to come.

REFERENCES

AASHTO (2010) Highway Safety Manual. American Association of State Highway and Transportation Officials, Washington, D.C.

Aryuyuen. S., & Bodhisuwan, W. (2013). Negative binomial-generalized exponential (NB-GE) distribution. *Applied Mathematical Sciences*, *7*, 1093–1105.

Barbour, A. D., Holst, L., & Janson, S. (1992). *Poisson approximation*. New York, NY: Clarendon Press.

Bowonrattenaset, P., & Budsaba, K. (2011). Some properties of the three-parameter crack distribution. *Thailand Statistician*, *9*(2), 195–203.

Cameron, A. C., & Trivedi, P. K. (1998). *Regression analysis of count data*. Cambridge, UK: Cambridge University Press.

Cordeiro, G. M., Hashimoto, E. M., & Ortega, E. M. M. (2014). The McDonald Weibull model. *Statistics*, *48*, 256–278.

Couturier, D.-L., & Victoria-Feser, M.-P. (2010). Zero-inflated truncated generalized Pareto distribution for the analysis of radio audience data. *Annals of Applied Statistics*, *4*, 1824–1846.

da Costa, J. O., Freitas, E. F., & Pereira, P. A. A. (2016). Collision prediction models with longitudinal data: A note on modeling collision frequency in road segments in Portugal. Paper presented at the *95th Annual Meeting of the Transportation Research Board*, Washington, D.C.

Davis, G. A. (2004). Possible aggregation biases in road safety research and a mechanism approach to accident modeling. *Accident Analysis and Prevention*, *36*(6), 1119–1127.

Dong, C., Richards, S. H., Clarke, D. B., Zhou, X., & Ma, Z. (2014). Examining signalized intersection crash frequency using multivariate zero-inflated Poisson regression. *Safety Science*, *70*, 63–69.

Elvik, R., & Mysen, A. B. (1999). Incomplete accident reporting: Meta-analysis of studies made in 13 countries. *Transportation Research Record*, *1665*, 133–140.

Fox, J.-P. (2013). Multivariate zero-inflated modeling with latent predictors: Modeling feedback behavior. *Computational Statistics and Data Analysis*, *68*, 361–374.

Francis, R.A., Geedipally, S.R., Guikema, S.D., Dhavala, S.S., Lord, D., & Larocca, S. (2011). Characterizing the performance of the Conway-Maxwell-Poisson generalized linear model. *Risk Analysis*, *32*(1), 167–183.

Geedipally, S. R., Lord, D., & Dhavala S. S. (2012). The negative binomial-Lindley generalized linear model: Characteristics and application using crash data. *Accident Analysis and Prevention*, *45*(2), 258–265.

Geedipally, S.R., Lord, D., & Dhavala, S.S. (2014). A caution about using the Deviance Information Criterion while modeling traffic crashes. *Safety Science*, *62*, 495–498.

Geedipally, S.R., Lord, D., & Park, B.-J. (2009). Analyzing different parameterizations of the varying dispersion parameter as a function of segment length. *Transportation Research Record*, *2103*, pp. 108–118.

Ghitany, M. E., Atieh, B., & Nadarajah, S. (2008). Lindley distribution and its application. *Mathematics and Computers in Simulation*, *78*, 39–49.

Gourieroux, C. A., & Visser, M. (1986). A count data model with unobserved heterogeneity. *Journal of Econometrics*, *79*, 247–268.

Guo, J., & Trivedi, P. (2002). Flexible parametric models for long-tailed patent count distributions. *Oxford Bulletin of Economics and Statistics*, *64*, 63–82.

Gupta, R. D., & Kundu, D. (1999). Generalized exponential distributions. *Australian and New Zealand Journal of Statistics*, *41*(2), 173–188.

Gupta, R., Szczesniak, R.D., & Macaluso, M. (2015). Modeling repeated count measures with excess zeros in an epidemiological. *Annals of Epidemiology*, *25*, 583–589.

Hallmark, S. L., Qiu, Y., Pawlovitch, M., & McDonald, T. J. (2013). Assessing the safety impacts of paved shoulders. *Journal of Transportation Safety and Security*, *5*(2), 131–147.

Han, S.-R., Guikema, S. D., & Quiring, S. M. (2009). Improving the predictive accuracy of hurricane power outage forecasts using generalized additive models. *Risk Analysis*, *29*, 1443–1453.

Hauer, E. (1997). *Observational before-after studies in road safety*. London: Pergamon Publications.

Hauer, E. (1999). Safety in highway design standards. *Proceedings of the 2nd International Symposium of Highway Geometric Design*, San Antonio, TX.

Hauer, E. (2015). *The art of regression modeling in road safety*. Switzerland: Springer.

Hauer, E., & Hakkert, A. S. (1988). Extent and some implications of incomplete accident reporting. *Transportation Research Record*, *1185*, 1–10.

Hauer, E., Ng, J. C. N., & Lovell, J. (1988). Estimation of safety at signalized intersections. *Transportation Research Record*, *1185*, 48–61.

Hickman, J. J. (1976). The common structures of statistical models of truncation, sample selection and limited variables and a simple estimator for such models. *Annals of Economic and Social Measurement*, *5*, 575–492.

Hilbe, J. M. (2011). *Negative binomial regression* (2nd ed.). Cambridge: Cambridge University Press.

Jovanis, P. P., & Chang, H. L. (1986). Modeling the relationship of accidents to miles traveled. *Transportation Research Record*, *1068*, 42–51.

Khazraee, S. H., Saez-Castillo, A. J., Geedipally, S. R., & Lord, D. (2015). Application of the hyper-Poisson generalized linear model for analyzing motor vehicle crashes. *Risk Analysis*, *35*, 919–930.

Kim, D., & Washington, S. (2006). The significance of endogeneity problems in crash models: An examination of left-turn lanes in intersection crash models. *Accident Analysis and Prevention*, *38*, 1094–1100.

Kumara, S. S. P., & Chin, H. C. (2005). Application of Poisson underreporting model to examine crash frequencies at signalized three-legged intersections. *Transportation Research Record*, *1908*, 46–50.

Lambert, D. (1992). Zero-inflated poisson regression, with an application to defects in manufacturing. *Technometrics*, *34*(1), 1–14.

Levine, N., & Lord, D. (2013). CrimeStat IV: Chapter 15 – *OLS regression modeling* (Part V: *Spatial modeling II*). Washington, D.C.: Ned Levine & Associates and the National Institute of Justice.

Lindley, D. V., (1958). Fiducial distributions and Bayes' theorem. *Journal of the Royal Statistical Society. Series B* (Methodological), *20*(1),102–107.

Lord, D. (2006) Modeling motor vehicle crashes using poisson-gamma models: Examining the effects of low sample mean values and small sample size on the estimation of the fixed dispersion parameter. *Accident Analysis & Prevention*, *38*(4), 751–766.

Lord, D., & Bonneson, J. A. (2005) Calibration of predictive models for estimating the safety of ramp design configurations. *Transportation Research Record*, *1908*, 88–95.

Lord, D., & Geedipally, S. R. (2011). The negative binomial-Lindley distribution as a tool for analyzing crash data characterized by a large amount of zeros. *Accident Analysis and Prevention*, *43*, 1738–1742.

Lord, D., Geedipally, S. R., & Guikema, S. (2010). Extension of the application of Conway-Maxwell-Poisson Models: Analyzing traffic crash data exhibiting under-dispersion. *Risk Analysis*, *30*, 1268–1276.

Lord, D., Geedipally, S. R., Persaud, B. N., Washington, S. P., van Schalkwyk, I., Ivan, J. N., ..., Jonsson, T. (2008). *Methodology for estimating the safety performance of multilane rural highways* (NCHRP Web-Only Document 126). Washington, D.C.: National Cooperation Highway Research Program.

Lord. D., & Mannering, F. (2010). The statistical analysis of crash-frequency data: A review and assessment of methodological alternatives. *Transportation Research Part A*, *44*, 291–305.

Lord, D., Washington, S. P., & Ivan, J. N. (2005). Poisson, Poisson-gamma and zero inflated regression models of motor vehicle crashes: Balancing statistical fit and theory. *Accident Analysis and Prevention*, *37*(1), 35–46.

Ma, J. (2009). Bayesian analysis of underreporting Poisson regression model with an application to traffic crashes on two-lane highways (Paper #09-3192). Paper presented at the *88th Annual Meeting of the Transportation Research Board*, Washington, D.C.

Mannering, F. L., & Bhat, C. R. (2014). Analytic methods in accident research: Methodological frontier and future directions. *Analytic Methods in Accident Research*, *1*, 1–22.

Mannering, F. L., Shankar, V., & Bhat, C. R. (2016). Unobserved heterogeneity and the statistical analysis of highway accident data. *Analytic Methods in Accident Research*, *11*, 1–16.

Miaou, S.-P. (1994). The relationship between truck accidents and geometric design of road sections: Poisson versus negative binomial regressions. *Accident Analysis and Prevention*, *26*, 471–482.

Miaou, S.-P., & Song, J. J., 2005. Bayesian ranking of sites for engineering safety improvements: Decision parameter, treatability concept, statistical criterion and spatial dependence. *Accident Analysis and Prevention*, *37*(4), 699–720.

Mitra, S., & Washington, S. (2007). On the nature of over-dispersion in motor vehicle crash prediction models. *Accident Analysis and Prevention*, *39*, 459–468.

Mitra, S. & Washington, S. (2012). On the significance of omitted variables in intersection crash modeling. *Accident Analysis and Prevention*, *49*, 439–448.

Neldelman, J., & Wallenius, T. (1986). Bernoulli trials, Poisson trials, surprising variance, and Jensen's inequality. *The American Statistician*, *40*, 286–289.

NHTSA. (2011). *The crash outcome data evaluation system (CODES) and applications to improve traffic safety decision-making*. Washington, DC: National Highway Transportation Safety Administration. Retrieved from http://www-nrd.nhtsa.dot.gov/Pubs/811181.pdf

Oh, J., Lyon, C., Washington, S. P., Persaud, B. N., & Bared, J. (2003). Validation of the FHWA crash models for rural intersections: Lessons learned. *Transportation Research Record, 1840*, 41–49.

Ortega, E. M. M., Cordeiro, G. M., & Kattan, M. W. (2012). The negative binomial–beta Weibull regression model to predict the cure of prostate cancer. *Journal of Applied Statistics, 39*(6), 1191–1210.

Park, B.-J., & Lord, D. (2009). Application of finite mixture models for vehicle crash data analysis. *Accident Analysis and Prevention, 41*(4), 683–691.

Park, B.-J., Lord, D., & Hart, J., (2010). Bias properties of Bayesian statistics in finite mixture of negative regression models for crash data analysis. *Accident Analysis and Prevention, 42*(2), 741–749.

PEO. (1997). *Highway 407 safety review*. Toronto: Professional Engineers of Ontario.

Poormeta, K. (1999). On the modelling of overdispersion of counts. *Statistica Neerlandica, 53*(1), 5–20.

Saboor, A., Kamal, M., & Ahmad, M. (2015). The transmuted exponential–Weibull distribution with applications. *Pakistan Journal of Statistics, 31*, 229–250.

Saengthong, P., & Bodhisuwan, W. (2013). Negative binomial-crack (NB-CR) distribution. *Journal of Pure and Applied Mathematics, 84*, 213–230.

Savolainen, P. T., Mannering, F. L., Lord, D., & Quddus, M. A. (2011). The statistical analysis of highway crash-injury severities: A review and assessment of methodological alternatives. *Accident Analysis & Prevention, 43*, 1666–1676.

Shankar, V., Milton, J., & Mannering, F. L. (1997). Modeling accident frequency as zero-altered probability processes: An empirical inquiry. *Accident Analysis & Prevention, 29*, 829–837.

Shaon, M. R. R. & Qin, X. (2016). Use of mixed distribution generalized linear models to quantify safety effects of rural roadway features. *Transportation Research Record, 2583*, 134–141.

Shaon, M. R. R., Qin, X., Shirazi, A., Lord, D., & Geedipally, S. (2017). *Development of a random parameters negative binomial-lindley generalized linear model to analyze over-dispersed data*. Working paper, Zachry Department of Civil Engineering, Texas A&M University, College Station, TX.

Shirazi, M., Dhavala, S. S., Lord, D., & Geedipally, S. R. (2017). A methodology to design heuristics for model selection based on the characteristics of data: Application to investigate when the negative binomial Lindley (NB-L) is preferred over the negative binomial (NB). *Accident Analysis and Prevention*, forthcoming. Retrieved from http://dx.doi.org/10.1016/j.aap.2017.07.002

Shirazi, M., Lord, D., Dhavala, S. S., & Geedipally, S. R. (2016). A semiparametric negative binomial generalized linear model for modeling over dispersed count data with a heavy tail: Characteristics and applications to crash data. *Accident Analysis & Prevention, 91*, 10–18.

Spiegelhalter, D. J., Best, N. G., Carlin, B. P., & Van Der Linde, A. (2002). Bayesian measures of model complexity and fit. *Journal of the Royal Statistical Society - Series B, 64*, 583–639.

St-Aubin, P., Saunier, N., Miranda-Moreno, L., & Ismail, K. (2013). Use of computer vision data for detailed driver behavior analysis and trajectory interpretation at roundabouts. *Transportation Research Record, 2389*, 65–77.

Vangala, P. (2015). *Negative binomial-generalized exponential distribution: Generalized linear model and its applications*. Master of Science dissertation, Zachry Department of Civil Engineering, Texas A&M University, College Station, TX.

Vangala, P., Lord, D., & Geedipally, S. R. (2015). Exploring the application of the negative binomial-generalized exponential model for analyzing traffic crash data with excess zeros. *Analytic Methods in Accident Research, 7*, 29–36.

Watson, A., Watson, B., & Vallmuur, K. (2015). Estimating under-reporting of road crash injuries to police using multiple linked data collections. *Accident Analysis and Prevention, 83*, 18–25.

Xu, J., & Sun, L. (2015). Modeling of excess zeros issue in crash count analysis. *Journal of Jilin University (Engineering and Technology Edition), 45*, 769–775.

Ye, F., & Lord, D. (2011). Investigating the effects of underreporting of crash data on three commonly used traffic crash severity models: Multinomial logit, ordered probit and mixed logit models. *Transportation Research Record, 2241*, 51–58.

Ye, F., & Lord, D. (2014). Comparing three commonly used crash severity models on sample size requirements: Multinomial logit, ordered probit and mixed logit models. *Analytic Methods in Accident Research, 1*, 72–85.

Young, W., Sobhani, A., Lenné, M. G., & Sarvi, M. (2014). Simulation of safety: A review of the state of the art in road safety simulation modelling. *Accident Analysis and Prevention, 66*, 89–103.

Zamani, H., & Ismail, N. (2010). Negative binomial-Lindley distribution and its application. *Journal of Mathematics and Statistics, 6*(1), 4–9.

Zamani, H., Ismail, N., & Faroughi, P. (2014). Poisson-weighted exponential univariate version and regression model with applications. *Journal of Mathematics and Statistics, 10*, 148–154.

Zha, L., Lord, D., & Zou, Y. (2015). The Poisson inverse Gaussian (PIG) generalized linear regression model for analyzing motor vehicle crash data. *Journal of Transportation Safety and Security, 8*(1), 18–35. doi:10.1080/19439962.2014.977502

Zou, Y., Lord, D., Zhang, Y., & Peng, Y. (2013). Comparison of Sichel and negative binomial models in estimating empirical Bayes estimates. *Transportation Research Record, 2392*, 11–21.

Zou, Y., Wu, L., & Lord, D. (2015). Modeling over-dispersed crash data with a long tail: Examining the accuracy of the dispersion parameter in negative binomial models. *Analytic Methods in Accident Research, 5–6*, 1–16.

Zou, Y., Zhang, Y., & Lord, D. (2012) Application of finite mixture of negative binomial regression models with varying weight parameters for vehicle crash data analysis. *Accident Analysis and Prevention, 50*, 1042–1051.

CHAPTER 15

CRASH SEVERITY METHODS

John N. Ivan and Karthik C. Konduri

ABSTRACT

Purpose – *This chapter gives an overview of methods for defining and analysing crash severity.*

Methodology – *Commonly used methods for defining crash severity are surveyed and reviewed. Factors commonly found to be associated with crash severity are discussed. Approaches for formulating and estimating models for predicting crash severity are presented and critiqued. Two examples of crash severity modelling exercises are presented and findings are discussed. Suggestions are offered for future research in crash severity modelling.*

Findings – *Crash severity is usually defined according to the outcomes for the persons involved. The definition of severity levels used by law enforcement or crash investigation professionals is less detailed and consistent than what is used by medical professionals. Defining crash severity by vehicle damage can be more consistent, as vehicle response to crash forces is more consistent than that of humans. Factors associated with crash severity fall into three categories – human, vehicle/equipment and environmental/road – and can apply before, during or after the crash event. Crash severity can be modelled using ordered, nominal or several different types of mixed models designed to overcome limitations of the ordered and nominal approaches.*

Safe Mobility: Challenges, Methodology and Solutions
Transport and Sustainability, Volume 11, 325–350

ISSN: 2044-9941/doi:10.1108/S2044-994120180000011017

Two mixed modelling examples demonstrate better prediction accuracy than ordered or nominal modelling.

Research Implications – *Linkage of crash, roadway and healthcare data sets could create a more accurate picture of crash severity. Emerging statistical analysis methods could address remaining limitations of the current best methods for crash severity modelling.*

Practical Implications – *Medical definitions of injury severity require observation by trained medical professionals and access to private medical records, limiting their use in routine crash data collection. Crash severity is more sensitive to human and vehicle factors than environmental or road factors. Unfortunately, human and vehicle factor data are generally not available for aggregate forecasting.*

Keywords: Haddon Matrix; KABCO; AIS; ISS; MAIS; injury; fatality; crash; proportional odds; multinomial models; partial proportional odds; generalised ordered logit

INTRODUCTION

This chapter is the fourth of six chapters that collectively represent state of the practice methodologies for both understanding and predicting the safety performance of transport networks. This chapter describes how to proceed when examining and understanding severity is of prime importance and is a complement to the appropriate management of cross-sectional crash data (Chapter 12), the management of crash data observed at the same transport network locations over time (Chapter 13), a focus on crash data that contain a relatively large number of zeros (numerous sites recording no crashes in a particular time period) (Chapter 14), methods for identifying high-risk sites or blackspots (Chapter 16) and finally methods for using surrogate measures of crashes to examine risk (Chapter 17).

While preservation of the planet and its resources for use by future generations is an important element of sustainability, maintaining quality of life or livability for residents is also important to keep our cities sustainable. According to Woodcock (2000, p. 6), 'Sustainability in urban design seeks to establish a sense of place by enhancing the public domain. This may be evidenced by an effective public transport network, safe streets, equity of access to retail and services as well as traditional commons such as parks'.

Woodcock identifies one of the key metrics for evaluating a sustainable transportation design as a 'reduced injury rate'. Similarly, the World Business Council for Sustainable Development (WBC), in its 'Sustainable Mobility Project', defines deaths and serious injuries by any travel mode as an indicator of sustainability (WBC, 2004).

Consequently, crash severity – the consequences of crashes and their costs, economic as well as social – is an important consideration in assessing the sustainability of transportation systems. This chapter discusses the state-of-the-art crash severity analysis methods. First, approaches for defining crash severity are introduced. Then, factors associated with crash severity and methods for measuring and defining them are discussed. After that, statistical methods for analysing factors associated with crash severity are presented. Two examples of crash severity modelling are presented. The chapter ends with a summary and conclusions.

DEFINING CRASH SEVERITY

KABCO

The KABCO scale is used in the United States by road agencies to define personal injury severity in crash reports filed by investigating police officers. Each person involved in a crash is assigned a degree of severity using this scale, and the most injured occupant defines the severity of the crash. The scale was defined by the National Safety Council and includes five levels defined by the National Highway Traffic Safety Association in the 4th (current) edition of the Model Minimum Uniform Crash Criteria (MMUCC4; DOT, 2012): K = Killed; A = Suspected Serious Injury; B = Suspected Minor Injury; C = Possible Injury; O = No Injury. MMUCC3 (the previous edition) and ANSI give the following definitions (ANSI, 2007; DOT, 2008): K = Killed; A = Incapacitating Injury; B = Non-Incapacitating Injury; C = Possible Injury; O = No Injury.

Because the KABCO severity is assigned by an investigating police officer, rather than a health professional, this definition of severity is imprecise and subject to the training that is given to investigating officers as well as interpretation by officers of what they see at the scene. For example, Level C has sometimes been defined as 'non-visible injury', which opens up the possibility of serious internal injuries being coded as C rather than A or B. It is noted that an injury that turns out to be fatal may be coded as A, B or C up to 30 days after the crash.

Abbreviated Injury Scale

The Abbreviated Injury Scale (AIS) was developed by the Association for the Advancement of Automotive Medicine (Gennarelli & Wodzin, 2005) to code injury types and injury severity based upon an in-hospital clinical assessment. The AIS is favoured over KABCO by public health professionals and epidemiologists. Coders trained specially on the AIS lexicon assign a separate code to each individual anatomical injury using existing medical records, with a severity score ranging from 1 (minor) to 6 (life threatening).

AIS results in a vector of codes addressing all major body regions. For comparison across individuals or from crash to crash, it is necessary to combine the vector of severity scores for any patient into a single metric. One way to do this is the Maximum AIS, which is the maximum AIS (MAIS) across all body regions. A MAIS of 3 or greater (MAIS 3+) is commonly used as the cut-off for defining serious injury (e.g., IRTAD, 2011).

However, the MAIS does not distinguish between patients with several serious injuries to different body regions and those with more localised injury. As an alternative, Baker, O'Neill, Haddon and Long (1974) developed the Injury Severity Score (ISS). The ISS is the sum of the squares of the most severe AIS scores in three different body regions. The highest possible ISS is 75, which indicates a fatality or no chance of survival. An ISS cut-off of 16+ has been used to define seriously injured occupants. The New ISS is similar to the ISS, but is computed as the sum of squares of the three most severe injuries, regardless of body region (Osler, Baker, & Long, 1997).

It is generally accepted that either MAIS3+ or ISS16+ is a more consistent indicator of a serious injury crash than say K and A crashes on the KABCO scale (Flannagan, Mann, & Rupp, 2013). Even for grading different levels of injury severity, the two AIS measures are still preferred to KABCO, not only because they can distinguish among more levels of severity, but even with the increased numbers of levels the differences among them are more consistent than among the levels of the KABCO scale since they are coded by clearly defined medical outcomes interpreted by medical professionals, rather than subjective assessments by public safety officers without specific medical training.

Vehicle Damage

Vehicle damage has been used as an alternative indicator to characterise crash severity. This measure is based on the destruction/deformation of the vehicles involved in a crash and thus may be more objective than personal injury severity as it can be readily seen and consistently measured. Qin, Wang and Cutler (2013a) suggest that the crashworthiness of the vehicle is the primary defence of the vehicle occupants; therefore, severe damage to the vehicle will often be associated with high human casualties as well. Huang, Chor and Haque (2008) and Quddus, Noland and Chin (2002) also analysed crash severity using vehicle damage as an indicator.

Vehicle damage is likely to be more consistent with respect to the manner and intensity of the collision than injury to the persons involved. Personal injury severity is known to vary with characteristics of the persons involved, such as age, physical infirmity, use of alcohol, seating position and restraint use. Huang et al. (2008) employed a hierarchical Bayesian analysis to account for variation in observed personal injury severity for individuals in the same vehicle and the same crash. They found that for individuals in the same vehicle for a given crash, the vehicle-level effect explained as much of the differences among observed personal injury severities as the individual characteristics. This suggests that the vehicle damage severity can be a good surrogate for the personal injury severity.

FACTORS ASSOCIATED WITH CRASH SEVERITY

Haddon Matrix

The Haddon Matrix, developed by William Haddon in 1970, is a commonly used paradigm for injury prevention (Haddon, 1999). The Matrix categorises factors associated with the occurrence of an injury or death on two dimensions: (1) whether they are related to the human, vehicles and equipment or the environment and (2) whether they were relevant before, during or after the event. Using this framework, the analyst can identify potential factors contributing to the occurrence and severity of the event that can help to determine appropriate interventions to mitigate the occurrence and consequences of similar events in the future. Table 1 shows a typical Haddon Matrix for a vehicle crash. The following sections discuss factors associated with crash severity in each of the columns of the table.

Human Factors

As noted in the previous section, human factors account for much of the variation in crash severity outcomes. These factors include both the physical

Table 1. A Typical Haddon Matrix for a Vehicle Crash.

Phase	Human Factors	Vehicle/Equipment Factors	Environmental and Road Factors
Pre-crash	Information Attitudes Impairment Police enforcement	Roadworthiness Lighting Braking Speed management	Road design and road layout Speed limits Pedestrian facilities
Crash	Use of restraints Impairments	Occupant restraints Other safety devices Crash-protective design	Crash-protective roadside objects
Post-crash	First-aid skills Access to medics	Ease of access Fire risk	Rescue facilities Road congestion

Source: http://en.wikipedia.org/wiki/Haddon_Matrix

characteristics and the behaviour of the individuals involved in a crash and can be related to either the occurrence of the crash or its severity, or both. Table 2 lists human factors frequently found to be associated with crash severity; as noted, some are also related to crash occurrence.

Table 2. Human Factors Related to Crash Severity.

Factor	Association with Crash Severity	Source(s)
Age	Seniors and children are more susceptible to serious injuries than young or middle-aged adults.	Qin et al. (2013a); Zonfrillo, Elliott, Flannagan and Durbin (2011)
Alcohol use	Significant amounts of alcohol in the blood make body systems frailer.	Jensen (1999)
Seating position	Occupants in each seating position in a vehicle sustain different injuries depending on how the vehicle is impacted.	Liu, Lai and Ma (2012); Viano & Parenteau (2010)
Choice of restraint use	Vehicle occupants using the restraints provided (e.g., seatbelts and child seats) are less likely to sustain serious injuries or cause injuries to others in the vehicle.	Zonfrillo et al. (2011); Nukenine and Daniel (2011)
Pedestrians	For pedestrians property-damage-only is usually not a possible outcome; otherwise no crash would be recorded.	Islam, Serhiyenko, Ivan, Ravishanker and Garder (2014)

Vehicle/Equipment Factors

The intensity of crash severity outcomes is directly related to the laws of physics. More specifically, it is the magnitude of forces exerted against the vehicles and persons involved in a crash that explain the severity of the crash. In turn, the magnitude of force is derived from the rate of deceleration that occurs in a crash and the protection afforded by the vehicle.

Consequently, vehicle/equipment factors that affect the rate of deceleration and absorption of forces are related to crash severity. These factors include the manner of collision, the size of the vehicles involved, the bumper height of the vehicles involved, the braking capabilities of the vehicles involved and the travel speed of the vehicles involved before the crash.

Table 3 discusses these factors briefly. Speed is an especially critical factor in the case of crashes between vehicles and pedestrians. Rosén, Stigson and Sander (2011) analysed knowledge about impact speed and pedestrian crash severity.

Table 3. Vehicle and Crash Factors Associated with Crash Severity.

Factor	Association with Crash Severity	Source(s)
Manner of collision	How and where each vehicle was impacted will affect how much damage the vehicle sustains and, in turn, how severely the occupants are injured.	Funk, Cormier and Manoogian (2012)
Size of vehicle	Heavy vehicles such as trucks and buses have greater mass and thus result in greater impact on any vehicles with which they collide; occupants of the heavier vehicle will often have less severe injuries than those in the lighter vehicle.	AlEassa, Eid and Abu-Zidan (2013); Qin, Wang and Cutler (2013b)
Bumper height	When vehicles that collide have bumpers at different heights, the bumpers may not function as intended for dissipating the force of the impact.	Mizuno and Arai (2010)
Braking capability	Vehicles with better performing brakes (e.g., ABS) may enable drivers to avoid collisions or reduce severity of collisions by reducing the impact speed.	Kaplan and Prato (2012)
Driving speed	Higher driving speeds provide drivers with less time to process information and react. Furthermore, higher driving speeds usually result in higher collision speeds and an attendant increase in crash severity and crash fatality rates.	Nilsson (2004); Elvik, Christensen and Amundsen (2004); Rosén et al. (2011)

They found that above 75 km/h (about 46.6 miles/h) the pedestrian is more than 50% likely to be killed. The impact speed must be below 30 km/h (about 18.6 miles/h) to have a negligible probability of the pedestrian being killed.

Consequently, speed at time of collision is obviously strongly associated with the severity of a crash. Unfortunately, this information is usually not available to analysts, as it must be estimated by an expert at the scene of the crash and the extra effort to do this is rarely taken, except in cases of commercial carriers or litigation via crash reconstruction.

Environmental and Road Factors

In addition to the humans and vehicles involved, where the crash occurs (environmental and road factors) can also explain the variation in its severity. Road factors pertain to characteristics of the roadway and immediate roadside design that contribute to a crash occurrence or severity. Environmental factors are related to conditions other than the road that were unique to the time and location of the crash. Table 4 lists some common environmental and road factors and how they are related to crash severity.

Table 4. Environmental and Road Factors Associated with Crash Severity.

Factor	Association with Crash Severity	Source(s)
Road surface	Slippery conditions reduce braking distance and may cause impact speed to be higher.	Milton, Shankar and Mannering (2008); Caliendo, Maurizion and Parisi (2007)
Inclement weather	Drivers may choose lower speeds in inclement weather to compensate, though this effect is not consistent for all types of precipitation or conditions.	Hayat, Debbarh, Antoniou and Yannis (2013); Theofilatos and Yannis (2014); Yu and Abdel-Aty (2014)
Roadside conditions	An unforgiving roadside, for example, a sharp drop off or a steep cliff-face, is likely to result in a vehicle roll-over in the event of a run-off-road crash.	Peng, Geedipally and Lord (2012); Schrum, De Albuquerque, Sicking and Reid (2014); Stephan and Newstead (2014)
Rural location	Crashes occurring at locations far from an emergency room or trauma centre are more likely to end up with fatalities, as the critical response time window may be lost en route.	Clark, Winchell and Betensky (2013); Chen et al. (2009)

MODELLING CRASH SEVERITY

Ordered Versus Un-Ordered Severity Outcomes

Crash severity outcomes are discrete in nature and, further owing to the inherent ordering of the outcomes, crash severity is commonly treated as an ordered (or ordinal) outcome variable. For example, when crash severity is reported on the KABCO scale (as defined earlier), the outcomes are discrete in nature. Further, the outcome levels can be ordered based on the nature of injury represented by each outcome level, from most extreme to least extreme.

Linear regression is not appropriate for modelling such a variable, since each incremental increase in severity level does not imply the same increase in magnitude of severity. Instead, ordered model formulations have been commonly used because they acknowledge the inherent ordering in the crash severity outcomes (Kockelman & Kweon, 2002). Another option is to treat crash severity as a nominal variable (Savolainen, Mannering, Lord, & Quddus, 2011). In this case, it is assumed that each level of the variable is not only discrete, but also independent, and the outcome levels are assumed to be mutually exclusive and the order of the values is not relevant. The following subsections discuss model formulations that treat crash severity as an ordered variable, a nominal variable and as a mixture of ordered and nominal variables.

Ordered Outcome Models

In ordered outcome models crash severity is treated as an ordinal variable and the inherent ordering in the outcome levels is explicitly acknowledged. The model formulation assumes that for a given crash *i*, the observed crash severity outcome (y_i) is a function of an underlying latent propensity value (y_i^*). Subsequently, the underlying propensity is mapped to the observed crash severity outcomes and is specified as a function of observed exogenous variables as shown below:

$$y_i^* = \beta' X_i + \varepsilon_i \tag{1}$$

where X_i represents the vector of explanatory variables predicting the crash severity outcome, representing any of the factors discussed above. Assuming that *K* predictors are used, X_i is a $K \times 1$ vector providing values of each of the explanatory variables for crash occurrence *i* and β is the corresponding

$K \times 1$ vector of coefficients associated with the explanatory variables. ε_i represents a random error term, which captures the portion of the propensity that cannot be explained, or that is not observed, or that is not measured correctly.

Assuming there are $j = 1,2,\ldots,J$ levels of outcomes for crash severity, the mapping between the underlying propensity value (y_i^*) and the observed crash severity outcome (y_i) is shown in Eq. (2):

$$y_i = j \text{ if } \mu_{j-1} < y_i^* \leq \mu_j \tag{2}$$

where μ_s represents the threshold value for the underlying propensity with the index s ranging from 0 to J. Further, the threshold values are ascending, with the increasing value of index s as $\mu_0 < \mu_1 < \ldots < \mu_J$. The boundary threshold values are assumed to satisfy the conditions: $\mu_0 = -\infty$ and $\mu_J = \infty$ so that the propensity value expressed on a continuous scale can be fully mapped to the J levels of crash severity outcomes.

With the above as preliminaries for the ordered outcome models, the probability of observing an outcome j for crash occurrence i is given as

$$P(y_i = j) = \Phi\left(\mu_j - \beta' X_i\right) - \Phi\left(\mu_{j-1} - \beta' X_i\right) \tag{3}$$

where $\Phi(\cdot)$ is the cumulative probability value for the distribution assumed for the error term (ε_i). When the error term (ε_i) is assumed to follow a standard logistic distribution, the ensuing ordered outcome model form is referred to as an ordered logit (OL) model, and when the error term is assumed to follow a standard normal distribution the resulting model form is referred to as the ordered probit model.

The β and μ_s coefficients can be estimated using the maximum likelihood estimation technique. For a sample data set with N observations of crashes, the likelihood function can be expressed as shown in Eq. (4):

$$L = \prod_{i=1}^{N} \prod_{j=1}^{J} P(y_i = j)^{I_{i,j}} \tag{4}$$

where, $I_{i,j}$ is an indicator variable which is equal to 1 if the crash severity outcome (y_i) is equal to j and equal to a value of 0 if the crash severity outcome is not equal to j.

A key assumption in the ordered outcome model formulation described above is that the impact of the different factors on the underlying propensity is assumed to be the same across all crash severity outcomes. This assumption is also referred to as the parallel lines assumption (because the coefficients

β representing the slope of the line for the underlying propensity function are the same across all choice outcomes for a given explanatory variable) or the proportional odds (PO) assumption (because the cumulative odds ratios for any two crash severity outcomes are constant across outcome categories). This assumption is inappropriately restrictive in many crash severity modelling applications, because some factors may have a different impact (both in terms of the magnitude and sign) across different levels of crash severity (Savolainen et al., 2011).

Nominal Outcome Models

In the multinomial probability model formulation, explanatory variables are allowed to influence levels of crash severity outcome differently, thus overcoming the PO assumption of the ordered outcome model formulation. However, in this modelling approach, the order inherent in the levels of crash severity outcomes is ignored and the levels are treated as independent of one another. The roots of multinomial probability models can be traced back to the application of random utility theory in behavioural economics. In the transportation domain, this approach is used to model activity and travel choices made by individuals.

Following the notation introduced for ordered outcome models (wherein i represents the crash instance, and y_i represents the observed crash severity outcome for crash i), the probability that a given crash i will result in a crash severity outcome j is defined by a propensity function, U_{ij}, as defined in the following equation:

$$U_{ij} = \beta_j X_{ij} + \varepsilon_{ij} \tag{5}$$

where X_{ij} is a vector of explanatory variables, β_j is the vector of coefficients associated with the explanatory variables for outcome j and ε_{ij} is the error term. Subsequently, the equation for the probability of observing outcome j for crash i can be derived by estimating the probability that the underlying propensity value for outcome j is greater than or equal to the propensity values for all other outcome values. Assuming the set $\overline{J}$ represents the set of possible crash outcomes 1,2,…,J, the probability of observing outcome j is as shown in Eq. (6):

$$P(y_i = j) = P\left(U_{ij} \geq U_{ik} \forall\, k \in \overline{J},\, k \neq j\right) \tag{6}$$

Substituting Eq. (5) in Eq. (6) and specifying the distribution of the error term will lead to an operational multinomial probability model. When the error terms are assumed to be normally distributed, then the multinomial probability model is referred to as the multinomial probit model (MNP). Alternatively, when the error terms are assumed to be independent and identically distributed, and follow a Type-1 generalised extreme value (Gumbel) distribution, the multinomial logit model (MNL) results. Further, based on the treatment of the β_j coefficients in the multinomial probability models, mixed multinomial probability models have been proposed in the literature that allow for coefficients to vary across crash instances. The mixed-model formulations attempt to capture additional heterogeneity in crash severity outcomes due to varying effects of explanatory variables (Milton et al., 2008).

The MNL variant of the multinomial probability model has been more commonly used to model crash severity compared to the MNP model, because the former has a closed form probability expression (see Eq. (7)):

$$P(y_i = j) = \frac{e^{U_{ij}}}{\sum_{k=1}^{J} e^{U_{ik}}} \tag{7}$$

Similar to ordered outcome models, the maximum likelihood estimation technique can be used to estimate the β_j coefficients for all crash severity outcomes. For a sample data set with N crash observations, the likelihood function for the sample is shown below:

$$L = \prod_{i=1}^{N} \prod_{j=1}^{J} P(y_i = j)^{I_{i,j}} \tag{8}$$

where, $I_{i,j}$ is an indicator variable which is equal to 1 if the crash severity outcome (y_i) is equal to j and equal to a value of 0 if the crash severity outcome is not equal to j.

Even though the multinomial probability model is able to overcome the PO assumption by accommodating different coefficient values for explanatory variables across levels of crash severity outcome, it also suffers from limitations for modelling crash severity accurately. The multinomial probability model treats crash severity as a nominal variable, which may lead to inefficient estimates because the model formulation does not account for the additional information contained in the sample data set due to the inherent ordering in the crash severity outcomes. Further, there is a risk of overfitting the data, potentially compromising the model's ability to predict crash severity outcomes on an out-of-sample data set. Lastly, a key assumption in the MNL model formulation is the independent and identical distribution assumption of the error terms. If there are common unobserved factors that affect the underlying propensities for different crash severity outcomes, then this assumption is violated and the inferences drawn based on the MNL coefficient estimates will be incorrect. In instances where this assumption is violated, alternative model formulations, such as the mixed logit or MNP model that relax this assumption by accounting for unobserved heterogeneity, are more appropriate (Mannering & Bhat, 2014). However, these model formulations suffer from model estimation challenges and estimation routines for these model formulations are not readily available in commonly used statistical packages. For example, in the MNP model, while there are no restrictive assumptions about the error terms, the probability expression does not have a closed form solution and, depending on the number of levels of crash severity, one may experience difficulties in estimating the model coefficients. Selection of a final model form must be based on a careful evaluation of alternative model formulations. In evaluating the alternative model formulations, a combination of engineering judgement, coefficient and model significance measures, and diagnostic tests can be utilised.

Alternative Model Formulations

More recently, researchers have adopted alternate model formulations for modelling crash severity outcomes in an effort to overcome limitations of the ordered outcome and multinomial probability modelling frameworks (as described above). These alternate model formulations attempt to extend the ordered outcome modelling framework by relaxing the PO assumption while avoiding the issues of overfit and heterogeneity when employing the multinomial probability models.

Two extensions of the ordered outcome models, namely the partial PO (PPO) model (Mooradian, Ivan, Ravishanker, & Hu, 2013) and the generalised ordered model (Srinivasan, 2002), are being increasingly used in crash severity modelling. There are a number of other approaches that address various limitations of using nominal and ordered modelling frameworks for injury severity analysis. Mannering and Bhat (2014) provide a detailed account of the state of methodological research for modelling of injury severity. In the PPO model, the PO assumption is relaxed by allowing coefficients associated with a subset of explanatory variables to vary across levels of the crash severity outcomes. On the other hand, in the generalised ordered logit (GOM) model, the PO assumption is implicitly relaxed by parameterising the threshold values as a linear combination of explanatory variables. Even though the conceptual approaches to relaxing the PO assumption are different in the PPO model and the generalised ordered model, Eluru and Yasmin (2015) establish that the modelling approaches are mathematically identical. Therefore, only the PPO model is discussed here and the readers are referred to Eluru and Yasmin (2015) for a comprehensive discussion of the similarity between the two modelling formulations.

The formulation of the PPO model closely follows the ordered outcome model formulation. It proceeds by defining an underlying propensity value y_i^* for each crash occurrence i and, subsequently, the propensity values are mapped to the observed outcomes y_i. However, unlike the specification of y_i^* in the ordered outcome model, the PPO model allows for the coefficients to be fixed or vary across levels of injury severity:

$$y_{ij}^* = \beta' X_i + \gamma_j' Z_i + \varepsilon_i \tag{9}$$

where X_i represents the vector of explanatory variables for which the PO assumption holds, and Z_i is the vector of explanatory variables for which the PO assumption is relaxed. β and γ_j represent the coefficients associated with the corresponding vectors of explanatory variables X_i and Z_i respectively. ε_i is the random error term; again a logistic or normal distribution may be selected. The expression for probability of observing an outcome j for crash occurrence i is

$$P(y_i = j) = \Phi\left(\mu_j - \beta' X_i - \gamma_j' Z_i\right) - \Phi\left(\mu_{j-1} - \beta' X_i - \gamma_{j-1}' Z_i\right) \tag{10}$$

where μ_s represents the threshold value with the index j ranging from 0 to J, and it maps the underlying propensity to the injury severity outcome level as shown in Eq. (2).

The coefficients β, γ_j and μ_s can be estimated using the maximum likelihood estimation procedure. In estimating a PPO model analysts can perform a Wald test, as described in Mooradian et al. (2013) to appropriately choose variables to include in the vectors X_i and Z_i.

The ordered outcome model is just a special case of the PPO model. When the slopes of the explanatory variables are all equal across levels of injury severity (i.e., $\gamma_j' = 0$), then the PPO formulation resembles the ordered outcome model formulation. Similarly, a PPO model that places all variables into the Z vector is 'comparable' to a multinomial probability model, although they are still not identical because the multinomial probability model treats the dependent variable as a nominal variable, ignoring the inherent ordering between injury severity outcomes. Based on the above comparisons among the different model formulations, the fit of a PPO model of injury severity will lie somewhere between the model fit of a restrictive ordered outcome model and that of a more unrestrained multinomial probability model of injury severity (Mooradian et al., 2013).

Other Issues

Observation and collection of crash data are inherently prone to sampling bias. Two common issues are systemic crash under-reporting, which usually affects lower-severity crashes the most, and unobserved heterogeneity in cases where multiple crashes are observed at the same location. The reader is referred to Ye and Lord (2011), Savolainen et al. (2011), Patil, Geedipally and Lord (2012) and Yasmin and Eluru (2013) for more extended discussion of these topics. There is also a growing body of literature on accommodating unobserved heterogeneity in addition to observed heterogeneity. For example, Eluru, Bagheri, Miranda-Moreno and Fu (2012) developed a latent segmentation-based OL model to model injury severity of drivers at highway–railway crossings. The highway–railway crossings are probabilistically assigned to different latent classes based on the crossing attributes and, within each latent class, an OL model of injury severity was estimated. Mannering and Bhat (2014) and Mannering, Shankar and Bhat (2016) provide a good review of the state of research aimed at modelling unobserved heterogeneity and other methodological issues for accident analysis. Additionally, there is also an issue of sample size for adequate statistical model estimation (Ye & Lord, 2014).

MODELLING EXAMPLES

In this section, two examples of mixed modelling are presented. It must be noted that the findings noted in this section are data specific. The examples must be used as a guide for model formulations to consider and understand the process to identify the best model. The analysis presented in these examples must be replicated for each case to ensure that the best model formulation is used.

Example 1

Mooradian et al. (2013) compared the performance of PPO models for crash severity to ordinal and multinomial approaches on both large-scale (200,000 crashes) and smaller-scale (20,000 crashes) sample sets. Using crash data from state roads in Connecticut from January 1995 to December 2009, they examined the three models based on three general criteria: model fit adequacy, covariate values and significance levels, and holdout prediction accuracy. The following covariates were used: indication of at-fault vehicle, indication of senior status (65 or more years old), access control (limited access or surface street), land use (urban or rural), weather (inclement or non-inclement) and crash type. Crash type was a categorical variable, grouped based on similarities of contributing factors to the crash (Zhang, Ivan, & Jonsson, 2007). Factor-level proportions for the entire data set can be found in Table 5.

A Wald test of PO was used to determine which predictor variables belonged to the subset q that rejects the PO assumption. This test takes the multinomial outcome variable and dichotomises it based on cumulative probability, using $P(S_n \geq j)$ and $P(S_n < j)$ for each j. Similar to the Independence of Irrelevant Alternatives test, this method simply determines whether the effect of a variable will remain the same across all 'cuts' of j. Fig. 1(a) and (b) illustrate this graphically. The test finds the empirical logits, where:

$$Emplogit = \text{In}\left[\frac{P(S_n \geq j)}{P(S_n < j)}\right] \tag{11}$$

for each cut of $S_n \geq j$ and $S_n < j$. These empirical logits are plotted across the support of the predictor variable and examined for parallelism. If there is significant deviation from parallelism, we can conclude that the variable will

Table 5. Factor-Level Proportions for Model Covariates.

Variable	Description	Proportion = 1
Fault	1 = in at-fault vehicle; 0 = not at fault	0.5086
Senior	1 = senior; 0 = non-senior	0.075
Urban land use	1 = urban; 0 = rural	0.9448
Limited access	1 = limited access; 0 = surface road	0.3135
Inclement weather	1 = inclement weather; 0 = no inclement weather	0.2136
Same direction collision	1 = same direction collision; 0 = otherwise	0.1535
Angle/turning collision	1 = angle/turning collision; 0 = otherwise	0.2148
Rear-end collision	1 = rear-end collision; 0 = otherwise	0.4499
Head-on, object collision	1 = head-on/object collision; 0 = otherwise	0.1552
Backing/parking collision	1 = backing/parking collision; 0 = otherwise	0.0186
Pedestrian crash	1 = pedestrian crash; 0 = otherwise	0.0053
Jackknife	1 = jackknife; 0 = otherwise	0.0006

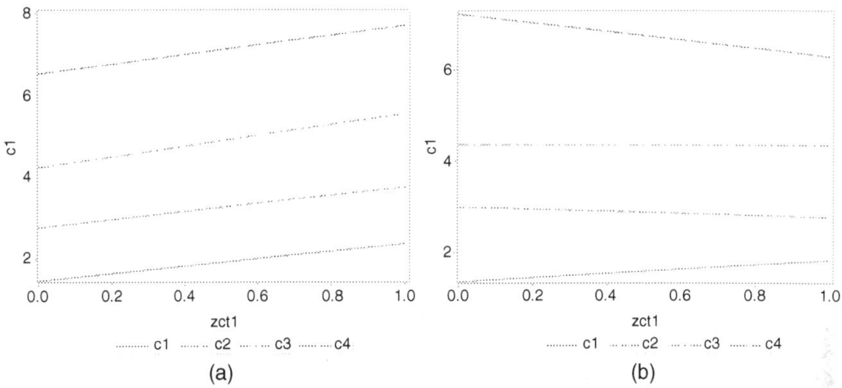

Fig. 1. Example of (a) No PO Rejection and (b) PO Rejection.

likely reject the PO assumption. The variables found to reject the PO assumption were the at-fault indicator, the urban land use indicator and two crash types (rear-end collisions and head-on/object collisions).

In order to compare the fit of the PPO model to the alternatives, models were run separately for each year of the data to provide an adequate sample for fit and to avoid fit problems that may come with different populations of crashes from different years. Log-Likelihoods, the Akaike Information Criterion (AIC) and the Bayesian Information Criterion (BIC) were computed for each of the 15 years. Table 6 shows the averages of each measure for the ordinal, PPO and multinomial models. The multinomial model has the best average fit, with the ordinal having the worst. As evidenced by the values for standard error, there are no significant differences between the Log-Likelihood, AIC and BIC among the three models.

Similar to the fit comparison, covariate results were compared in each of the 15 yearly data sets among the three model types. Table 7 lists whether a covariate is significant for some, all or none of the 15 years. The at-fault indicator, access control, weather and five out of seven of the crash types were significant in the majority of the models. The covariate effects with the greatest magnitude in the model were found to be same-direction and backing/parking collisions, which consistently lowered probabilities of injuries and fatalities, and pedestrian collisions, which dramatically increased the probability of injury and fatality. Overall, the PPO model shows similar coefficient values and significance levels to the multinomial model in variables that reject the PO assumption. When variables do not reject the PO assumption, the coefficients more closely resemble the ordinal model.

Table 6. Comparison of Log-Likelihood, AIC and BIC.

Criterion	Ordinal		Partial Proportional Odds		Multinomial	
	Mean	Standard error	Mean	Standard error	Mean	Standard error
LL	−95440.6	7920.791	−94149.621	7793.518463	−93585	7729.315
AIC	190913.3	15841.58	188355.241	15587.03693	187274	15458.63
BIC	191073.4	15842.44	188641.011	15587.03693	187794.4	15461.43

Holdout prediction was performed in a small (20,000 cases) and large (200,000) sample of the data set, with 10% of the data in each sample assumed unknown in the modelling process. Samples were randomly selected from the population using the SAS SURVEYSELECT procedure (Geisser, 1993). The comparison of predicted probabilities to the observed outcomes within the holdout data on an aggregate basis (Table 8) shows the predictive ability of each of the models. PPO has a fairly constant error percentage, at 0.73% for the small set and 0.71% for the large set. The ordinal data and the multinomial data change more dramatically from the small sample to the large sample, from 3.06% to 0.97% error for the ordinal models and 2.32% to 0.83% error for the multinomial model.

This study shows the following regarding choice of ordered, PPO or multinomial models:

- More covariates were found significant in the PPO model than either of the other two models. By using multiple coefficients only for variables that have significant changes between levels, the PPO model effectively accounts for PO violation without using unnecessary degrees of freedom.
- Holdout prediction was most accurate for the PPO model. This may indicate that the PPO model converges towards the overall proportions of injury severity levels more quickly than the ordinal or multinomial and may have improved aggregate prediction accuracy in smaller sample sizes.

Example 2

Yasmin and Eluru (2013) is another study that compared alternate discrete variable model formulations for modelling driver injury severity. In particular, the study estimated seven different models, including three models based on formulations that treat injury severity as an ordered outcome variable,

Table 7. Covariate Comparison of Model Results.

Model type	Ordinal				Partial Proportional Odds				Multinomial			
Level	Possible injury	Minor injury	Serious injury	Fatal	Possible injury	Minor injury	Serious injury	Fatal	Possible injury	Minor injury	Serious injury	Fatal
Intercept	All	All	All	All	All	All	All	All	All	All	All	All
Fault	All				All	All	All	Some	All	All	All	15
Senior	None				Some				All	Some	All	None
Land use	None				None	All	All	All	None	Some	All	Some
Access control	All				All				Some	Some	Some	All
Weather	All				All				None	All	Some	None
Same-direction collision	Some				All				All	All	All	Some
Angle/turning collision	None				None				All	None	None	Some
Rear-end collision	Some				Some	All	All	All	None	Some	Some	Some
Head-on, object collision	Some				Some	Some	Some	All	None	Some	Some	All
Packing/parking collision	All				All				All	All	All	All
Pedestrian crash	All				All				Some	All	Some	All
Jackknife	None				None				None	None	None	None

Note: Cell values indicate whether a covariate is significant to the 95% level in some, all, or none of the 15 models.

Table 8. Aggregate Holdout Prediction Mean Absolute Percentage Error Values.

Sample Size	Ordinal	Partial Proportional Odds	Multinomial
20,000	3.0614	0.7310	2.3188
200,000	0.9715	0.7161	0.8323

namely OL, GOL and mixed generalised ordered logit (MGOL) and four models that treat injury severity as a nominal outcome variable, namely ML, nested logit (NL), ordered generalised extreme value logit (OGEV) and mixed multinomial logit model (MMNL). Data from the General Estimates Systems from year 2010 was used in the study. The focus of the study was on modelling the injury severity of drivers of passenger cars. In explaining the injury severity, five types of explanatory variables were used, including driver characteristics, vehicle-related variables, roadway attributes, environmental factors and crash characteristics.

After final model estimations, the authors found that the NL and OGEV model formulations did not offer additional explanatory power when compared to the MNL model form. Therefore, they focused the comparative analysis on the remaining five model formulations. Table 9 provides a comparison of the likelihood ratio values for the final specifications using the different model formulations. Table 9 shows inferences drawn from performing a likelihood ratio test (note that this is a nested hypothesis test; that is, it can only be applied to compare alternate model formulations/specifications where one of the models is subsumed within the other model). Inspecting the log-likelihood values and by applying the hypothesis test, the authors

Table 9. Model Estimation Summary.

Model	Log-likelihood	Inference Drawn from the Nested Hypothesis Test–Likelihood Ratio Test
OL	−10617.51	–
GOL	−10517.83	GOL is preferred to OL (X^2 = 199.36, degrees of freedom = 21)
MGOL	−10506.97	MGOL is preferred to GOL (X^2 = 21.72, degrees of freedom = 5)
MNL	−10517.59	–
MMNL	−10508.76	MMNL is preferred to MNL (X^2 = 17.66, degrees of freedom = 4)

identified MGOL as the best ordered model formulation and MMNL as the best unordered model formulation for modelling injury severity.

To further identify the best model formulations among MGOL and MMNL, Yasmin and Eluru (2013) applied a non-nested hypothesis test proposed by Horowitz (1983). From the test, the authors found that the MGOL model offered the best fit. Further, the authors confirmed the inference drawn from the hypothesis test by inspecting model goodness-of-fit statistics, including the BIC and AIC – the lower value of the statistic indicating better model fit. The corresponding values for the MGOL model were 21531.31 and 21124.45 compared to values of 21591.33 and 21140.14 for the MMNL model respectively.

Yasmin and Eluru (2013) also performed a validation analysis using a hold-out sample to test whether the observation of superior model fit of the MGOL model when compared to the MMNL model also translated to better injury severity predictions. To this end, a number of disaggregate and aggregate measures were compared to assess the quality of predictions from the two models. Comparison of the predictions did not show any model clearly outperforming the other and both models performed well in matching actual shares.

Yasmin and Eluru (2013) also explored differences in model specification and predictions between the two formulations when the data suffers from under-reporting issues – a common challenge with traffic safety data sets. For this, they generated a data set by randomly selecting 50% of the crashes from the original estimation data set. To assess the impacts of under-reporting, the authors compared elasticity values from six variables. The percentage differences between the estimates from the full dataset and the under-reported dataset are shown in Table 10 for the MGOL and MMNL formulations. It can be seen that under-reporting appears to affect both model formulations. These observations hold even after calibrating the constants in the MGOL and MMNL models estimated using the under-reported sample to match the shares of the full sample.

Important takeaways from this study for the choice of model formulation for modelling injury severity include the following:

- There are a number of ordered and unordered model formulations that are capable of overcoming the PO assumption of the ordered outcome models. However, it appears that the model formulations that acknowledge the ordered nature of the outcome variable while relaxing the PO assumption offer the best statistical properties for model estimation.
- Under-reporting is a significant issue and this affects the coefficient estimates irrespective of model formulation.

Table 10. Percent Difference in Elasticity Estimates between Full Sample and Under-Reported Sample Using MGOL and MMNL Model Formulations.

Explanatory Variable		MGOL Model		MMNL Model	
	Constants Calibrated	Non-incapacitating injury	incapacitating/ fatal injury	Non-incapacitating injury	Incapacitating/ fatal injury
Male	No	34.44	40.74	35.00	51.25
	Yes	9.88	14.87	8.17	11.90
Age less than 25 indicator	No	23.26	15.31	5.85	7.49
	Yes	12.93	11.62	15.88	18.39
Passenger car indicator	No	19.60	10.65	14.89	36.47
	Yes	8.43	6.78	35.89	9.70
High speed limit indicator	No	17.23	2.97	26.15	25.54
	Yes	0.74	2.51	1.20	10.53
Snowy surface indicator	No	31.61	24.70	51.13	22.53
	Yes	23.99	21.96	30.46	14.94
Head-on collision indicator	No	75.99	20.30	57.83	12.46
	Yes	38.41	14.96	21.19	2.13

SUMMARY AND CONCLUSIONS

Road crash severity is important for assessing the sustainability of the transportation system. This chapter has provided an overview of crash severity modelling, including definitions of crash severity, factors affecting it and statistical approaches for estimating models to predict it. Two examples of crash severity model estimation are provided in which alternative approaches are applied and compared. It is noted that, as with any statistical modelling exercise, the specific findings from these examples apply only to the data sets used for the modelling. These examples illustrate some of the issues with crash severity modelling but do not necessarily indicate what can be expected universally for any data conditions.

Regarding definitions of severity, in the US the KABCO scale is most commonly used for severity modelling by road safety professionals because it is most readily available in crash reports. The MAIS scale is preferred by health professionals due to its greater accuracy; however, because it must be coded by trained medical professionals it is not available in crash data sets unless they are linked to healthcare data sets. Such linkage is not performed routinely due to privacy and linkage consistency issues. Many factors are known to be related to crash severity. Person- and crash-related variables are generally found to be most significantly associated in prediction models. Statistical prediction approaches must account for both the ordinal nature of the crash severity outcome variable and limitations that ordinal modelling approaches impose on variable associations that improperly restrict model predictions.

Future research in crash severity prediction modelling would do well to focus in two primary areas:

- Linkage of crash, roadway and healthcare data sets to combine the best and most accurate data from all of these data sources to get the most complete and accurate picture of the crash severity phenomenon.
- Exploration of new statistical estimation methods to address the ordinal nature of crash severity outcomes as well as independence from one another in their associations with certain predictive factors.

ACKNOWLEDGEMENTS

The authors are grateful for helpful input from Dr. Carol Flannagan of the University of Michigan Transportation Research Institute and from the editors and reviewers in the preparation of this chapter.

REFERENCES

AlEassa, E., Eid, H., & Abu-Zidan, F. (2013). Effects of vehicle size on pedestrian injury pattern and severity: Prospective study. *World Journal of Surgery, 37*(1), 136–140.

American National Standards Institute (ANSI). (2007). *Manual on classification of motor vehicle traffic accidents ANSI D16.1-2007*. Itasca, IL. Retrieved from http://www-nrd.nhtsa.dot.gov/Pubs/07D16.pdf. Accessible on August 2, 2007.

Baker, S., O'Neill, B., Haddon, W., Jr., & Long, W. (1974). The injury severity score: A method for describing patients with multiple injuries and evaluating emergency care. *Journal of Trauma, 14*(3), 187–196.

Caliendo, C., Maurizion, G., & Parisi, A. (2007). A crash-prediction model for multilane roads. *Accident Analysis and Prevention, 39*, 657–659.

Chen, H., Ivers, R., Martiniuk, A., Boufous, S., Senserrick, T., Woodward, M., ..., Norton, R. (2009). Risk and type of crash among young drivers by rurality of residence: Findings from the DRIVE Study. *Accident Analysis & Prevention, 41*(4), 676–682.

Clark, D., Winchell, R., & Betensky, R. (2013). Estimating the effect of emergency care on early survival after traffic crashes. *Accident Analysis & Prevention, 60*, 141–147.

Department of Transportation (DOT). (2008). *MMUCC guideline model minimum uniform crash criteria* (3rd ed.). DOT HS 810 957 Washington DC.

Department of Transportation (DOT). (2012). *MMUCC guideline model minimum uniform crash criteria* (4th ed.). DOT HS 811 631 Washington DC.

Eluru, N., Bagheri, M., Miranda-Moreno, L., & Fu, L. (2012). A latent class modelling approach for identifying vehicle driver injury severity factors at highway-railway crossings. *Accident Analysis & Prevention*, *47*, 119–127.

Eluru, N., & Yasmin, S. (2015). A note on generalized ordered outcome models. *Analytic Methods in Accident Research*, *8*, 1–6.

Elvik, R., Christensen, P., & Amundsen, A. (2004). *Speed and road accidents. An evaluation of the power model*. TOI report 740/2004. Retrieved from http://www.trg.dk/elvik/740-2004.pdf

Flannagan, C., Mann, N., & Rupp, J. (2013). Measuring serious injuries in traffic crashes. In *Transportation Research Board 92nd Annual Meeting*, Washington, DC, January, Paper No. 13-4780.

Funk, J., Cormier, J., & Manoogian, S. (2012). Comparison of risk factors for cervical spine, head, serious, and fatal injury in rollover crashes. *Accident Analysis & Prevention*, *45*, 67–74.

Geisser, S. (1993) *Predictive inference*. New York, NY: Chapman and Hall. ISBN 0-412-03471-9.

Gennarelli, T., & Wodzin, E. (2005). *Abbreviated injury scale 2005*. Des Plaines, IL: Association for the Advancement of Automotive Medicine.

Haddon, W. (1999). The changing approach to the epidemiology, prevention, and amelioration of trauma: the transition to approaches etiologically rather than descriptively based, *Injury Prevention, 5*, 231–236.

Hayat, R., Debbarh, M., Antoniou, C., & Yannis, G. (2013). Explaining the road accident risk: Weather effects. *Accident Analysis & Prevention, 60,* 456–465.

Horowitz, J. (1983). Statistical comparison of non-nested probabilistic choice models. *Transportation Science*, *17,* 319–350.

Huang, H., Chor, H., & Haque, M. (2008). Severity of driver injury and vehicle damage in traffic crashes at intersections: A Bayesian hierarchical analysis. *Accident Analysis & Prevention, 40*, 45–54.

International Road Traffic Accident Database Group (IRTAD). (2011). *Reporting on serious road traffic casualties: Combining and using different data sources to improve understanding of non-fatal road traffic crashes.* Retrieved from https://www.itf-oecd.org/sites/default/files/docs/road-casualties-web.pdf

Islam, M., Serhiyenko, V., Ivan, J., Ravishanker, N., & Garder, P. (2014). Explaining pedestrian safety experience at urban and suburban street crossings considering observed conflicts and pedestrian counts, *Journal of Transportation Safety & Security*, *6,* 335–355.

Jensen, S. (1999). Pedestrian safety in Denmark. *Transportation Research Record*, *1674,* 61–69.

Kaplan, S., & Prato C. (2012). Braking news: Link between crash severity and crash avoidance maneuvers. *Transportation Research Record*, *2280,* 75–88.

Kockelman, K. M., & Kweon, Y.-J. (2002). Driver injury severity: An application of ordered probit models. *Accident Analysis and Prevention*, *34*, 313–321.

Liu, Y., Lai, X., & Ma, C. (2012). Influence of vehicle model year and occupant seating position on serious injury risk in near-side collision. *Journal of Automotive Safety and Energy*, *3*(4), 313–318.

Mannering, F., & Bhat, C. (2014). Analytic methods in accident research: Methodological frontier and future directions. *Analytic Methods in Accident Research*, *1*, 1–22.

Mannering, F., Shankar, V., & Bhat, C. (2016). Unobserved heterogeneity and the statistical analysis of highway accident data. *Analytic Methods in Accident Research*, *11*, 1–16.

Milton, J., Shankar, V., & Mannering, F. (2008). Highway accident severities and the mixed logit model: An exploratory empirical analysis. *Accident Analysis and Prevention*, *40*(1), 260–266.

Mizuno, K., & Arai, Y. (2010). The effectiveness of matching front rail heights in car-to-car crashes. *International Journal of Crashworthiness*, 15(6), 571–579.

Mooradian, J., Ivan, J., Ravishanker, N., & Hu, S. (2013). Analysis of driver and passenger crash injury severity using partial proportional odds models. *Accident Analysis & Prevention*, *58*, 53–58.

Nilsson, G. (2004). *Traffic safety dimensions and the power model to describe the effect of speed on safety* (Bulletin 221). Lund, Sweden: Department of Technology and Society, Lund Institute of Technology.

Nukenine, S., & Daniel, J. (2011). Impact of belt use by rear-seat occupants on injury severity of belted drivers in New Jersey. *Transportation Research Record*, *2265*, 91–99.

Osler, T., Baker, S. P., & Long, W. (1997). A modification of the injury severity score that both improves accuracy and simplifies scoring. *The Journal of Trauma and Acute Care Surgery*, *43*(6), 922–926.

Patil, S., Geedipally, S., & Lord, D. (2012). Analysis of crash severities using nested logit model: Accounting for the underreporting of crash data. *Accident Analysis & Prevention*, *45*, 646–653.

Peng, Y., Geedipally, S., & Lord, D. (2012). Effect of roadside features on single-vehicle roadway departure crashes on rural two-lane roads. *Transportation Research Record*, *2309*, 21–29.

Qin, X., Wang, K., & Cutler, C. (2013a). Analysis of crash severity based on vehicle damage and occupant injuries. *Transportation Research Record*, *2386*, 95–102.

Qin, X., Wang, K., & Cutler, C. (2013b). Logistic regression models of the safety of large trucks. *Transportation Research Record*, *2392*, 1–10.

Quddus, M., Noland, R., & Chin, H. (2002). An analysis of motorcycle injury and vehicle damage severity using ordered probit models. *Journal of Safety Research*, *33*, 445–462.

Rosén, E., Stigson, H., & Sander, U. (2011). Literature review of pedestrian fatality risk as a function of car impact speed. *Accident Analysis and Prevention*, *43*, 25–33.

Savolainen, P., Mannering, F., Lord, D., & Quddus, M. (2011). The statistical analysis of highway crash-injury severities: A review and assessment of methodological alternatives. *Accident Analysis & Prevention*, *43*(5), 1666–1676.

Schrum, K., De Albuquerque, F., Sicking, D., & Reid, J. (2014). Correlation between crash severity and embankment geometry. *Journal of Transportation Safety & Security*, *6*(4), 321–334.

Srinivasan, K. (2002). Injury severity analysis with variable and correlated thresholds: Ordered mixed logit formulation. *Transportation Research Record*, *1784*, 132–141.

Stephan, K., & Newstead, S. (2014). Characteristics of the road and surrounding environment in metropolitan shopping strips: Association with the frequency and severity of single-vehicle crashes. *Traffic Injury Prevention*, *15*, S74–S80.

Theofilatos, A., & Yannis, G. (2014). A review of the effect of traffic and weather characteristics on road safety. *Accident Analysis and Prevention*, *72*, 244–256.

Viano, D., & Parenteau, C. (2010). Severe-to-fatal injury risks in crashes with two front-seat occupants by seat belt use. *Traffic Injury Prevention*, *11*, 294–299.

Woodcock, S. (2000). *Sustainability design guidelines for urban release areas: A masterplan approach for developers*. A report submitted to the Penrith City Council. Institute for Sustainable Futures, University of Sydney, Sydney.

World Business Council for Sustainable Development (WBC). (2004). *Mobility 2030: Meeting the challenges to sustainability. The sustainable mobility project, full report*. Retrieved from http://wbcsdservers.org/wbcsdpublications/cd_files/datas/business-solutions/mobility/pdf/Mobility2030-FullReport.pdf

Yasmin, S., & Eluru, N. (2013). Evaluating alternate discrete outcome frameworks for modeling crash injury severity. *Accident Analysis and Prevention, 59,* 506–521.

Ye, F., & Lord, D. (2011). Investigating the effects of underreporting of crash data on three commonly used traffic crash severity models: Multinomial logit, ordered probit and mixed logit models. *Transportation Research Record, 2241*, 51–58.

Ye, F., & Lord, D. (2014). Comparing three commonly used crash severity models on sample size requirements: Multinomial logit, ordered probit and mixed logit models. *Analytic Methods in Accident Research, 1,* 72–85.

Yu, R., & Abdel-Aty, M. (2014). Analyzing crash injury severity for a mountainous freeway incorporating real-time traffic and weather data. *Safety Science, 63,* 50–56.

Zhang, C., Ivan, J., & Jonsson, T. (2007). *Collision type categorization based on crash causality and severity analysis*. Paper No. 07-2454. Transportation Research Board Annual meeting, Washington, DC.

Zonfrillo, M., Elliott, M., Flannagan C., & Durbin, D. (2011). Association between weight and risk of crash-related injuries for children in child restraints. *Pediatrics, 128*, 1148–1152.

CHAPTER 16

DETECTING HIGH-RISK ACCIDENT LOCATIONS

Simon Washington, Amir Pooyan Afghari and Mohammed Mazharul Haque

ABSTRACT

Purpose – *The purpose of this chapter is to review the methodological and empirical underpinnings of transport network screening, or management, as it relates to improving road safety. As jurisdictions around the world are charged with transport network management in order to reduce externalities associated with road crashes, identifying potential blackspots or hotspots is an important if not critical function and responsibility of transport agencies.*

Methodology – *Key references from within the literature are summarised and discussed, along with a discussion of the evolution of thinking around hotspot identification and management. The theoretical developments that correspond with the evolution in thinking are provided, sprinkled with examples along the way.*

Findings – *Hotspot identification methodologies have evolved considerably over the past 30 or so years, correcting for methodological deficiencies along the way. Despite vast and significant advancements, identifying*

Safe Mobility: Challenges, Methodology and Solutions
Transport and Sustainability, Volume 11, 351–382

ISSN: 2044-9941/doi:10.1108/S2044-994120180000011018

hotspots remains a reactive approach to managing road safety – relying on crashes to accrue in order to mitigate their occurrence. The most fruitful directions for future research will be in the establishment of reliable relationships between surrogate measures of road safety – such as 'near misses' – and actual crashes – so that safety can be proactively managed without the need for crashes to accrue.

Research implications – *Research in hotspot identification will continue; however, it is likely to shift over time to both closer to 'real-time' crash risk detection and considering safety improvements using surrogate measures of road safety – described in Chapter 17.*

Practical implications – *There are two types of errors made in hotspot detection – identifying a 'risky' site as 'safe' and identifying a 'safe' site as 'risky'. In the former case no investments will be made to improve safety, while in the latter case ineffective or inefficient safety improvements could be made. To minimise these errors, transport network safety managers should be applying the current state of the practice methods for hotspot detection. Moreover, transport network safety managers should be eager to transition to proactive methods of network safety management to avoid the need for crashes to occur. While in its infancy, the use of surrogate measures of safety holds significant promise for the future.*

Keywords: Hotspots; blackspots; transport network management; safety management; surrogate safety measures

INTRODUCTION

This chapter is the fifth of six chapters that collectively represent state of the practice methodologies for both understanding and predicting the safety performance of transport networks. This chapter describes methods for identifying high-risk sites or blackspots on a transport network and is a complement to the appropriate management of cross-sectional crash data (Chapter 12), the management of crash data observed at the same transport network locations over time (Chapter 13), a focus on crash data that contain a relatively large number of zeros (numerous sites recording no crashes in a particular time period) (Chapter 14), methods for examining and understanding crash severity (Chapter 15) and finally methods for using surrogate measures of crashes to examine risk (Chapter 17).

Motor vehicle crashes claim millions of human lives and cost billions of monetary losses to society each year. To mitigate these human and financial losses, road transport authorities routinely assess transport network risk and apply crash mitigation strategies to increase safety performance. Budget constraints and scarcity of resources, however, often force governments to choose a sub-sample of locations for remedial treatment from among treatable transport network locations. Transport system locations (e.g., road segments, signalised intersections, etc.) experiencing relatively high crash counts are referred to as blackspots (also known in the literature as hotspots, sites with promise and accident-prone locations). The process whereby system safety managers identify potential blackspots across the transport system is called blackspot identification (BSI) – also known as network screening. As stated by Cheng and Washington (2005), 'the objective of hot spot identification is to identify transportation system locations that possess underlying correctable safety problems, and whose effect will be revealed through elevated crash frequencies relative to similar locations'.

Identification of blackspots is complicated by numerous factors, including crash severities (e.g., fatal versus injury crashes), random fluctuations of crashes from year to year, different exposure levels (traffic volumes), variation in geometric design and operational features (e.g., signal phasing, shoulder and median design), variation in weather (e.g., rain, wind, fog) and differences in driving populations (e.g., younger drivers, older drivers). Moreover, poor safety performance of a site is based on how a site performs relative to how it is expected to perform. This continuing need to correctly identify blackspots has led researchers to develop and test different BSI methodological approaches over the years. These approaches range from simple calculations of crash risk to advanced mathematical tools that account for complex mechanisms of crash occurrence. While the advanced BSI methodological approaches have been developed to address the complexities mentioned previously, the more simple methods are often applied and appeal to practitioners. Further work is needed to both illuminate the trade-offs across methodological approaches and make more complicated techniques accessible to practitioners.

Erroneous identification of blackspots has prompted the development of robust and quantitative measures to evaluate the performance of various BSI methodologies. These performance measures aim to quantify the ability of BSI approaches to correctly identify blackspots. Although there is extensive research on the development of various BSI methods, there is comparatively little research on comparing the performance across methods.

This chapter first discusses how blackspots arise. It then reviews in detail various BSI methodologies. Following this review is a discussion of what criteria are used to evaluate the performance of BSI methodologies. Future research directions in BSI methodology conclude the chapter.

WHAT ARE BLACKSPOTS?

Understanding how blackspots are conceptualised, and how and why they arise, is fundamental to BSI methodological development. The fundamental questions underpinning the development of BSI methods may be understood by considering the following questions:

- How is crash risk defined and what are acceptable measures of crash risk?
- What delineates between acceptable versus unacceptable safety performance or crash risk?
- What assumptions are made when applying BSI methods?

The first order of business is to establish a risk context. Risk can be measured in a variety of contexts – an individual, a mode, a city or a transport system. Traditionally, transport system managers are concerned with crash risk hotspots at transport network system locations. For example, a safety engineer might be focused on the safety of high-speed, limited-access roads. Thus, computation and comparison of crash risk along segments of these roads is a primary focus – enabling a manager to understand the range and distribution of risk observed across these locations. Typically a manager would examine risk across all segments of high-speed, limited-access roads within their jurisdiction. The focus of course could instead be on signalised intersections, rural roads, on-ramps, so on – with a similar focus shifting to risk on these transport system locations.

The next step in understanding the detection of high-risk crash locations is to quantify the relative risk of crash occurrence – measured in crashes per unit of exposure (e.g., traffic volumes, time, length of road, population, etc.). The most straightforward way is to count the total number of crashes at a transport system location as a measure of risk, and so the higher the crash count, the higher the risk of crash occurrence. State-of-the-practice modelling of crash risk (commonly referred to as count data modelling in the literature) models or predicts crash counts as a function of site characteristics, including vehicle exposure metrics.

Crash risk, as might be expected, is not constant over the transport network. We would not expect crash risk to be constant across segment types, because driving task requirements, vehicle driver behaviour, environment and roadside environments are fundamentally different across the transport network. Factors that are thought to influence the probability of crash occurrence are often referred to as crash-contributing factors. Some contributing factors are related to engineering design considerations (e.g., lane widths, median treatments), while others are related to the roadside environment (e.g., drop-offs, trees, slopes, etc.), surrounding traffic (e.g., percentage of trucks) and environmental conditions (e.g., glare, rain, ice, wind). At the centre of all this lies the driver, whose reactions to particular situations will determine whether an incident or accident occurs – thus driver factors also play a significant role. However, from the perspective of site-specific safety and crash risk – which is determined by a collection of crashes over time – driver attributes are not related to site-level safety unless obtained in a meaningful aggregate form. For example, a site with a large percentage of young male drivers might experience relatively high numbers of crashes.

Once factors that influence crash risk are determined, the profession applies a well-accepted concept of 'expected safety' of a site, pioneered by Ezra Hauer (Hauer, 1986; Stokes & Mutabazi, 1996). The notion is that the long-run safety performance of a site, corrected for factors known to influence safety, is the basis by which other site's safety performances should be compared. Sites that perform consistently worse than expected are deemed worthy of improvement – or at least worthy of inspection. A consistency criterion is applied to ensure that a site did not just randomly perform worse than expected, and so we look for sites that perform worse than average over several time periods of observation – typically on the scale of several years. The models used to estimate the expected safety performance of sites are generically referred to as safety performance functions (SPFs).

The practicalities involved with improving sites often dictate that only the worst-performing sites can be improved in any given budget cycle. Thus, safety improvements are often considered and applied to the top 1%, 5% or 10% of the worst-performing sites depending on a variety of factors, including available budget, cost of treatments, number of sites being considered, so on. In some jurisdictions costs of improvement are considered, and so benefits (i.e., cost savings associated with injuries and fatalities reduced) and costs, often in the form of benefit-to-cost ratios, are considered in the final allocation of public funds.

Driven largely by pragmatic considerations, safety managers see crash risk as a dual-state, with a site considered to be either risky (unsafe) or non-risky (safe). This artificially imposed dual state will vary across sites and depend on numerous factors, including the expected safety predicted by the SPF. Differences in safety performance of sites across jurisdictions, regions, states and countries can also be significant.

The SPFs developed to predict the expected safety of sites are used to screen networks and identify potential blackspots. This process – BSI – is a routine practice of safety managers around the globe. It is important to emphasise that this process identifies potential blackspots, as site visits are required to determine whether a potential blackspot can be treated cost-effectively with engineering or behavioural countermeasures, whether there is an unobserved spatial cause (e.g., local population of animals) or whether it is anomalous (random).

STATE OF THE PRACTICE IN BSI

Overview of BSI Approaches

The past four decades have seen significant development of BSI methodologies. For convenience, such approaches can be grouped into six categories: naïve, SPFs of expected crash frequencies, empirical Bayes' (EB) and Bayes', severity-based, spatial and continuous risk profile (CRP) approaches. While extensive improvements have been achieved and a variety of methods have been developed, defining a crash risk indicator and selection criteria to prioritise sites against the indicator have always been inseparable and key parts of every approach. Each approach has an underpinning philosophy and procedure for estimating relative crash risk. This section briefly introduces these approaches, while the following section delves into more technical detail.

Simple ranking of sites based on crash frequency, crash reduction potential[1] and rate quality control (RQC) are among the most common naïve approaches for identifying blackspots. The 'naïve' descriptor is used to signify that these approaches do not attempt to account for the expected safety performance of a site but simply assess safety performance usually in the most recent time period. SPFs have been developed as an improvement to the naïve classical approaches by capturing the concept of expected safety performance of a site – where a site's observed safety performance is compared to its expected safety performance. SPFs account for exogenous factors that influence safety, such as exposure to traffic, weather effects and design configurations, and serve as a universally accepted method for identifying high-risk locations.

The SPF has been modified to account for a site's crash history, thus correcting for potential regression to the mean (RTM) effects caused by non-random selection of sites. RTM is an effect whereby outlying observations of crash counts in any given period are likely to be less outlying – or closer to the mean crash count – in subsequent observation time periods. The EB approach has attracted universal attention due to the simplicity of correcting for potential RTM effects. In the EB approach, estimates of the means and variances of a site's crash history are used to adjust for possible RTM effects when computing expected crash counts. The Bayesian approach for addressing RTM makes use of the probability distribution of crash history to adjust for RTM effects, thus in some cases being more precise. The Bayesian approach is more complicated than the EB approach, and so often in practice the former method is employed.

The need to account for and differentiate among crash severities occurring across sites has led researchers to develop methods for incorporating crash severities into BSI procedures, including multivariate SPFs, mixed modelling and equivalent property damage only (EPDO) methods. The multivariate methods allow prediction of crash frequency by severity category, while the EPDO methods allow for direct accounting of crash severity in a single metric, similar to crash frequency.

Spatial relationships of various crash locations within a transport network arise from shared omitted spatial effects used in SPFs and from effects on safety that are spatial in nature. Animal-related collisions are an ideal example, as animals tend to congregate in habitats with defined spatial boundaries – thus their effects on road crashes are related to proximity to their habitat. Micro-climates can also create similar spatial effects on safety. With the development of various geographic information system (GIS) tools and spatial analytical capabilities, the spatial effects of sites can be accommodated without necessarily understanding the underlying causes. Nearest neighbour, spatial auto correlation and Kernel density estimation are the common spatial approaches that have been applied to capture the safety effects of location, without identifying specifically what factors give rise to those effects.

The vast majority of network BSI approaches break roads into segments and define intersections to consist of network features within 500 feet (or some other fixed length) of the intersection centreline. Crashes are then allocated to the appropriate road segments upon which they occurred, and analysis and SPFs are built around these segments. Monitoring crashes on a dynamic length of roadway segments and looking for crash frequency peaks in a continuous observance are the fundamental principles behind the development of CRP approaches. These approaches are computationally more complex but offer the ability to detect hotspots of differing segment lengths. While CRP approaches have been modified to identify blackspots proactively, they have not been implemented widely due to their complexity and general lack of software availability. As segments analysed change by small increments along a network, it is not easy to conduct a CRP analysis without the assistance of sophisticated software.

Fig. 1 illustrates a schematic of BSI approaches. The underlying assumptions and analytical process of each approach are also provided in the schematic.

Before BSI methodologies are described, it is important to discuss the context under which they are typically applied. BSI methodologies are analytical tools that enable a network to be screened for *potential* blackspots. These methods are typically applied using databases and computer analytics (e.g., spreadsheets, software, etc.) at jurisdictional headquarters. Sites are identified, or screened, as potential blackspots because there remains a need upon subsequent site visitation to substantiate an engineering or operational deficiency, as distinct from a random 'up' fluctuation in crashes. Often this distinction is determined through an on-site audit (through inspection of crash reports) and through engineering inspections. Because a transport network is typically large, both in terms of kilometres of roads and geographical spread, visiting sights to inspect them is an expensive and time-consuming task. If the jurisdiction is a state or province, distances can indeed be vast, requiring overnight stays by site-inspection crews.

Because of the significant resources involved, being as 'correct as possible' in identifying potential blackspots is critically important. There are two errors that can be made when applying BSI methods. The first is to identify a potential blackspot erroneously, while the second is a failure to identify a potential blackspot. The first error results in sending crews to sites that are not blackspots. The second involves failing to send crews to sites that are potential blackspots. Both errors are undesirable and result in misallocation of public resources. For these reasons, spending time to develop and/or apply the most accurate and reliable of BSI methods is a fruitful undertaking.

After sites are visited and verified as being 'treatable', engineering and/or operational improvements are identified to remedy the engineering and/or operational deficiency. Sometimes the deficiency may be behavioural in nature – in which case appropriate solutions would be considered. For example, distraction, fatigue, impaired driving and inattention are driver behaviours that may require a combination of targeted policies and enforcement to address. As is often the case, however, the agencies responsible for overseeing engineering improvements are not also responsible for implementing behavioural ones – thus coordination across agencies is required to facilitate effective treatment.

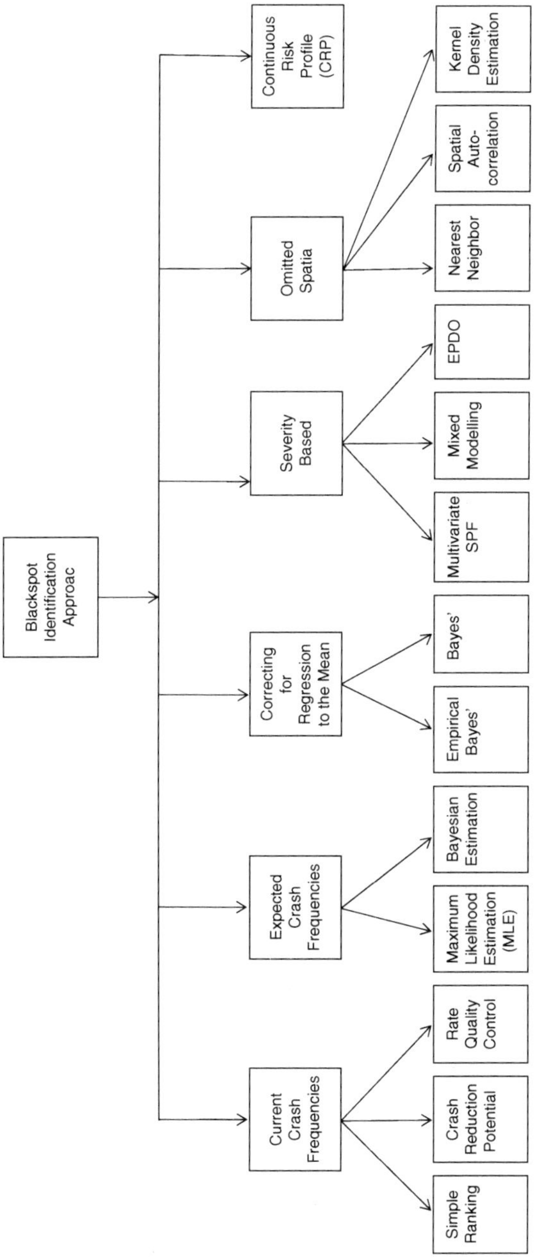

Fig. 1. A Schematic of BSI Approaches Discussed in the Literature.

Sites that are identified as blackspots are typically treated using 'blackspot' funds; hence, the ranking facilitates the treatment of sites most costly in terms of societal impact (Deacon, Zegeer, & Deen, 1975; Hauer, 1986). In practice, often costs of treatment are also considered, so cost–benefit ratios are calculated such that public monies are invested most wisely. This is typically done after sites have been identified as potential blackspots and confirmed through site inspections.

Naïve Classical Approaches

The most basic method that has been used to identify potential crash blackspots (from herein the word 'potential' is dropped when referring to blackspots) in a transport network is to use raw crash counts or crash frequencies as an indicator of risk for every location relative to a subset of similar locations (e.g., high-volume urban intersections, two-lane rural roads, etc.) and then rank them from highest to lowest crash frequency per unit time. In the case of road segments, lengths are often standardised (e.g., 1 mile) so that crash frequency comparisons are valid. While this approach is simple to apply, the use of crash counts neglects the opportunities for risk – or exposure to risk – reflected in part through traffic volumes. The desire to capture exposure to risk has led to the calculation of crash rates, which are simply the crash counts divided by a measure of average exposure at each site (Norden, Orlansky, & Jacobs, 1956). Smeed (1949) showed that the total number of crashes at sites is proportionate to the number of vehicles (Smeed, 1949), and this finding began the trend of using traffic volumes as measures of exposure in which to calculate crash risk. In current practice, average annual daily traffic (AADT) is generally accepted as the measure of exposure on road segments, whereas entering traffic volumes are used for intersections.

It is important to highlight that driver (and passenger) exposure to crash risk is complex and is not perfectly captured by AADT or entering volumes, as the calculation of risk outcomes is complex. For example, crashes on road segments can arise as either single-vehicle or multi-vehicle crashes. As exposure increases initially on a road segment, single-vehicle crash risk also rises. As exposure continues to increase, single-vehicle crash risk declines and multi-vehicle crash risk begins to increase. These and other complications lead to an imprecise measure of exposure used in the calculation of risk, despite its widespread acceptance and application in road safety. Adding further to the complexity of the relation between crash outcomes and exposure, Hauer (1997) fundamentally changed the state of the practice when he demonstrated that a non-linear relationship between exposure and crashes exists on many types of transport segments. This non-linearity means that comparing crash rates across sites is not an objective way of comparing crash risk, as sites with large traffic volumes are expected to have lower crash rates than similar sites with lower volumes. Fortunately, additional analytical tools were proposed and have been widely adopted to address this additional complexity (more is provided on this later).

In the late 1970s and early 1980s, researchers recommended comparing crash counts (or crash rate) at each location to the mean crash count (or rate) at similar, comparable sites, and referred to this approach as CRP, as discussed previously. The underlying principal of the crash reduction potential approach is that the safety performance of a site should be compared to its expected, long-run safety performance in order to determine whether or not it is a blackspot. Subsequently, high-risk locations were identified as locations with a crash reduction potential greater than zero (meaning a location's crash count is above the average) (Hauer, 1986, 1980; Maher & Mountain, 1988). This comparison between observed crash counts (or rate) of a location and mean crash count (or rate) of the reference population led to the initial understanding that a threshold criterion for selecting blackspots was ultimately insightful and gave rise to the notion that sites could perform both 'better' and 'worse' than expected with respect to safety. Later, other types of thresholds were applied to compare crash rates at various transport network locations with a predefined control rate to ensure the achievement of a certain safety level. This latter approach is referred to in the literature as the RQC method (Norden, Orlansky, & Jacobs, 1956; Stokes & Mutabazi, 1996).

A major deficiency with naïve, classical approaches is their ignorance of the RTM effect; in spite of crash counts tending towards their long-term means, there is not an appropriate mechanism in naïve classical approaches to correct for this effect.

Safety Performance Functions

A major development in BSI methodology was the introduction of crash prediction models or SPFs. SPFs are an extension of expected safety performance beyond accounting for the effects of exposure and typically account for geometric and operational features such as number of approach lanes and surrounding land use (e.g., urban).

In the mid-1980s researchers started using SPFs to evaluate the effects of treatments after implementation. The argument behind their application was that the observed crash count in the before observation period is not an accurate predictor of what safety 'would have been' after treatment, as it does not account for potential changes that could have occurred after treatment (e.g., weather, exposure, resurfacing, etc.). Thus, standard practice adopted the use of an expected value estimate for the crash frequency, correcting for known changes in the after period (Hauer, 1986). These estimates are based typically on cross-sectional data from the safety performance of similar sites, accounting for the effects of features known to influence safety. The use of SPFs in evaluation studies dramatically increased the interest in the statistical modelling of crashes. These SPFs overcame several limitations of naïve classical approaches in which the simple use of crash counts did not account for changes in roadway, environmental and generally exogenous factors that can influence crash occurrence (Hauer, 1997). Importantly, SPFs account for non-linear relationships between crashes and exposure (Hadayeghi, Shalaby, & Persaud, 2003; Hauer, 1997; Miaou & Lord, 2003) and capture and quantify the random and uncertain nature of crash occurrence.

One of the statistical features of SPFs is that models require assumptions about the distribution of crash counts. Crash counts at an individual transport network location (e.g., road segment or intersection) are generally and reasonably assumed to follow a Poisson distribution (Miaou & Lum, 1993a, 1993b). Because of unobserved heterogeneity across sites, crash data in aggregate (i.e., from multiple sites) are often over-dispersed relative to the Poisson distribution. As such, count data modelling techniques, like the Poisson regression model (when not over-dispersed) and negative binomial regression models (when over-dispersed), and a variety of extensions are routinely used to develop SPFs. Assuming that the total number of crashes (Y) in any fixed observation period (e.g., month, year, etc.) follows a Poisson distribution with parameter μ (expected mean crash count), the expected (mean) number of crashes for a site is used as an estimate of the long-run average of a site. Recently, some researchers have developed and applied statistical models of crashes that provide estimates of quantiles of the crash counts (e.g., 50th percentile or median, 95th percentile) and found that a quantile regression analysis is able to identify different thresholds of crashes that may correspond more seamlessly with BSI approaches (Washington, Haque, Oh, & Lee, 2014). Researchers have also examined influences on μ of the effects of a variety of engineering factors (e.g., roadway geometric characteristics, speed limits), weather (e.g., rainy days), traffic (e.g., percentage of heavy vehicles) as well as exploring various model functional forms relating explanatory variables to the mean number of crashes (Anastasopoulos & Mannering, 2009; Geedipally & Lord, 2008; Kim, Washington, & OH, 2006; Lee & Mannering, 2002;

Lyon, Oh, Persaud, Washington, & Bared, 2003; Miaou & Lum, 1993a; Milton & Mannering, 1998; Oh, Lyon, Washington, Persaud, & Bared, 2003; Shankar, Mannering, & Barfield, 1995). In common state-of-the-practice applications of statistical models, crash counts are considered to have a log linear relationship with their causal factors. Two of the most commonly used SPFs are shown in Eqs. (1) and (2):

$$\text{Intersections: } \mu = F_1^{\alpha_1} F_2^{\alpha_2} e^{(\alpha_3 X_3 + \alpha_1 X_4 + \alpha_5 X_5 + \cdots)} \tag{1}$$

$$\text{Road segments: } \mu = L^{\beta_0} V^{\beta_1} e^{(\beta_2 X_2 + \beta_3 X_3 + \beta_4 X_4 + \cdots)} \tag{2}$$

where μ is the expected mean of crash counts of a site, F_1 and F_2 are traffic volumes for the major and minor approaches at intersections, V is the traffic volume on a road segment, L is the length of the road segment, X_i are additional explanatory variables and α_i, β_i are estimated regression parameters. As stated previously, often these models are negative binomial models or a close derivative. Two common estimation techniques to obtain parameters in the SPFs are maximum likelihood estimation (MLE) and the Bayesian estimation.

The MLE method seeks parameter estimates that maximise the likelihood (or probability) of observing the sample under the assumed distributional assumptions. In the BSI terminology, this is equivalent to selecting model parameters that maximise the likelihood of μ being equal to Y. Parameters are assumed to be approximately normally distributed (or t-distributed)

Another method for the estimation of regression parameters is through the application of Bayes' theorem and sampling-based estimation. Bayes' theorem enables the incorporation of prior information in parameter estimation. According to Bayes' theorem, the posterior probability of crashes is correlated with the likelihood of observed crashes as well as prior information about the predicted crash mean:

$$P[\mu|Y] = \frac{P[Y|\mu]\pi[\mu]}{\int [m(y)]dy} \tag{3}$$

where $P[\mu|Y]$, $P[Y|\mu]$ and $\pi[\mu]$ are referred to as the posterior, the likelihood and the prior, respectively, and $\int m(y)dy$ is the marginal distribution of observed crash data.

Bayes' theorem is often operationalised using Markov Chain Monte Carlo (MCMC) sampling (Washington, Karlaftis, & Mannering, 2003). MCMC offers a significant advantage in the application of Bayes' methods in that complicated likelihood functions and posteriors can be considered in model estimation more readily than using maximum likelihood or similar approaches. While the details of MCMC are available in Washington et al. (2003), the basic notion is that samples are randomly drawn from prior distributions and combined with the likelihood (i.e., data) to obtain sampling-based estimates from the posterior. Of course, there is no such thing as a 'free lunch' and so MCMC is generally less efficient than maximum likelihood methods for obtaining parameter estimates, thus, usually yielding larger prediction intervals (equivalent to confidence interval in MLE) around parameter estimates. There are also interpretive advantages of the Bayesian approach – especially when informative priors are used – but these advantages are often overlooked in favour of the computational advantages.

In pursuing robust and state-of-the-art models for predicting expected safety performance, much research has focused on the statistical characteristics or performance of models, addressing topics including the commonly observed over-dispersion of crash data, a preponderance of zeroes, unobserved spatial and temporal heterogeneity, outcome correlation and modelling of crash severity to name a few. Interested readers should consult Lord and Mannering (2010), Savolainen et al. (2011) and Mannering and Bhat (2014) for extensive reviews of the evolution of statistical models of crash data, their shortcomings, methodological frontiers and future directions.

The outcome of SPFs – whether estimated via MLE or Bayesian estimation – is the predicted mean crash count of a transport network location, often interpreted to reflect the unobserved true safety of the site. The observed crash count at the same site is an indicator of how the site is performing with respect to safety (in a given observation period) relative to how it *should* perform, as predicted by an SPF (Hauer, 1996, 1997). This method, sometimes referred to as 'excess crash counts', aims to identify as blackspots those sites with significantly lower predicted compared to observed crash counts. Other studies have compared the observed crash count of a site with the predicted mean crash count of a sample or comparison group of sites which are similar to the crash location (McGuigan, 1981, 1982). The differences between these two approaches really depend on how many covariates are included in the SPF for predicting expected safety – a small number of predictors (e.g., AADT only) will usually include a large number of similar or comparison sites, whereas a large number of predictors (e.g., shoulder width, median type, lane width, etc.) will limit the comparison sites to very few, 'similar' sites. Another selection criterion has been developed in the Bayesian estimation of SPFs in which a predefined threshold value is set in determining the posterior probabilities. Blackspots are then identified as those locations with higher posterior probability of exceeding a threshold value. This method is referred to as the Bayesian probability of excess (Heydecker & Wu, 2001).

Empirical Bayesian Approach

Classical or frequentist models of crash risk (i.e., estimation of SPFs) attempt to capture the uncertainty in crash occurrence within and across sites by fitting probability distributions to these crashes. One aspect of this uncertainty is that 'excess crashes' (i.e., observed crashes exceed expected) in any period are likely to 'regress to the mean'. In other words, sites that appear to perform worse than average in an observation period are likely to improve in the subsequent period as a statistical artefact. The more outlying a site is with respect to expected safety performance, the larger the anticipated RTM effect in the subsequent observation period. So predicting how a site should perform in the future should account for RTM. This fundamental notion of combining information from comparison sites and the variance of crashes 'within' a site led to the development of the EB approach, which is the most common BSI approach for accurately predicting the future safety performance of a site (Hauer, 2003). The EB approach is an analytically simplified version of Bayesian estimation.

Parameter estimation in the EB approach is accomplished by simplifying the Bayesian rule into a proportion relationship between the posterior and the product of likelihood and prior:

$$P[\mu|Y] \sim P[Y|\mu]P[\mu]$$

The EB estimator is a weighted sum of the predicted mean value and the observed mean and variances of observations such that:

$$\mathrm{EB}(\mu) = w_1 y + w_2 \mu' \tag{5}$$

$$w_1 = \frac{\mu'}{\left(\mu' + \frac{1}{k}\right)} \quad \text{and} \quad w_2 = \frac{1}{k\left(\mu' + \frac{1}{k}\right)}$$

where μ' is the predicted crash count, y is the observed (empirical) crash count, k is a constant for a given model that is estimated during the SPF calibration process and w_1 and w_2 are weights calculated based on the mean and variance of the SPF estimate (Persuad, Lan, Lyon, & Bhim, 2010). Parameters of the EB model are achieved using MLE.

Common criteria for identifying blackspots when using the EB approach are the potential for improvement (PFI) and the critical EB (EB_c) estimate methods. The PFI makes use of the difference between the EB estimate and the predicted crash mean (outcome of the SPF) for each site and ranks sites according to decreasing PFI. Sites with large, positive PFIs are identified as the most correctable 'potential' blackspots (Montella, 2010). The second criterion makes use of the difference between the EB estimate of every location and the mean EB estimate of a reference population, which is fairly similar to the location as the critical EB_c estimate, where the locations with positive EB_c are identified as blackspots (Cheng & Washington, 2008; Hauer, 1992, 1996; Hauer & Persaud, 1984; Persaud et al., 1999).

Hauer and Persaud (1984), Persaud (1986), Higle and Witkowski (1988), Persaud and Hauer (1984) and other researchers have argued that the EB is superior compared to traditional approaches for identifying blackspots. Subsequently, researchers have applied the EB approach and compared it to the performance of other approaches (Cheng & Washington, 2005; Congdon, 2007; Elvik, 2008; Hauer, 1997; Hauer, Kononov, Allery, & Griffith, 2002; Mitra & Washington, 2007; Park & Lord, 2007; Persaud, Lyon, & Nguyen, 1999; Washington et al., 2003). All of these researchers conclude that the EB approach outperforms alternative approaches both in terms of the statistical goodness-of-fit on data and identifying crash blackspots. As a result, the EB remains the most widely used approach for identifying blackspots.

Notwithstanding, some researchers have identified flaws with the EB approach. Hauer (1997) and Carlin and Louis (2000) argue that the use of crash history twice during the EB approach can lead to erroneous results. They indicate that the crash history is first used to calibrate the mean function and once again to determine the prior probability of crash mean (Carlin & Louis, 2000; Hauer, 1997). It is better to use a reference group to estimate the mean function and history for prior. Miaou and Lord (2003) also demonstrate a shortcoming of the EB approach in capturing 'uncertainty' between covariates and safety. They argue that the point estimates of covariates in SPFs are fixed after model calibration, and the uncertainty is merely presented by an over-dispersion term. However, the calibration process is accomplished via crash data which are subject to uncertainty, and thus point estimates should also be subject to uncertainty (Huang, Chin, & Haque, 2009). Congdon (2003) and Hauer (1997) point to the weakness of the EB approach in capturing time-variant correlations among crashes from year to year.

As stated previously, interpretative advantages and computational enhancements associated with Bayesian models have made these methods appealing to researchers and academics. One of the important advantages of Bayesian estimation is its capability in handling computations of complex distributions and mixtures of distributions. Crash data are usually over-dispersed and one solution to capture over-dispersion is to allow randomness (ε_0 which follows a probability distribution) among the crash means across observed sites. As crashes are Poisson-distributed, incorporating a random distribution into the mean function complicates regression model likelihood functions. The Bayesian methodology, combined with advances in simulation and computational techniques, such as MCMC simulation, has facilitated the use and testing of a wide range of distributions in crash modelling. Consequently, the choice of the distribution for ε_0 is much less restricted (Washington et al., 2003), and more complicated likelihood functions are estimable using this approach.

The Bayesian method directly estimates the posterior distribution by integrating the prior and data likelihoods relative to the marginal distribution (computing the integration in the denominator in Eq. (3)). This approach accommodates a wide range of distributions to be considered as prior information and distribution of observed data. The Bayesian approach has several advantages over the EB method, including the incorporation of probability distributions rather than point estimates of means and variances. As expected, a Bayesian analysis reveals a distribution of outcomes (posterior distribution) rather than point estimates (Gelman et al., 2003; Miaou & Lord, 2003). Another advantage of a Bayesian analysis is interpretation. The posterior distribution yields the probabilities of model parameters given the observed data, rather than the probability of observing the data given the model parameters – the conditional probability obtained from classical estimation techniques. This latter advantage is often lost on practitioners and researchers, despite it being a fundamental argument found in the early statistics literature, where Bayesian and classical statisticians often debated the merits of Bayesian estimation. Finally, MCMC estimation of Bayesian models often results in inflated standard errors compared to classical estimation techniques and thus are, in general, less efficient (Congdon, 2007).

Some studies have compared the performance of the Bayesian and the EB approaches to identify blackspots and conclude that the Bayesian approach significantly outperforms the EB approach (Huang et al., 2009), while others (Miaou & Lord, 2003) have found that the two approaches are comparable and thus it may not be worth the extra effort to adopt a Bayesian analysis as it complicates modelling. The Bayesian approach is recommended in cases involving relatively small numbers of comparison groups for the SPF calibration (Persaud et al., 2010). Even though the Bayesian approach provides a more flexible platform for parameter estimation and model calibration, the EB approach remains appealing due to the ease of implementation. The ultimate decision on estimation approach taken is thus a function of sample size, complexity of assumed theoretical distributions, complexity of functional form of SPF and importance of the analysis.

Severity-Based Approaches

Incorporating crash severity into SPF estimation has been another focus of BSI methodological development. The motivation for incorporating severity into BSI methodology is that the societal impact and harm are measured not by the number of crashes but by severity of crashes and their differential costs and impact on society. Such severity levels range between PDO crashes, where no one is injured, and fatalities where lives are lost. As such, it is argued that severity should be taken into account when identifying the most potentially harmful network locations. Thus, it is argued that methods that ignore severity are not able to identify hotspots that are most harmful to society (Deacon et al., 1975). Accordingly, numerous studies have examined and incorporated crash severity in the context of BSI procedures. These studies can be categorised into three main groups including multivariate SPFs, mixed models and EPDO.

The first category is similar to the SPF approach, with a difference being that crash severities are separated and their counts considered simultaneously in a multivariate regression analysis (Park & Lord, 2007). In this approach the risk indicators (predictors of crash counts) are similar but different for each severity class – differing both in estimates of parameters and the subset of included predictors.

The second category combines crash severity probabilities (predicted by multinomial logit or mixed logit models) with the total predicted crash count at every site (obtained from count models) to obtain estimates of crash frequency by severity category. Although traditional severity analysis of crashes focuses on the probability of different crash severities across individuals, some recent studies have used site-specific covariates in a multinomial logit (Congdon, 2003) and a mixed logit platform (Gelman et al., 2003) to derive the proportions of certain injury-severity categories (PDO, minor injury, major injury and fatalities) across sites. The difference between such approaches and recent site-level logit modelling methodologies is that these site-specific predicted proportions of injury crashes have the potential to differentiate among crash counts of different severity levels at each site. Since total crash count is still used in these models to determine different portions of certain severities, this second category is applicable to any BSI approach which yields a total crash count indicator of the risk, including the EB approach. In a third approach incorporating crash severity, the count of crashes for different severities is converted to an EPDO count by applying weights based on average cost ratios by severity. For example, the average fatal crash may cost 1,000 times more than the average PDO crash and thus is weighted by a factor of 1,000 – or is equivalent to 1,000 PDOs. Subsequently, the weighted sum of all types of crashes at each site is used as the measure of crash risk and forms the basis of SPF modelling. This approach, not surprisingly, is referred to as the

EDPO approach (Oh, Washington, & Lee, 2010). The selection threshold criteria for the EPDO approach are referred to as the excess EPDOs (Washington et al., 2014). The expected EPDO of a site is compared to its observed EPDO, and the results are ranked accordingly. The sites with the largest excess EPDOs are identified as blackspots. Such a calculation has the same shortcomings as naïve classical approaches (i.e., not accounting for RTM). An alternative approach is to calculate EB estimates distinctly for each severity level, generate PFIs (as stated later in the chapter) and then apply EPDO weights to the PFIs of different levels. Ultimately, the weighted sum of all PFIs at each site may then be used to identify blackspots (National Research Council, 2010).

Accounting for Unobserved Spatial Effects

Because SPFs often suffer from omitted variables, the unobserved effects of spatial influences are often statistically significant in crash models. For example, collisions with wild animals occur in locations with close proximity to animal populations (e.g., deer, kangaroos, etc.). It is difficult to include a variable that could explain the increase in crash frequency caused by proximity to wild animals, and so inclusion of a spatial variable will often pick up unobserved spatial effects. Other unobserved spatial effects include, but are not limited to, micro-climates (wind, rain, icing, etc.), glare and land use effects not captured by other variables (e.g., number of vehicles entering a parking lot). Proximity to drinking establishments and high pedestrian activity zones may also be partially captured through spatial effects.

The general approach for estimating unobserved spatial effects is to spatially search for clusters of locations with similar effects on crash frequency prediction (or severity). The nearest neighbour and the Kernel density estimation (or grid-based) methods were initially introduced to perform spatial analysis (Levine, Kim, & Nitz, 1995). Such methods were capable of identifying high crash count sites within a certain distance from a centroid. However, they are based on point coordinates and do not take into account attributes of crash locations. To address this drawback, spatial statistical auto correlation techniques were introduced to test for spatial correlation between various crash locations and their specific attributes (Mitra, 2009). Moran's Index, Geary's ratio, local indicators of spatial associate, time lag between events, network distance weighted clustering and K-means clustering are among such tools that have been applied to evaluate the possible auto correlation of crash locations, grouping them based on their similarity or dissimilarity in crash occurrence and their contributing factors (Mitra, 2009). Ultimately, crash sites which reveal strong correlation within close proximity to one another are clustered together and identified as blackspots. For additional detail on spatial analysis tools, the reader is referred to Anderson (2009), Mitra (2009) and Prasannakumar, Vijith, Charutha and Geetha (2011).

Although the spatial approach represents a supplemental path for identifying crash blackspots based on their spatial attributes, the approach in practice is a band-aid solution to address the omission of important variables that are correlated across space. More specifically, it makes no sense that a spatial coordinate would itself have an effect on crashes – it is an observed effect that resides at that spatial coordinate that is captured by the coordinate.

Continuous Risk Profile Approach

Important assumptions underlying most BSI methods are that crashes occur independently at each individual transport network location, crash rates are constant over each individual segment for an observation period (e.g., 1 year), and the causes of crash occurrence are independent of each other at different network locations (Chung & Ragland, 2009). These assumptions may be violated on roadway segments in certain fairly common circumstances. For example, the risk of crash occurrence at a particular location could be influenced by a variety of factors at other potential locations, such as upstream congestion or a downstream crash. These secondary crashes are spatially correlated on the transport network. Another assumption that arises as a consequence of the way methodologies are applied is that blackspots are fixed-length locations in accordance with segment lengths used for analysis. So, for example, in a blackspot analysis that employs 1-km segment lengths, blackspots that are associated with point locations (e.g., a poorly placed utility pole) and blackspots that arise from a design deficiency over considerable distance (such as lack of a median barrier) are too short and too long, respectively, for conventional BSI approaches to efficiently identify.

These limitations have led to the development of 'moving-window' approaches, including the sliding average window approach (Kwon, Park, Chung, & Yeo, 2012), the CRP approach (Chung, Jang, Madanat, & Washington, 2011) and the peak searching algorithm approach (Hauer, Kononov, Allery, & Griffith, 2002). In these approaches, the aim is to monitor crash rates/expected frequency over a fixed/dynamic length of roadway segments called 'windows' as shown in Fig. 2(a). In the first two approaches – sliding average window and CRP – crash rates at each location (e.g., d) are averaged over the desired road segment within the window. The window is then moved from the beginning (d_{start}) to the end (d_{end}) of the roadway segment and the methodology 'looks for' crash peaks. Subsequently, as depicted in Fig. 2(b), a predetermined threshold is utilised and

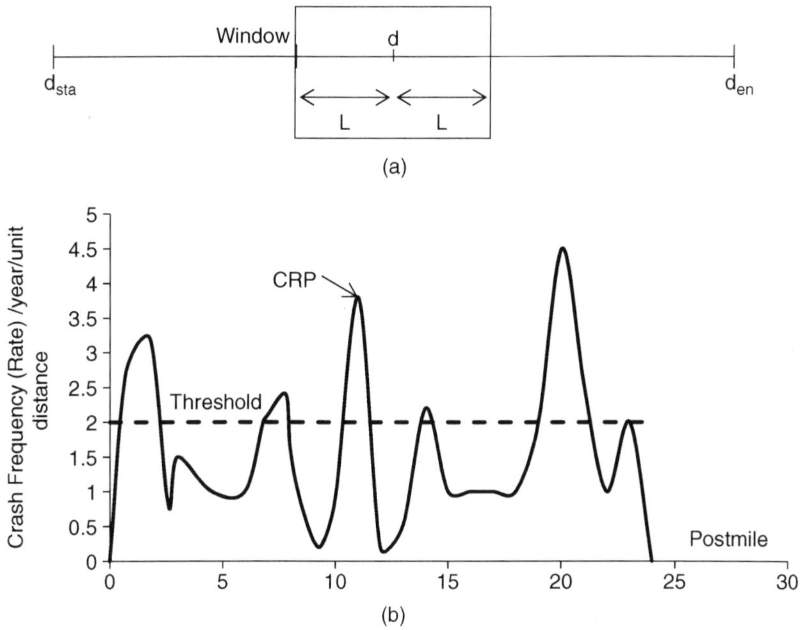

Fig. 2. The CRP Approach to Identify Blackspots: (a) Hypothetical Roadway Segment with the Fixed Window for Assessing CRP and (b) CRP for the Hypothetical Roadway Segment. *Source*: Chung & Ragland (2009).

locations of the network, which show higher crash rates than the threshold, are identified as blackspots and flagged for further investigation. The naïve use of crash rates in these two approaches, however, does not account for the RTM effect and also ignores the non-linear relationship between crashes and traffic volume.

A third approach (i.e., the *peak searching algorithm*) makes use of expected crash frequency (the output from SPFs) and thus addresses this shortcoming. In addition, the precision of network screening is enhanced by comparing the coefficient of variation (rather than expected frequency) at each site with a predetermined threshold. Those sites with a coefficient of variation more than the threshold are then identified as blackspots – yielding two simultaneous BSI thresholds (Task, 2002).

EVALUATION CRITERIA

With so many alternative BSI methods detailed in the literature there is a need for objective criteria and a framework to compare their performance.

Several challenges confronting researchers wishing to compare approaches are immediately obvious. First, BSI methods are used to screen for potential safety issues, and thus the actual (true) status of sites with empirically observed crash counts is not truly known. Second, false positives (FPs) and false negatives (FNs) are the predominant measures reported in the literature to measure BSI performance, but these measures are limited in insight, as they ignore the continuous nature of predicted and observed crash counts and ranking data. Not surprisingly, there is a history of literature around the development of methods for evaluating the performance of BSI approaches.

Early studies comparing the performance of the EB approach with other approaches used false identification of blackspots (Hauer & Persaud, 1984; Persaud & Hauer, 1984) – meaning FPs and FNs, or type-I and type-II errors. These errors provide two metrics for comparing BSI approaches and prove to be quite limited regarding deep insight. For example, sites are identified as 'safe' or 'unsafe' because they meet or exceed a threshold, for example, are in the 'top X%' of unsafe sites in a region. In this definition a site just above a threshold may be quite similar in safety performance to a site just below the threshold – yet is labelled as 'safe' and 'unsafe', respectively. Clearly, greater clarity can be provided to distinguish between such sites.

Responding to the dearth of statistical measures for BSI performance evaluation, new measures were introduced to capture the greater complexity associated with BSI methods (Cheng & Washington, 2008). Five new metrics were introduced: *false identification test*, the *site consistency test*, the *method consistency test*, the *total rank differences test* and the *Poisson mean differences test*.

False Identification Test

This simple test was first used in early comparisons of BSI approaches. A simple test is accomplished by computing the percentage of FPs and FNs identified by a method. FPs are sites that are *truly* safe but are wrongly considered as blackspots. In contrast, FNs are sites that are *truly* unsafe but are considered as safe. Recall that the distinction between 'safe' and 'unsafe' is merely its ranking in relation to other sites and its position relative to a threshold. Often FPs and FNs are reported as percentages – the percentage of sites incorrectly identified as unsafe, or incorrectly identified as safe. In contrast, the percentage of correct identifications consists of correct positives and negatives identified by the BSI methodology (Elvik, 2007). Nevertheless, the false identification test has been applied in a number of safety studies examining BSI methods (Cheng & Washington, 2005; Persaud & Hauer, 1984;

Washington & Cheng, 2005). As stated previously, a correct, objective application of the test requires that *truly* safe and hazardous sites are known a priori, which requires the use of experimentally controlled, simulated data. For details on how a simulation experiment might be conducted and these identifications be calculated, see Cheng and Washington (2005).

Site Consistency Test

The site consistency test measures the efficiency of the BSI method in identifying a network location (site) that reveals consistently poor performance over subsequent periods of time. The fundamental logic behind this test is that in the absence of any changes and in case of a location being identified as a blackspot, that location should be inferior in terms of safety performance in subsequent periods. Alternatively, we would not expect a site to vacillate between 'unsafe' and 'safe' in subsequent observation periods, and it should produce a consistent number of crashes. Using this simple logic, the test compares the total sum of observed crashes in time period $i+1$ in the 'top X%' of unsafe sites identified by the BSI method j in time period i, with the total sum of observed crashes in time period $i+1$ in the 'top X%' of unsafe sites identified by other BSI approaches in time period i. The larger this total sum of observed crashes is, the more consistent the BSI method is. Consequently, the BSI approach which reports a larger total sum of observed crashes in time period $i+1$ is the one that has been capable of consistently identifying sites with poor performance in the absence of any changes and thus is the superior method.

Method Consistency Test

In contrast to the site consistency test, the method consistency test aims to measure the number of blackspot sites commonly identified in two consecutive time periods. The logic of this test is that roadway characteristics of a transport network are typically under similar operating conditions (e.g., similar traffic volume, drivers' population, etc.) in two consecutive time periods. Thus, their safety performance would be expected to remain relatively unchanged in the absence of any treatment. Consequently, a consistent BSI method is one that identifies a similar set of blackspots in two consecutive time periods. In the method consistency test the intersection of blackspots in two consecutive time periods is compared across various BSI methods. The BSI method which yields the higher number of intersections in common is the most consistent method.

Total Rank Differences Test

While the two previous methods make use of counts of blackspots, this test takes into consideration the rank of each site in the final top α per cent list of blackspots. On the basis of similar assumptions as previous tests (i.e., no treatment applied and roadway characteristics remain unchanged), the total rank difference test determines the sum of rank differences between the period i and $i+1$. The BSI method which has the smaller sum of differences is more consistent and outperforms the other methods.

Poisson Mean Differences Test

One important disadvantage of the false identification test is that it computes a metric based on a binary outcome (blackspot or not) based on underlying continuous data. There is a variance component – captured by the estimated Poisson mean plus extra variation – that can be used to determine how well a method predicts the underlying (true) mean. For instance, one method may misidentify a site with a very small variation from the true mean, while the same site may have a larger mean using another method. The Poisson mean differences test measures the sum of the absolute differences between the true Poisson means and predicted Poisson means. Similar to the false identification test, this test requires that hazardous sites are known a priori via data simulation.

To overcome the fundamental and vexing problem of not knowing the true status of sites, several research teams have conducted complex simulation experiments, whereby synthetic crash data were simulated from sites with assumed underlying a priori 'blackspot status' (Cheng & Washington, 2005; Higle & Hecht, 1989; Maher & Mountain, 1988). This approach enables the objective comparison of BSI performance across a variety of metrics and systematic comparison under a variety of experimental conditions. Montella (2010) used this approach and conducted a comparative analysis of several BSI approaches, including crash frequency, EPDO, crash rate, proportion method, EB estimate of total crash frequency, EB estimate of severe-crash frequency and PFI. His experiments showed the outperformance of the EB method for identifying priority investigation locations. The proportion method, on the other hand, performed the poorest among all BSI methods. Nevertheless, simulating crash data under experimental conditions remains the only truly objective way of comparing the performance of BSI approaches.

RESEARCH GAPS AND LIKELY FUTURE DIRECTIONS

Despite extensive methodological enhancements and progress made during the past four decades, there remain significant opportunities for further development of BSI methodologies. In particular, little attention has been paid to questioning and possibly revising the theory underlying BSI methodologies; to date, much of the work has been empirically motivated. Identifying gaps in the literature is critically important, as a comprehensive examination of the underlying theory of BSI approaches will help to identify where useful, new research contributions can be made both in basic and applied research applications. The following areas represent a partial list of opportunities for further research in BSI methodology:

1. While BSI approaches typically utilise crash counts (either observed, predicted or a combination of these two) as the measure of risk, there seems to be a great potential in other measures of crash risk. For example, there is extensive research in the literature in the area of crash severity analysis, which aims to determine various causal factors contributing to different crash severities. Although these models have been partially utilised in the BSI procedures (in severity-based approaches), the potential of explicitly utilising discrete probabilities of crash severities to identify blackspots has been generally neglected. Recently, some studies have investigated a BSI methodology using a categorical analysis (Ferreira & Couto, 2012, 2015). However, there is a significant need to test and compare the efficiency of such methods against current approaches using objective criteria.
2. Gaining considerable attention are surrogate measures of safety (see Chapter 17). Use of these measures in BSI approaches has been hampered by lack of theory and evidence linking surrogate measures to actual crash experience; however, this lack of evidence is being addressed. As studies linking surrogate measures to actual crashes and the nature of these relationships are determined, opportunities to use surrogate measures to identify blackspots will present themselves.
3. Just-in-time crash prediction is of significant interest, as the risk of crash occurrence may be affected by (and thus predicted by) short-term variations in transport system operating conditions. Potential contributing factors and interactions between facility design parameters, traffic flow, speeds, lighting and weather conditions may assist in the prediction of heightened crash risk (Abdel-Aty & Pemmanaboina, 2006; Hossain & Muromachi, 2012; Lee, Hellinga, & Saccomanno, 2003). Current BSI approaches, however, are targeted to support reactive rather than proactive

safety actions. More work is needed to evolve and demonstrate viability and value of these methods, which in principal rely on crash prediction.

4. Integration of data mining and clustering techniques has led to new insights regarding how groupings and sub-samples of sites systematically differ in crash risk. The concept of clustering was introduced into road safety in the late 1990s by Karlaftis and Tarko (1998), revealing the importance of grouping observations based on common features, such as socioeconomic, traffic and infrastructure characteristics. Later studies have confirmed these findings and illustrate the need to analyse sub-populations. This process of identifying homogeneous clusters of data as subsets of the main population based on minimising dissimilarities within clusters and maximising dissimilarities between is likely to remain a topic of future research (de Oña, López, Mujalli, & Calvo, 2013).
5. Latent class models apply statistical criteria to evaluate the appropriate number of clusters and, more importantly, membership probabilities assigned to observations that allow different distributions for different variables, in contrast with traditional clustering techniques that assign an observation to one cluster (Depaire, Wets, & Vanhoof, 2008; Eluru, Bagheri, Miranda-Moreno, & Fu, 2012, Ma & Kockelman, 2006; Malyshkina & Mannering, 2010; Park & Lord, 2009; Park, Lord, & Lee, 2014; Peng & Lord, 2011; Riviere, Lauret, Ramsamy, & Page, 2006; Xie, Lord, & Zhang, 2007; Zou, Zhang, & Lord, 2014). Interpretations of such models will need to be tied to theories of crash occurrence and may lead to new insights.
6. The relationship between exposure and crashes is complex and depends upon context; however, it is often modelled quite simply as a linear or exponential function. Exposure at intersections, for example, involves turning movements, queuing and gap acceptance – all of which are commonly omitted in accounting for the effects of exposure on crash prediction. Additional work is needed to articulate theories around this relationship and support them empirically.
7. Current BSI methodologies make use of subjective thresholds to distinguish between 'high-risk' and 'non-high-risk' sites. These typically 1%, 2% and 5% thresholds are subjective, because they are dictated by available safety investment resources, and because a given threshold is dependent upon the sample of sites. For example, remediating the 'worst' site in a sample will lower the threshold – thus potentially moving a previously 'safe' site above the new threshold. Is it theoretically justifiable that a site that was considered safe previously is now considered unsafe simply because the worst sites were improved? Moreover, how can subjective thresholds across sites best be considered for optimal safety investment? Is a top-2% two-lane rural highway site less safe than a top-2% motorway segment?

One solution to this conundrum is a prospective cost-effective analysis proposed by Hauer (2000), in which costs are determined based on particular project types and benefits are proportional to either expected crash frequency or expected excess crash frequency (Hauer, 2000). It has been proposed that such a cost–benefit analysis should be accomplished for the most disaggregated categorisation possible, such as area type (urban/rural), intersection or roadway type, traffic levels, so on. (Task, 2002). Because public funds are used to remediate sites – and efficient investment is sought – the application and definition of thresholds remains the subject of potential research.

8. Crashes arise from different categories of underlying processes or factors as they relate to the transport network. For example, engineering features will contribute to crashes at specific network locations, whereas behavioural factors may be more related to time of day and driver attributes. Animal-involved collisions may be linked to surrounding environment factors, as would be micro-climates. These various distinct sources of crash causes will give rise to crashes in different ways from the perspective of a network location. Crash counts and the associated probability distributions that arise from such processes have traditionally been modelled as a single process – with a single distribution used to capture the sum of these crashes. Opportunities exist to postulate and model these latent distributions more precisely for potentially improved understanding of crash variability across sites. Recent research has shown that separating distinct sources of crash causes reveals more information about blackspots and thus may lead to a more efficient and precise BSI (Afghari, Haque, Washington, & Smyth, 2016).
9. Current BSI procedures require a high-risk site to be 'fully developed' prior to detection. In other words, a site must be a location with significant crash occurrence – including injuries and sometimes deaths – before it is identified as 'high-risk'. Early efforts have recognised that this extremely undesirable pretext requires a solution, a method that can detect 'infant' blackspots. The premise here is that detecting blackspots earlier may serve the public good. Proactive detection of blackspots is in its infancy and remains an important area of future research.
10. Most methodologies for blackspot detection rely on segmentation of transport networks using somewhat arbitrary criterion – such as a homogenous number of lanes and posted speeds. This segmentation involves two assumptions that may be questionable: (1) blackspots are best identified as segments of 1-km lengths and (2) the transition between segments (e.g., transition between two- and four-lane road or change in posted speed limit) is not a source of contribution to a blackspot.

The notion that a blackspot may require remedial treatment along a significant length of road (e.g., rumble strips or median barrier) suggests that the safety performance along various lengths of road should be accommodated in a blackspot methodology. Moreover, the transition from a four-lane to two-lane road or reduction in speed limits may be a location of increased crash risk, yet will not be detected by conventional BSI methodologies. Both of these assumptions are worthy of exploration through research.

11. Finally, studies that enable the development of relationships between crashes and surrogate measures are needed to develop the next generation of proactive safety investment tools. Understanding, for example, how time-to-collision and post-encroachment time metrics relate to crashes at sites will enable the justification of safety investments prior to a crash occurring. This research is in its infancy and is likely to transform how safety investments are allocated across transport networks.

NOTE

1. Not to be confused with continuous risk profile.

REFERENCES

Abdel-Aty, M. A., & Pemmanaboina, R. (2006). Calibrating a real-time traffic crash-prediction model using archived weather and ITS traffic data. *IEEE Transactions on Intelligent Transportation Systems*, *7*(2), 167–174.

Afghari, A. P., Haque, M. M., Washington, S. P., & Smyth, T. (2016). Bayesian latent class safety performance function for identifying motor vehicle crash black spots. *Transportation Research Record: Journal of the Transportation Research Board*, *2601*, 90–98.

Anastasopoulos, P. C., & Mannering, F. L. (2009). A note on modeling vehicle accident frequencies with random-parameters count models. *Accident Analysis & Prevention*, *41*(1), 153–159.

Anderson, T. K. (2009). Kernel density estimation and K-means clustering to profile road accident hotspots. *Accident Analysis & Prevention*, *41*(3), 359–364.

Carlin, B. P., & Louis, T. A. (2000). *Bayes and empirical Bayes methods for data analysis*. Boca Raton, FL: CRC Press.

Cheng, W., & Washington, S. P. (2005). Experimental evaluation of hotspot identification methods. *Accident Analysis & Prevention*, *37*(5), 870–881.

Cheng, W., & Washington, S. (2008). New criteria for evaluating methods of identifying hot spots. *Transportation Research Record: Journal of the Transportation Research Board*, *2083*(1), 76–85.

Chung, K., Jang, K., Madanat, S., & Washington, S. (2011). Proactive detection of high collision concentration locations on highways. *Transportation Research Part A: Policy and Practice*, *45*(9), 927–934.

Chung, K., & Ragland, D. R. (2009). *The continuous risk profile approach for the identification of high collision concentration locations on congested highways* (pp. 463–480). Transportation and Traffic Theory 2009: Golden Jubilee.

Congdon, P. (2003) *Applied bayesian modelling*. West Sussex, England: John Wiley & Sons.

Congdon, P. (2007). *Bayesian statistical modelling* (Vol. 704). West Sussex, England: John Wiley & Sons.

Deacon, J. A., Zegeer, C. V., & Deen, R. C. (1975). Identification of hazardous rural highway locations. *Transportation Research Record: Journal of the Transportation Research Board*, *543*, 16–33.

de Oña, J., López, G., Mujalli, R., & Calvo, F. J. (2013). Analysis of traffic accidents on rural highways using latent class clustering and Bayesian networks. *Accident Analysis & Prevention*, *51*, 1–10.

Depaire, B., Wets, G., & Vanhoof, K. (2008). Traffic accident segmentation by means of latent class clustering. *Accident Analysis & Prevention*, *40*(4), 1257–1266.

Eluru, N., Bagheri, M., Miranda-Moreno, L. F., & Fu, L. (2012). A latent class modeling approach for identifying vehicle driver injury severity factors at highway-railway crossings. *Accident Analysis & Prevention*, *47*, 119–127.

Elvik, R. (2007). *State-of-the-art approaches to road accident black spot management and safety analysis of road networks*. Oslo, Norway: Transportøkonomisk institutt.

Elvik, R. (2008). Comparative analysis of techniques for identifying locations of hazardous roads. *Transportation Research Record: Journal of the Transportation Research Board*, *2083*(1), 72–75.

Ferreira, S., & Couto, A. (2012). Categorical modeling to evaluate road safety at the planning level. *Journal of Transportation Safety & Security*, *4*(4), 308–322.

Ferreira, S., & Couto, A. (2015). A probabilistic approach towards a crash risk assessment of urban segments. *Transportation Research Part C: Emerging Technologies*, *50*, 97–105.

Geedipally, S. R., & Lord, D. (2008). Effects of varying dispersion parameter of Poisson-gamma models on estimation of confidence intervals of crash prediction models. *Transportation Research Record: Journal of the Transportation Research Board*, *2061*(1), 46–54.

Gelman, A., Carlin, J. B., Stern, H. S., Dunson, D. B., Vehtari, A., & Rubin, D. B. (1996). *Bayesian data analysis*. Boca Raton, FL: Chapman & Hall/CRC Texts in Statistical Science.

Hauer, E. (1980). Bias-by-selection: Overestimation of the effectiveness of safety countermeasures caused by the process of selection for treatment. *Accident Analysis & Prevention*, *12*(2), 113–117.

Hadayeghi, A., Shalaby, A. S., & Persaud, B. N. (2003). Macrolevel accident prediction models for evaluating safety of urban transportation systems. *Transportation Research Record: Journal of the Transportation Research Board*, *1840*(1), 87–95.

Hauer, E. (1986). On the estimation of the expected number of accidents. *Accident Analysis & Prevention*, *18*(1), 1–12.

Hauer, E. (1992). Empirical Bayes approach to the estimation of "unsafety": The multivariate regression method. *Accident Analysis & Prevention*, *24*(5), 457–477.

Hauer, E. (1996). Identification of sites with promise. *Transportation Research Record: Journal of the Transportation Research Board*, *1542*(1), 54–60.

Hauer, E. (1997). *Observational before/after studies in road safety. Estimating the effect of highway and traffic engineering measures on road safety*. Oxford: Elsevier Science Ltd.

Hauer, E. (2000). *Working Paper 10. Screening: Methods and software.* Unpublished paper prepared for Task Force Developing Safety Analyst. Retrieved from http://www. roadsafetyresearch.com

Hauer, E. (1997). *Observational before after studies in road safety*. Bingley, England: Emerald Group Publishing Limited

Hauer, E., Kononov, J., Allery, B., & Griffith, M. (2002). Screening the road network for sites with promise. *Transportation Research Record: Journal of the Transportation Research Board, 1784*(1), 27–32.

Harwood, D. W., Torbic, D. J., Richard, K. R., & Meyer, M. M. (2010). Safety Analyst: Software tools for safety management of specific highway sites. Report FHWA-HRT-10-063, U.S. Federal Highway Administration.

Hauer, E., & Persaud, B. N. (1984). *Problem of identifying hazardous locations using accident data*. Washington, DC: Transportation Research Board.

Heydecker, B., & Wu, J. (2001). Identification of sites for road accident remedial work by Bayesian statistical methods: An example of uncertain inference. *Advances in Engineering Software, 32*(10), 859–869.

Higle, J. L., & Hecht, M. B. (1989). A comparison of techniques for the identification of hazardous locations. *Transportation Research Record: Journal of the Transportation Research Board, 1238*, 10–19.

Higle, J. L., & Witkowski, J. M. (1988). Bayesian identification of hazardous locations (with discussion and closure). *Transportation Research Record: Journal of the Transportation Research Board, 1185*, 24–36.

Hossain, M., & Muromachi, Y. (2012). A Bayesian network based framework for real-time crash prediction on the basic freeway segments of urban expressways. *Accident Analysis & Prevention, 45,* 373–381.

Huang, H., Chin, H. C., & Haque, M. (2009). Empirical evaluation of alternative approaches in identifying crash hot spots. *Transportation Research Record: Journal of the Transportation Research Board, 2103*(1), 32–41.

Karlaftis, M. G., & Tarko, A. P. (1998). Heterogeneity considerations in accident modeling. *Accident Analysis & Prevention, 30*(4), 425–433.

Kim, D.-G., Washington, S., & Oh, J. (2006). Modeling crash types: New insights into the effects of covariates on crashes at rural intersections. *Journal of Transportation Engineering, 132*(4), 282–292.

Kwon, O. H., Park, M. J., Chung, K., & Yeo, H. (2012). Comparing the performance of sliding moving window, peak searching, and continuous risk profile methods for identifying high collision concentration locations. *Transportation Research Board 91st Annual Meeting,* Washington D.C., No. 12-2293.

Lee, C., Hellinga, B., & Saccomanno, F. (2003). Real-time crash prediction model for application to crash prevention in freeway traffic. *Transportation Research Record: Journal of the Transportation Research Board, 1840*, 67–77.

Lee, J., & Mannering, F. (2002). Impact of roadside features on the frequency and severity of run-off-roadway accidents: An empirical analysis. *Accident Analysis & Prevention, 34*(2), 149–161.

Levine, N., Kim, K. E., & Nitz, L. H. (1995). Spatial analysis of Honolulu motor vehicle crashes: I. Spatial patterns. *Accident Analysis & Prevention, 27*(5), 663–674.

Lord, D., & Mannering, F. (2010). The statistical analysis of crash-frequency data: A review and assessment of methodological alternatives. *Transportation Research Part A: Policy and Practice, 44*(5), 291–305.

Lyon, C., Oh, J., Persaud, B., Washington, S., & Bared, J. (2003). Empirical investigation of interactive highway safety design model accident prediction algorithm: Rural intersections. *Transportation Research Record: Journal of the Transportation Research Board, 1840*(1), 78–86.

Ma, J., & Kockelman, K. (2006). Crash frequency and severity modeling using clustered data from Washington state. In *Proceedings of IEEE Intelligent Transportation Systems Conference, 2006*. Toronto, Ontario: Canada.

Maher, M., & Mountain, L. (1988). The identification of accident blackspots: A comparison of current methods. *Accident Analysis & Prevention, 20*(2), 143–151.

Malyshkina, N. V., & Mannering, F. L. (2010). Zero-state Markov switching count-data models: An empirical assessment. *Accident Analysis & Prevention, 42*(1), 122–130.

Mannering, F. L., & Bhat, C. R. (2014). Analytic methods in accident research: Methodological frontier and future directions. *Analytic Methods in Accident Research, 1*, 1–22.

McGuigan, D. (1981). The use of relationships between road accidents and traffic flow in "black-spot" identification. *Traffic Engineering & Control, 22*(HS–032), 669.

McGuigan, D. (1982). Non-junction accident rates and their use in 'black-spot' identification. *Traffic Engineering & Control, 23*(HS-033), 020.

Miaou, S.-P., & Lord, D. (2003). Modeling traffic crash-flow relationships for intersections: Dispersion parameter, functional form, and Bayes versus empirical Bayes methods. *Transportation Research Record: Journal of the Transportation Research Board, 1840*(1), 31–40.

Miaou, S.-P., & Lum, H. (1993a). Modeling vehicle accidents and highway geometric design relationships. *Accident Analysis & Prevention, 25*(6), 689–709.

Miaou, S.-P., & Lum, H. (1993b). Statistical evaluation of the effects of highway geometric design on truck accident involvements. *Transportation Research Record: Journal of the Transportation Research Board, 1407*, 11–23.

Milton, J., & Mannering, F. (1998). The relationship among highway geometrics, traffic-related elements and motor-vehicle accident frequencies. *Transportation, 25*(4), 395–413.

Mitra, S. (2009). Spatial autocorrelation and Bayesian spatial statistical method for analyzing intersections prone to injury crashes. *Transportation Research Record: Journal of the Transportation Research Board, 2136*(1), 92–100.

Mitra, S., & Washington, S. (2007). On the nature of over-dispersion in motor vehicle crash prediction models. *Accident Analysis & Prevention, 39*(3), 459–468.

Montella, A. (2010). A comparative analysis of hotspot identification methods. *Accident Analysis & Prevention, 42*(2), 571–581.

National Research Council (2010). *Highway safety manual*. Washington, DC: American Association of State Highway and Transportation Officials.

Norden, M., Orlansky, J., & Jacobs, H. (1956). Application of statistical quality-control techniques to analysis of highway-accident data. *Highway Research Record, 117*, 17–31.

Oh, J., Lyon, C., Washington, S., Persaud, B., & Bared, J. (2003). Validation of FHWA crash models for rural intersections: Lessons learned. *Transportation Research Record: Journal of the Transportation Research Board, 1840*(1), 41–49.

Oh, J., Washington, S., & Lee, D. (2010). Property damage crash equivalency factors to solve crash frequency-severity dilemma. *Transportation Research Record: Journal of the Transportation Research Board, 2148*(1), 83–92.

Park, B.-J., & Lord, D. (2009). Application of finite mixture models for vehicle crash data analysis. *Accident Analysis & Prevention, 41*(4), 683–691.

Park, B.-J., Lord, D., & Lee, C. (2014). Finite mixture modeling for vehicle crash data with application to hotspot identification. *Accident Analysis & Prevention*, *71*, 319–326.

Park, E. S., & Lord, D. (2007). Multivariate Poisson-lognormal models for jointly modeling crash frequency by severity. *Transportation Research Record: Journal of the Transportation Research Board*, *2019*(1), 1–6.

Peng, Y., & Lord, D. (2011). Application of latent class growth model to longitudinal analysis of traffic crashes. *Transportation Research Record: Journal of the Transportation Research Board*, *2236*(1), 102–109.

Persaud, B. N. (1986). Safety migration, the influence of traffic volumes, and other issues in evaluating safety effectiveness: Some findings on conversion of intersections to multiway stop control. *Injury*, *73*, 81.

Persaud, B. N., & Hauer, E. (1984). Comparison of two methods for de-biasing before-and-after accident studies (Discussion and Closure). *Transportation Research Record: Journal of the Transportation Research Board*, *975*, 43–49.

Persaud, B. N., Lan, B., Lyon, C., & Bhim, R. (2010). Comparison of empirical Bayes and full Bayes approaches for before–after road safety evaluations. *Accident Analysis & Prevention*, *42*(1), 38–43.

Persaud, B. N., Lyon, C., & Nguyen, T. (1999). Empirical Bayes procedure for ranking sites for safety investigation by potential for safety improvement. *Transportation Research Record: Journal of the Transportation Research Board*, *1665*(1), 7–12.

Prasannakumar, V., Vijith, H., Charutha, R., & Geetha, N. (2011). Spatio-temporal clustering of road accidents: GIS based analysis and assessment. *Procedia-Social and Behavioral Sciences*, *21*, 317–325.

Riviere, C., Lauret, P., Ramsamy, J. M., & Page, Y. (2006). A Bayesian neural network approach to estimating the energy equivalent speed. *Accident Analysis & Prevention*, *38*(2), 248–259.

Savolainen, P. T., Mannering, F. L., Lord, D., & Quddus, M. A. (2011). The statistical analysis of highway crash-injury severities: A review and assessment of methodological alternatives. *Accident Analysis & Prevention*, *43*(5), 1666–1676.

Shankar, V., Mannering, F., & Barfield, W. (1995). Effect of roadway geometrics and environmental factors on rural freeway accident frequencies. *Accident Analysis & Prevention*, *27*(3), 371–389.

Smeed, R. J. (1949). Some statistical aspects of road safety research. *Journal of the Royal Statistical Society. Series A (General)*, *112*(1), 1–34.

Stokes, R. W., & Mutabazi, M. I. (1996). Rate-quality control method of identifying hazardous road locations. *Transportation Research Record: Journal of the Transportation Research Board*, *1542*(1), 44–48.

Washington, S. P., & Cheng, W. (2005). *High risk crash analysis*. Report FHWA-AZ-05-558, Arizona Department of Transportation, Phoenix: Arizona.

Washington, S. P., Karlaftis, M., & Mannering, F. (2003). *Statistical and econometric techniques for transportation data analysis*. New York, NY: CRC/Chapman Hall Press.

Washington, S. P., Haque, M. M., Oh, J., & Lee, D. (2014). Applying quantile regression for modeling equivalent property damage only crashes to identify accident blackspots. *Accident Analysis & Prevention*, *66*, 136–146.

Xie, Y., Lord, D., & Zhang, Y. (2007). Predicting motor vehicle collisions using Bayesian neural network models: An empirical analysis. *Accident Analysis & Prevention*, *39*(5), 922–933.

Zou, Y., Zhang, Y., & Lord, D. (2014). Analyzing different functional forms of the varying weight parameter for finite mixture of negative binomial regression models. *Analytic Methods in Accident Research*, *1*, 39–52.

CHAPTER 17

SURROGATE MEASURES OF SAFETY

Andrew P. Tarko

ABSTRACT

Purpose – *This chapter overviews surrogate measures of safety to help better understand the related challenges and opportunities. The chapter is meant to serve as a primer for practitioners looking for alternative methods of evaluating safety where crashes are lacking or are insufficient.*

Approach – *The historical perspective and the current state-of-the-art thinking are presented in an organised manner with a focus on fundamental concepts, traffic measurement techniques and estimation of the relationships between surrogate events and collisions.*

Findings – *An analysis of the published research and its findings indicates that traffic conflicts are the most promising surrogates. They enable evaluation of the safety implications of a wide range of road and traffic conditions. The required ecological consistency between conflicts and collisions can be ensured by sufficient nearness of conflicts to collisions. Several methods of estimating the relationship between conflicts and crashes are discussed. Behavioural measures of safety are also discussed. Although easier to measure than conflicts, behavioural measures should be used with caution.*

Safe Mobility: Challenges, Methodology and Solutions
Transport and Sustainability, Volume 11, 383–405

ISSN: 2044-9941/doi:10.1108/S2044-994120180000011019

Research on surrogate measures of safety may provide a basis for improving microsimulation models as tools of safety evaluation.

Practical implications – *Current changes in vehicle and road instrumentation affect safety at a rate that exceeds the efficiency of the traditional crash-based methods of safety analysis. Accurate and quick measurement of safety with surrogate measures offers a viable solution. They are also a necessary condition of gaining a better understanding of safety and finding more effective solutions for safety problems.*

Keywords: Traffic conflicts; crash surrogates; risky events

INTRODUCTION

This chapter is the final of six chapters that collectively represent state of the practice methodologies for both understanding and predicting the safety performance of transport networks. This chapter describes methods for using surrogate measures of crashes to examine risk and is a complement to the appropriate management of cross-sectional crash data (Chapter 12), the management of crash data observed at the same transport network locations over time (Chapter 13), a focus on crash data that contain a relatively large number of zeros (numerous sites recording no crashes in a particular time period) (Chapter 14), methods for examining and understanding crash severity (Chapter 15) and methods for identifying high-risk sites or blackspots on a transport network (Chapter 16).

Previous chapters have presented specific research methods and the knowledge gained by applying these methods to crash data. While the value of crash data is undisputable, there is also a considerable amount of concern about the downsides of crash data, such as the low quality of the data, the difficulty in tracking the sequence of events leading to a crash and, consequently, the limitations of crash data for both acquiring new knowledge and evaluating new safety improvement methods.

One of the problems with crash data is the focus on crash events while neglecting successful events that do not lead to crashes. Crash-focussed data are particularly useful in exploring the risk of injury and fatality given a crash, but they do not support research exploring the risk of a crash itself particularly well where *risk* is defined in both cases as the probability that exposure to a hazard (crash or being on a road) leads to a negative consequence (Ropeik, 2002). Current safety studies compensate for missing information about the events between crashes with roughly assessed or highly aggregate quantities, such as the overall exposure and the average risk, during considerably long periods of months or years. The fallacy of this approach, which is a necessity rather than a choice, was discussed by Davis (2004).

Data highly aggregated with limited or non-existent representation of causality cannot support timely and accurate estimation of safety. The need for a quick and accurate estimation of safety has been amplified with the growing presence of modern technology on roads and in vehicles that quickly change road safety in a way not always sufficiently controlled or even understood. Traditional crash-based methods of acquiring safety knowledge cannot keep up with these changes. There is an upside to this situation however. Modern technologies may help collect safety-related data at and between crashes that may be utilised to provide a never-before available insight to the risk of a crash and the risk of a severe outcome if a crash occurs. The crash risk is used in its probabilistic meaning when applied to short periods. Alternatively, the high risk of crashes can be detected for certain conditions or locations by analysing the frequency and severity of crashes over longer periods by using representative periods.

The earliest mentioning of alternative methods of viewing and analysing safety that do not rely solely on crash data can be found in Perkins and Harris (1968). Thus, the exploration of alternative measures of safety, called surrogate measures of safety to emphasise the need to replace or supplement crash data, has been carried for at least 50 years. This has resulted in various concepts and a variety of definitions, sometimes inconsistent with each other. It is impossible to present all of them, their chronology and impact in a way that would acknowledge all who contributed to the area, especially when only a single chapter is devoted to this topic and not an entire book. Because of that, this chapter focusses on surrogate events and related concepts that are internally consistent and have a good chance, in the judgement of the author, to be useful in the emerging area of autonomous vehicles and computer microsimulation for safety analysis. To make the reader aware that surrogate events are not the only type of surrogate measures, behaviour-related surrogate measures are briefly covered. In summary, the primary objective of this chapter is to introduce surrogate measures of safety as being viable and of growing importance, while the ambition of giving credit to all the documented efforts is not pursued.

Considerable research efforts in the 1970s and 1980s established fundamentals of traffic conflicts but they did not deliver a practical method of measuring safety. The recently observed revival of this research area is motivated by the growing need for timely safety measurement and the new opportunities afforded by modern technologies. Furthermore, the improved statistical methods have raised hope again for applying the traffic conflicts technique to deliver quick measurements and gain a better understanding of safety. At least three benefits of traffic conflicts and other surrogate measures of safety are widely understood:

1. Detecting the excessive risk of crashes on the road.
2. Improving the knowledge of conditions leading to a crash or increasing the crash probability.
3. Estimating the effectiveness of the experimental and existing countermeasures.

Finding a way to achieve these benefits will lead to effective crash avoidance systems in modern vehicles, well-estimated safety relationships and efficient methods for evaluating safety improvements.

As already mentioned, this chapter focusses on the leading surrogate measures of safety in the mainstream research that could become the basis for practical methods and tools of safety analysis. Although the driver perspective is presented in the discussion as vital and difficult to ignore, this chapter is intended for transportation engineers and therefore traffic flow is the dominant perspective of the discussion; surrogate measures of safety are presented as potential tools for evaluating road and traffic control improvements.

A brief discussion of surrogate events follows this introduction and identifies traffic conflicts as the most promising surrogates of crashes. Then, the historical context and recent proposals related to traffic conflicts are presented, as well as the challenges of observing traffic conflicts in naturalistic driving and roadside studies. Correct representation and estimation of the relationship between conflicts and crashes are critical, and the various current models and the prospect of their estimation are discussed. To close the chapter, the frequently proposed or used surrogate measures and computer simulation are also discussed in separate sections, followed by a chapter summary.

HISTORICAL PERSPECTIVE OF TRAFFIC CONFLICTS

The concept of a dangerous interaction is defined in De Silva (1942). Although the term 'traffic conflict' was used for dangerous interactions by Klebelsberg (1964), its first application to identifying safety-related problems is attributed to the Detroit General Motors Laboratory (Perkins & Harris, 1968). At that time, the meaning was somewhat wider than today and included not only interactions between vehicles but also traffic violations

(Campbell & King, 1970; Perkins & Harris, 1968). For example, entering an intersection on a red signal was a traffic conflict even when no actual interaction between vehicles took place. Hayward (1971) narrowed the meaning of traffic conflicts to dangerous interactions where at least one of the involved drivers had to change speed or direction to avoid collision. He estimated the times to collision (TTC) repeatedly updated during observed conflicts. At each calculation instant, the current distance between vehicles was divided by the current speed difference. The vehicles' positions were recorded on a 16-mm film and the TTC calculated with a computer program. He considered an event a conflict if the TTC was shorter than 1 s.

Since the late 1970s, many authors in the United States and Europe have contributed to developing the traffic conflicts technique, including producing standards and national methods (Chin & Quek, 1997; Hauer, 1978; Hayward, 1971; Hydén, 1987; Sayed & Zein, 1999; Shinar, 1984; Svensson, 1998; van der Horst, 1982, 1990; van der Horst & Kraay, 1986; van der Parker & Zegeer, 1989; Zegeer & Deen, 1977). The first Workshop on Conflict Techniques in Oslo in 1977 brought a formal definition of traffic conflicts: 'A traffic conflict is an observable situation in which two or more road users approach each other in space and time to such an extent that there is a risk of collision if their movements remain unchanged'. This definition was further modified in the Dutch method, DOCTOR (van der Horst & Kraay, 1986), by adding a condition for a 'recordable' crash warranted with a considerable probability of personal injury or material damage.

Several authors have tried to connect traffic conflicts with crashes with mixed results (Hauer & Garder, 1986; Migletz, Glauz, & Bauer, 1985; Williams, 1981). Another line of research attempted to improve the technique by revisiting the measurement of the conflict's nearness to a collision (Hayward, 1971; Hydén, 1987; Minderhoud & Bovy, 2001; Svensson & Hydén, 2006; van der Horst, 1990; Vogel, 2003). The past work on measuring traffic conflicts to reduce or eliminate human observers and estimating conflict–collision relationships is discussed in separate sections later in this chapter. A comprehensive overview of the work on traffic conflicts and other surrogate measures can be found in Zheng, Ismail and Meng (2014c).

Campbell, Joksch and Green (1996) introduced the concept of precipitating events, such as driver error, that trigger a sequence of events potentially leading to a crash. This idea was later used by Davis, Hourdos, Xiong, and Chatterjee (2011) in a more formalised manner. Tarko (2012) considered inclusion of near-crash events without evasive manoeuvres among surrogate events. These events are not acceptable and are most likely caused by road users' errors. Furthermore, certain causes of crashes (or precipitating events), such as a driver having a heart attack or a vehicle's mechanical failure, offer little chance to apply evasive manoeuvres, thus the corresponding traffic conflicts with observable evasions may be quite infrequent.

AETIOLOGY OF TRAFFIC CONFLICTS AND CRASHES

Many surrogate measures of safety have been proposed and used in road and traffic engineering. It is widely accepted that a good surrogate measure of safety must meet three conditions:

1. It properly captures the effect of road and traffic changes.
2. It is correlated with the crashes affected by these changes.
3. It is practical.

Properly defined traffic conflicts meet all three of the following conditions: (1) traffic conflicts are affected by traffic, road and other factors; (2) although a conflict is not a crash, a chain of events connects the two events that constitutes the crash causality and, furthermore, traffic conflict and crash occurrences share many traffic and road factors; and (3) traffic conflicts happen more frequently than crashes and current measurement techniques allow their observation with a reasonable level of accuracy.

Fig. 1 postulates two general sources of aetiological consistency between good surrogates and crashes: association and causality. A causal relationship between a crash and its surrogate is denoted by the arrows. Surrogate events and crashes are correlated through common conditions they share (B Conditions in Fig. 1). A specific surrogate event or measure is useful for studying the safety effect of a certain treatment only if the treatment applies to (changes) B Conditions that affect both the surrogate and the crashes. Thus, a surrogate that shares many conditions also supports the evaluation of many treatments.

Surrogate events, to be adequate, must be related to crashes through a common cause. For instance, confusing road design may cause both driver error (precipitation event) and a consequent situation that requires an evasion to correct the error. At this critical moment the situation may turn into a conflict or a crash. From this perspective, a conflict and a crash belong to the same class of risky events brought about by the same cause. Thus, a traffic conflict and a crash tend to be similar by manner of interaction (right-angle, rear-end, roadway departure), type of road users involved (vehicles, trucks, pedestrians) and other event attributes.

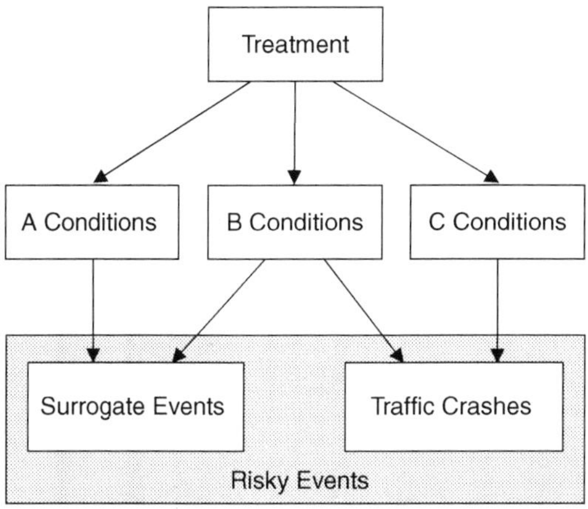

Fig. 1. Aetiology of Surrogate Events and Crashes.

A Conditions affect the frequency of surrogate events but do not affect the frequency of crashes. If a traffic conflict is incorrectly defined by allowing TTC to be too long, some intentional events acceptable by drivers will be misclassified as undesirable conflicts. For example, an efficient training of drivers may increase drivers' confidence and their acceptance of shorter TTC without increasing the frequency of crashes.

On the other hand, increased frequency of steering failures in vehicles (C Conditions) does not have to lead to an increased frequency of observable conflicts if most of the steering failures end up as crashes. A more interesting case is a slippery pavement downstream of the spot where drivers start braking in response to a hazard. If the traffic conflict is defined through a TTC measured at the beginning of the evasive manoeuvre (Hydén, 1987), then the frequency of detected traffic conflicts will remain unchanged while the frequency of crashes will increase due to unsuccessful evasive manoeuvres. Applying the minimum TTC, on the other hand, would detect the change in the minimum value, helping to detect the increased frequency of conflicts.

The aetiological consistency of conflicts and collisions promotes a wide range of B Conditions and a narrow range of A and C Conditions (Fig. 1). The aetiological consistency requires that conflicts and crashes belong to the same class of risky events. Thus, the manner of conflict (e.g., right-angle conflict) is consistent with the manner of crash (right-angle crash between the same traffic movements as in a conflict). Furthermore, a traffic conflict must be the result of a driver error as most crashes are. To claim that the observed conflict is the result of an error the conflict must be a traffic interaction that is not acceptable to the involved drivers. The closeness of a collision and the quickness of the evasive manoeuvre are often the only available evidence to decide that the interaction is a conflict if the drivers are not interviewed.

The quickness of a driver's response to a traffic event has been used as a sole criterion to detect a traffic conflict in naturalistic driving, where it is easier to observe the behaviour of a subject vehicle than its interaction with other vehicles (Bagdadi, 2013; Dozza & González, 2012; Wu & Jovanis, 2013). This attempt generated a large number of false positives, because drivers' responses were inadequate for the traffic situation or difficult to interpret. Drivers were responding rapidly to the close proximity of other vehicles around them even if they were moving along non-conflicting trajectories. Such behaviour indicates that a close proximity of another vehicle is not desirable and it is the effect of a driver's error. In fact, a rapid and not fully controlled response to an error in close proximity of another vehicle may lead to a collision even if the original trajectories were not conflicting. On the other hand, right-angle collisions recorded by Japanese police at an intersection in Tokyo included a large number of cases without any attempts from drivers to avoid the crash (Ueyama, 1997). The lack of awareness of the hazard is the explanation. Thus, cases of 'near-misses' without evasive manoeuvres are the result of human errors (Tarko, 2012). In other words, a close proximity between road users in traffic is the manifestation of an operational failure, particularly if a potential collision may lead to a severe outcome.

It can be concluded from this discussion that properly defined traffic conflicts or near-crashes have the potential to serve as comprehensive surrogate measures of safety that would be useful in a wide range of road, traffic, human conditions and treatment studies. This statement is particularly true for studies addressing the risk and rate of crashes in order to prevent crashes from occurring. Use of traffic conflicts to study the severity of crashes may be much more challenging than studying the crash risk or frequency.

MEASURES OF CRASH PROXIMITY

The closeness (or proximity) to a collision can be measured in time via TTC or post-encroachment time (PET). It can also be measured in space as the distance between two road users or between a road user and an obstruction. The selection of a measure of proximity affects the traffic conflict's ability to reflect certain safety factors and to be applicable to certain types of risky events.

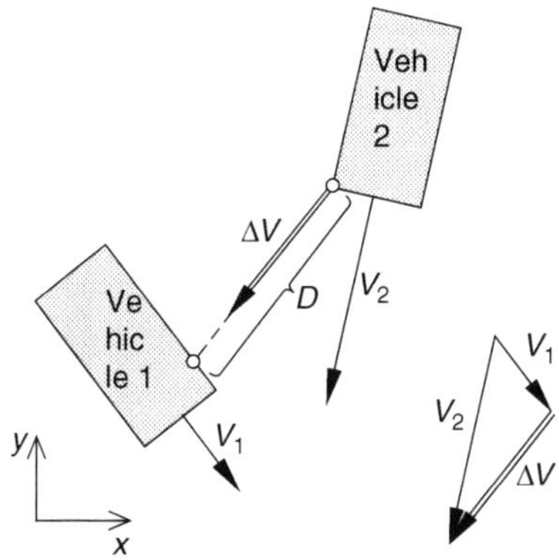

Fig. 2. Time to Collision TTC = $D/\|\Delta V\|$ with Consideration of the Point of Physical Contact at Collision of Vehicles (adapted from Hayward, 1971).

Hayward (1971) calculated the TTC at a sequence of instants by dividing the distance between two vehicles on (x,y) plane at a certain instant with the instantaneous speed difference (Fig. 2). It was assumed that the speeds and the directions of travel observed at the instant remain unchanged. Hayward recommended the minimum TTC reached during the interaction as a measure of crash proximity. Hydén (1987), on the other hand, proposed the proximity to crash to be measured with the time to accident (TTA) calculated at the moment when the evasive action is initiated. This measure requires detecting the beginning of the evasive manoeuvre, which may not always be obvious, and it does not reflect the effectiveness of the evasive manoeuvre on the crash proximity. TTC and TTA are not applicable to risky events without evasive manoeuvres.

PET involves two elements: (1) the conflict area and (2) the order in which two vehicles pass the conflict area. PET for a right-angle conflict is measured between the time when the first vehicle leaves the conflict area and the time when the second vehicle enters the conflict area (Fig. 3). In the car-following case, PET is equivalent to a time headway and it requires that the second vehicle moves faster than the first. PET cannot be applied to collisions with stopped vehicles or fixed obstructions. On the other hand, it does apply to risky events without evasive manoeuvres.

Hayward proposed the TTC calculation for a general case of two vehicles based on the speed difference in the (x,y) plane (Fig. 2). The differential speed ΔV must point out at the other vehicle to ensure existence of the TTC value. Otherwise, the two vehicles miss each other in the hypothetical collision area

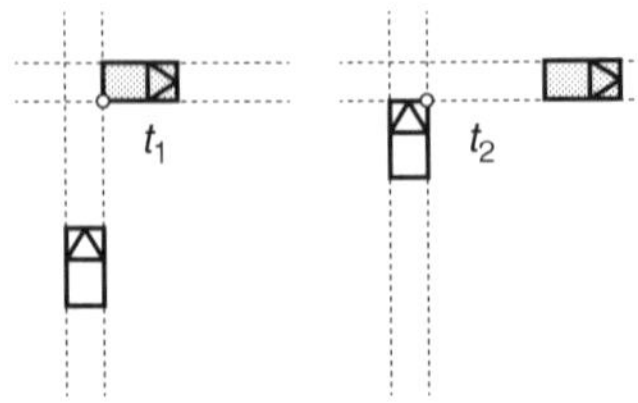

Fig. 3. Post-Encroachment Time ($t_2 - t_1$).

and PET can be calculated instead. Calculation of PET in such a general case is even more complex than TTC. PET applies to near-crashes without evasive manoeuvres, but it is not applicable to near-crashes with fixed obstructions.

Vehicles' paths and speeds are important for proper interpretation of the collision closeness and the severity of potential collision. In other words, the speeds of vehicles during a surrogate event must be sufficiently high to constitute a dangerous outcome of the potential collision and to indicate that the event resulted from human error. Some studies combine the proximity with the speeds of the vehicles to arrive at measures that combine the closeness to the potential collision and the severity of the outcome of the potential crash (Hydén, 1987; van der Horst & Kraay, 1986). The following section discusses surrogate events with a focus on crash proximity, taking the assumption that the speeds of the interacting vehicles warrant a sufficiently serious hazard. This and other measures of proximity are discussed in Gettman, Pu, Sayed and Shelby (2008) and Laureshyn, Svensson and Hydén (2010).

OBSERVING TRAFFIC CONFLICTS

Observing how traffic progresses from normal situations to high-risk interactions between vehicles can reveal events and their circumstances that direct traffic development towards or away from a collision. Such observations under similar conditions for a sufficiently long period may help identify the key factors of the pre-collision process and evaluate the risk of a crash.

An inherent problem with observing traffic to analyse safety was noted by Hydén (1987). Observing severe traffic conflicts (close to a crash, almost crash) ensures aetiological consistency between the observed events and the corresponding collisions, but this benefit comes at the cost of a dramatically increased observation time. Songchitruksa and Tarko (2006) concluded that observing surrogate events to estimate the frequency of crashes at an intersection with extreme value statistics may require several weeks. This data collection period is much longer than the 1 or 2 days recommended by existing traffic conflict techniques that offer already estimated conflict–collision relationships based on previous studies at many similar intersections and crash data collected over several years. The downside of the existing traffic conflict techniques is their lack of applicability to new road designs or after considerable changes in overall road safety.

Research on traffic observation techniques for studying safety focusses on two general types of observations: in-vehicle and roadside. The in-vehicle method (naturalistic driving) is centred on individual drivers and their behaviours, and interactions with other drivers along the route. Sufficiently complete and accurate measurements are possible if the vehicle is properly instrumented. Even then, the most reliable observations are of the participating drivers inside instrumented vehicles, but measuring their interactions with the road and other vehicles is much more challenging (Hallmark et al., 2011; Hunt, Vandervalk, & Snyder, 2011). For that reason, some authors have investigated the possibility of detecting traffic conflicts solely based on a driver's behaviour (Dozza & González, 2012; Wu & Jovanis, 2013). The results indicate that the physical proximity of interacting vehicles, whether in time or space, is important for correctly interpreting driver behaviour.

The early roadside observation techniques resorted to manual recording of traffic conflict observations (Sayed & Zein, 1999; van der Horst & Kraay, 1986; Zegeer & Deen, 1977). The subjective classification of traffic interactions as conflicts was inaccurate and inconsistent (Grayson, Hydén, Kraay, Muhlrad, & Oppe, 1984; Shinar, 1984). To reduce subjectivity, Hayward (1971) introduced the TTC measured by an observer equipped with a monitor and 18-mm film. A similar technique but with a more advanced video technology was used by Hydén (1987), and video-based TTC measurements have advanced significantly since then. The most advanced methods today reduce the role of human observers but do not eliminate them completely (Saunier & Sayed, 2008; Saunier, Sayed, & Ismail, 2010; Sayed, Ismail, Zaki, & Autey, 2012; Sayed, Zaki, & Autey, 2013). In some studies the traditional definition of traffic conflict with extrapolated speeds along known paths was expanded to the spatial distributions of the paths (Saunier & Sayed, 2008).

Only severe interactions leave no doubt of human error, making these interactions aetiologically consistent with crashes. This premise led to observing rare events requiring a long observation time as already mentioned. Any technique that requires human supervision seems impractical. Fortunately, there are new generations of sensors, including LiDAR and light field cameras, which promise better accuracy and easier automation of measurements.

Roadside observations utilise sensors positioned at optimal locations. These sensors allow tracking multiple vehicles and vulnerable vehicles with similar accuracy but only within the range determined by the position and capabilities of the roadside sensors. Another limitation is the lack of personal information about observed objects. Thus, many important variables are unavailable or difficult to observe (e.g., driver's age, driving experience, personality traits and current psychological states, as well as the technical conditions of the pavement and vehicle).

The naturalistic driving and roadside observation techniques have intrinsic weaknesses that are difficult to overcome. Integrating the two types of observations collected at multiple sources might be the answer.

THE CONFLICT–COLLISION RELATIONSHIP

Many research efforts have been focussed on conforming and estimating the relationship between traffic conflicts and crashes. The SHRP2 program deployed more than 2,000 vehicles for almost 1 year to collect safety data. Fewer than 200 crashes (Hauer & Garder, 1986; Migletz et al., 1985; Williams, 1981) were observed during that period. It is clear that collisions must be supplemented with conflicts or other relevant surrogates to gain better insight into the safety conditions. One important question is the relationship between conflicts and collisions. The progress in estimating the relationship between collisions and conflicts is shown in Table 1. The models are listed in the order they are discussed here, which does not follow the order of their publication.

The Type 1A model in Table 1 was proposed by Hauer (1982) and later estimated by Migletz et al. (1985) and Hydén (1987) based on an observed number of collisions and conflicts at multiple locations. The focus was on the coefficient K, which was assumed to be fixed. The data themselves were the most difficult challenge. Collisions are so infrequent that it takes dedicated agencies a long period of time to record a sufficient number of these events; it is believed that a considerable number still remain unreported. The details of a traffic event leading to a crash are typically poorly documented. On the other hand, traffic conflicts are expensive and troublesome to observe, thus only much shorter observation periods are practical. Consequently, crash period T is much longer than conflict period t and conditions Y of crashes are different from conditions X of conflicts. The strong heterogeneity of these conditions – mostly aggregated or unknown for crashes – combined with the typically small samples might be the main cause of, at best, mixed results of the first attempts to estimate the conflict–collision relationship. Another

Table 1. Models of Conflict–Collision Relationship.

Type	Model Form	References
Type 1A	**Accidents**$(T, Y) = K($**Conflicts**(t, X)	Hauer (1982) Migletz et al. (1985) Hydén (1987)
Type 1B	**Accidents**$(T, Y) = \sum_s K_s \cdot$ **Conflicts**$_s(t, X)$	Hauer & Garder (1986) Sayed & Zein (1999)[a] Guo, Lauer, Hankey & Dingus (2010)
Type 2	**Accidents**$(T, Y) = K(Y, X) \cdot$ **Conflicts**(t, X)	Wu & Jovanis (2012)[b] El-Basyouny & Sayed (2013)
Type 3	**Accidents**$(t, X) = \sum_s R_s(X)$ **Conflicts**$_s(t, X)$	Davis et al. (2011) Campbell et al. (1996) Tarko & Songchitruksa (2005) Songchitruksa & Tarko (2006)
Type 4	**Accidents**$(t, X) = R(X) \cdot$ **Conflicts**(t, X)	Tarko (2012) Jonasson & Rootzén (2014), Zheng, Ismail & Meng (2014a) Zheng, Ismail & Meng (2014b)
Type 5	$\frac{\textbf{Accidents}(T_A, Y_A)}{\textbf{Accidents}(T_B, Y_B)} = \frac{\textbf{Conflicts}(t_A, X_A)}{\textbf{Conflicts}(t_B, X_B)}$	Sacchi, Sayed & deLeur (2013)

Notation: **Accidents**(T, Y) = expected number of accidents during period T and under conditions Y; **Conflicts**(t, X) = number of conflicts during period t (much shorter than T) and under conditions X; **Conflicts**$_s(t, X)$ = number of conflicts of severity s during period t and under conditions X; K = conflict-crash conversion coefficient; K_s = conflict-crash conversion coefficient for conflicts with severity s; $R_s(X)$ = risk (probability) of crash given risky event of severity s; $R(X)$ = risk (probability) of a crash given risky event; T_B, T_A = period of counting crashes before and after improvement, respectively; t_B, t_A = period of counting conflicts before and after improvement, respectively.

[a]The model's form is linear Accidents$(T, Y) = K_0(Y) + K_1(Y) \cdot$Conflicts$(t)$, but the authors have estimated different models for different conditions Y.

[b]The model is probabilistic and it estimates the probability of a crash given conflict conditions (true positive).

problem is that the coefficient K should be assumed dependent on conditions (Y, X). The Type 1B model, proposed by Hauer and Garder (1986), is an extension of the early version by considering the conflict severity s.

The Type 2 model assumes that the relationship between the collisions and conflicts may be different for different conditions. It was estimated with count models, and the conflict rates are included among the explanatory variables (El-Basyouny & Sayed, 2013; Guo et al., 2010; Sayed & Zein, 1999). Conditions X and Y may still be inconsistent.

The Type 3 model involves the concept of conflicts as precipitating events with different probabilities of 'precipitation' – collision (Campbell et al., 1996). In this case, a conflict is not a crash and a crash is not a conflict. Formulation of this model was based on an understanding of the crash occurrence mechanism and the model may be considered causal. The model estimates the expected number of crashes in the same period when the precipitating events occur. The mechanism of a collision is well defined, with contributing probabilities of precipitation $R_s(X)$ under various conflict severities s and conditions X. Davis et al. (2011) initiated this line of research by reconstructing accidents and applying the counterfactual technique to 'simulate' various possible alternative hypothetical developments to estimate the risk of crash as a function of various conditions. This perspective allows considering events aetiologically inconsistent with collisions because the irrelevance of such events is represented in this model with negligible probability of collision. To some extent, this perspective was proposed by Svensson and Hydén (2006) in their proposition of considering events that are not conflicts. This model may be difficult to estimate if the observation period is relatively short. Precipitating events with a high probability of a collision are infrequent and they generally do not appear in a short period. This difficulty may lead to underestimation of the number of crashes.

The Type 4 model defines collisions as being the same type of events as conflicts, assuming the continuity of severity s as it was in the original approach to traffic conflicts presented in Glauz and Migletz (1980), Hydén (1987) and Svensson (1998). This perspective makes estimation of the model easier for short periods without underestimation thanks to extrapolating any observed events into infrequent and severe events including collisions. Campbell et al. (1996) and Songchitruksa and Tarko (2006) proposed using extreme value statistics. Tarko (2012) then presented traffic conflicts as statistical exceedances above a certain threshold of crash proximity s. Similar to the Type 3 model, the Type 4 model estimates the number or probability of crashes for the same period when the conflicts were observed. Unlike the Type 3 model, however, the observed events must be aetiologically consistent with collisions and thus should be severe conflicts. The most exciting ability of these models is the possibility of estimating them without crash data (Songchitruksa & Tarko, 2006; Tarko, 2012; Tarko & Songchitruksa, 2005). Recently, Zheng et al. (2014a) confirmed that the exceedances-based estimation is more efficient than that based on extreme values. The same authors used exceedances to estimate the link between the observed conflicts and crashes according to the framework represented by the Type 4 model (Zheng et al., 2014b). The number of crashes predicted in the model may be different from the number reported by police due to the under-reporting problem and possible inconsistency in the collision definitions applied by police and assumed in the model.

Although the Type 5 model does not have the appeal of the Type 3 and Type 4 models, it is discussed here for its potential usefulness in estimating the so-called crash modification factors that express the relative changes in safety caused by certain treatments. Sacchi et al. (2013) proposed observing traffic conflicts in conditions (Y_B, X_B) before a treatment and in conditions (Y_A, X_A) after implementation of the treatment to estimate the corresponding crash modification factor. In this setting, the change in safety may be attributed to the treatment. Other adjustments may be necessary through control sites. The potential weakness of this model is an implicit assumption that the treatment does not change the relationship between conflicts and collisions. The latter assumption can be defended in cases where the treatment does not obviously affect the risk of failure of the evasive manoeuvre and only affects the probability of the conflict; indeed, it might be in the case of an improved ramp angle as reported by Sacchi et al. (2013), which obviously improved the visibility of vehicles on the major road while its effect on the braking manoeuvre to avoid the collision could be neglected. On the other hand, the relationships of the Type 5 model cannot be applied to study improvements that impact the effectiveness of evasive manoeuvres, such as increasing the pavement friction.

SPEED SELECTION AND OTHER BEHAVIOURAL MEASURES

Although traffic conflicts are the most comprehensive surrogate measure of safety, there are other surrogate measures that may be convenient for studies with objectives limited to confirming if a treatment improves safety or selecting the most effective treatment among several alternatives. In such cases, other measures such as speed selection, lane changing, gap acceptance, etc., might be a suitable choice.

The connection between speed and safety is unquestionable although its nature and mechanism are still the subject of research and debate. The large majority of experts agree that high speed and its variance are associated with the increased probability of a crash if all other conditions remain constant. Treatments that change a driver's speed selection without changing the physical conditions of the road and its environment include speed limits, speed limit enforcement, economic incentives for driving within the speed limit and education about the risk of driving at excessive speed.

The speed–safety relationship becomes complex if a treatment affects the driving environment. Changes to the driving environment that are noticeable to the driver can affect the driver's risk perception and may cause speed adjustments. For example, drivers reduce their speeds on sharp horizontal curves, but they may 'undercompensate' for the added risk; the actual crash risk on a curve thus is still higher than on a straight segment although the speed is lower. This mechanism reverses the expected speed–safety relationship. On the other hand, the idea behind traffic calming is to change the road in a way that increases the perception of hazard or the discomfort of driving at high speed to the point that drivers 'over-compensate' for the added risk and the road actually becomes safer. In this case, the speed reduction and increase in safety are observed simultaneously.

It is also possible that the changes in the driving environment are not noticeable to the driver, even if they indeed affect the driver's safety. For example, modifying traffic signal coordination in order to reduce arrival time at the next intersection when stopped vehicles are still present there may reduce rear-end collisions. This improvement is not obvious to drivers and thus it may not affect their perception of risk; drivers will not be tempted to reduce their speed or the distance to other vehicles in front of them. In this case, any behaviour-based measure cannot serve as a surrogate measure of safety.

Fig. 4 presents the connection between behaviour, environment and safety. A good behavioural measure, such as free-flow speed, is a convenient surrogate measure of safety if the studied treatment does not affect the road and other components of the driving environment. Otherwise, the relationship between speed and safety is unclear; therefore, relying solely on driver behaviours may be misleading.

Strong statistical evidence of the relationship between the mean speed and safety was provided by Nilsson (2004), who proposed a power model to describe the relationship between the relative change in the mean speed and the crash frequency. This model was re-estimated with meta-analysis by Elvik, Christensen and Helene Amundsen (2004) and then Elvik (2013).

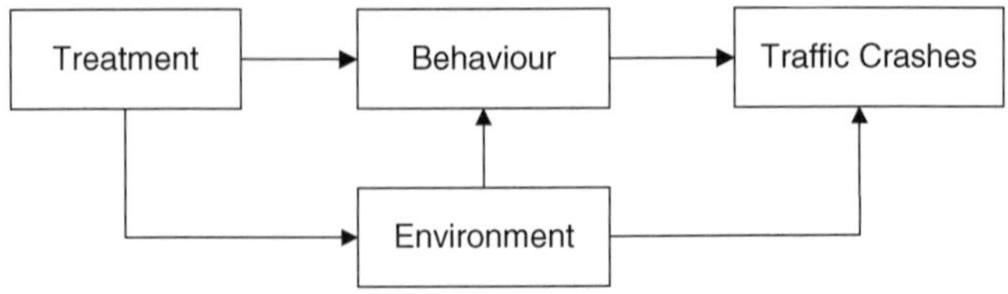

Fig. 4. Behavioural Measures as Surrogate Measures of Safety.

The authors emphasised that the relationship holds when confounding factors are controlled for. Keeping the cofounding factors of speed, such as road geometry or weather, constant is consistent with the preceding discussion of the conditions required for the validity of the speed–safety relationship. The most recent adjustment of the model was described in Elvik (2013). The meta-analysis applied to a large number of studies pointed to the exponential function rather than to the power function as a better representation of the speed–safety relationship. Nevertheless, the estimated relationships varied strongly across individual studies; in many cases the relationship was even reversed (negative exponent). The presence of not-fully controlled conditions external to drivers could be the source of this result, together with many other potential causes. Many authors have statistically connected detector-measured traffic characteristics with the risk of collision (e.g., Abdel-Aty, Hassan, Ahmed, & Al-Ghamdi, 2012; Abdel-Aty, Uddin, Abdalla, Pande, & Hsia, 2004; Hossain & Muromachi, 2011; Hourdos, Garg, Michalopoulos, & Davis, 2006; Oh, Oh, & Ritchie, 2005; Pande & Abdel-Aty, 2006; Xu, Tarko, Wang, & Liu, 2013). Unlike the already discussed long-term impact of the average speed on safety, this relationship is short term, and it reflects the safety impact of dynamic interactions between vehicles manifested via temporal and spatial speed variability.

COMPUTER SIMULATION OF SAFETY

Existing computer simulation models have been developed for studying traffic performance at individual roads and in road networks. They are indispensable in estimating speeds, delays and other performance measures for new road systems. These models adequately represent typical, or standard, traffic events that frequently occur, are observed and are well understood. The essence of traffic safety lies in the dangerous and infrequent interactions that are difficult to observe and study, and their nature therefore is not as well understood as the other more frequent events. Crashes are not intended to occur in the existing computer simulation models. The near-miss events may occur incidentally, but very little can be learned about safety on a simulated road.

Gettman et al. (2008) and Archer (2005) investigated the potential use of microscopic simulation to model traffic conflicts. The results from recent attempts to use the existing simulation tools, such as VISSIM, indicate an evident correlation between simulated interactions and traffic conflicts observed in the field and even collisions reported by police (e.g., Gettman et al., 2008; Huang, Liu, Yu, & Wang, 2013; Ozbay, Yang, Bartin, & Mudigonda, 2008). With the limitations of the current simulation models the reported correlation might be introduced by the exposure. Roads with heavy traffic tend to have more crashes and traffic interactions than roads with low traffic regardless of whether these interactions are simulated or observed.

The progress in measurement techniques and gaining a better understanding of dangerous interactions engendered by on-going research on traffic conflicts should allow proper simulation of more and more risky events. This knowledge is currently scant but will grow with the progress in analysing naturalistic driving and roadside measurements. With the growing knowledge of dangerous interactions, the existing models will be supplemented with components that properly represent the possibility of driver errors and the ensuing response to the near-collision situations under various conditions. The initial simulation-based models may combine simulation of traffic conflicts and the statistical components of the risk of collision conditioned on these conflicts. With the growing understanding of near-collision interactions more severe conflicts could be simulated up to the point of eliminating the statistical component.

CONCLUSIONS

In general, the surrogate measures of safety can be classified into several categories that include conflicts, behaviours, specific conditions and derivatives thereof. The most comprehensive measure is traffic conflicts, if properly defined and observed, because their occurrence is caused and their probability is affected by most of the conditions (factors) that affect the occurrence and probability of collisions. A properly defined traffic conflict is the one caused by human error. The closer the conflict is to the corresponding collision, the more aetiologically consistent the two are. The minimum TTC observed during the conflict process seems to be an appropriate measure of collision nearness in most cases. The minimum distance to collision with simultaneous consideration of the speeds and travel directions of the interacting vehicles might provide the most universal measure of collision nearness. The observation period to collect a sufficient number of conflicts is much longer than the period recommended in the existing traffic conflict techniques. Maturing measurement technologies will help achieve a practical method of observing traffic conflicts in extended periods without expensive human involvement.

Among several methods of estimating the connection between conflicts and collisions, the count or probabilistic models seem to promise the quickest road to practical results. To fulfil this promise, these models must include traffic conflict's other variables that affect the probability of the collision given the conflict. The biggest hurdle in estimating these models is the necessity of using crash data with their disadvantages and much longer observation periods necessary to observe traffic conflicts.

Exceedance models are the only models that can be estimated without crash data. They return the probability of collision given the conflict and, together with the known frequency of conflicts, they return the expected number of crashes during the period when conflicts were observed. The road to practical results is longer than the traditional count and probabilistic models. More research is needed to confirm the usefulness of these models and to advise the best data collection and model estimation methods.

Recent studies point to traffic conflicts as potentially useful for estimating the so-called crash modification factors. These factors are essential in predicting the safety benefits of various road and traffic control countermeasures. This approach may produce biased results if the evaluated treatment affects the probability of collision given the conflict.

Behaviour-based surrogate measures of safety, such as speeds or car-following headways, are useful if the studied treatment does not change the road and its environment, but rather directly changes driver behaviour. Otherwise, the risk compensation behaviour may skew the results considerably. Power and recently exponential relationships between speed and safety have been reported in the literature (Elvik, 2013).

Simulating road safety has advanced gradually at a pace determined by the growing knowledge of driver behaviour and performance during near-collision events. As this knowledge continues to grow existing models will be supplemented with components that properly represent the possibility of driver errors and responses to near-collision situations under various conditions. It seems that the road to simulating safety without resorting to statistical relationships will be a long one. Hybrid models that include simulation and statistical components offer a less remote and more practical solution. The current surrogate measures of safety research may deliver the two elements of hybrid models: improved simulation of conflicts and a statistical bridge linking conflicts with collisions.

REFERENCES

Abdel-Aty, M., Hassan, H., Ahmed, M., & Al-Ghamdi, A. (2012). Real-time prediction of visibility related crashes. *Transportation Research Part C*, *24*, 288–298.

Abdel-Aty, M., Uddin, N., Abdalla, F., Pande, A., & Hsia, L. (2004). Predicting freeway crashes based on loop detector data using matched case-control logistic regression. *Transportation Research Record*, *1897*, 88–95.

Archer, J. (2005). *Indicators for traffic safety assessment and prediction and their application in micro-simulation modelling: A Study of urban and suburban intersections*. Dissertation, Lund University.

Bagdadi O. (2013). Assessing safety critical braking events in naturalistic driving studies. *Transportation Research Part F, 16*, 117–126.

Campbell, K., Joksch, H. C., & Green, P. E. (1996). *A bridging analysis for estimating the benefits of active safety technologies.* UMTRI-96-18 Final Report. University of Michigan, Transportation Research Institute, Ann Arbor, MI.

Campbell, R. E., & King, L. E. (1970). The traffic conflicts technique applied to rural intersections. *Accident Analysis and Prevention, 2*(3), 202–221.

Chin, H., & Quek, S. (1997). Measurement of traffic conflicts. *Safety Science, 26*, 169–185.

Davis, G. A. (2004). Possible aggregation biases in road safety research and a mechanism approach to accident modeling. *Accident Analysis and Prevention, 36*(6), 1119–1127.

Davis, G. A., Hourdos, J., Xiong, H., & Chatterjee, I. (2011). Outline for a causal model of traffic conflicts and crashes. *Accident Analysis and Prevention, 43*, 1907–1919.

De Silva, H. R. (1942). *Why we have automobile accidents*. New York, NY: Wiley.

Dozza, M., & González, N. P. (2012). Recognizing safety-critical events from naturalistic driving data. *Procedia-Social and Behavioral Sciences, 48*, 505–515.

El-Basyouny, K., & Sayed, T. (2013). Safety performance functions using traffic conflicts. *Safety Science, 51*, 160–164.

Elvik, R. (2013). A re-parameterisation of the power model of the relationship between the speed of traffic and the number of accidents and accident victims. *Accident Analysis and Prevention, 50*, 854–860.

Elvik, R., Christensen, P., & Helene Amundsen, A. (2004). *Speed and road accidents: An evaluation of the power model.* Oslo – Norge: TØI, Transportøkonomisk Institutt.

Gettman, D., Pu, L., Sayed, T., & Shelby, S. (2008). *Surrogate safety assessment model and validation: Final report.* Final Report: FHWA-HRT-08-051, USDOT, FHWA.: Turner-Fairbank Highway Research Center, McLean, VA.

Glauz, W. D., & Migletz, D. J. (1980). *Application of traffic conflicts analysis at intersections.* Report No. NCHRP 219. Transportation Research Board, Washington, D.C.

Grayson, G. B., Hydén, C., Kraay, J. H., Muhlrad, N., and Oppe, S. 1984. *The Malmo study: A calibration of traffic conflict techniques*. Report No. R-84-12. Institute for Road Safety Research, Leidschendam.

Guo, F., Klauer, S. G., Hankey, J. M., & Dingus, T. A. (2010). Near crashes as crash surrogate for naturalistic driving studies. *Transportation Research Record: Journal of the Transportation Research Board, 2147*, 66–74.

Hallmark, S., Yihsu, Y., Boyle, L., Carriquiry, A., Tian, Y., & Mudgal, A. (2011). *Evaluation of data needs, crash surrogates, and analysis methods to address lane departure research questions using naturalistic driving study data.* SHRP 2 Report S2-S01E-RW-1. Transportation Research Board, Washington, D.C.

Hauer, E. (1978). *Traffic conflict surveys: Some study design considerations.* TRRL Supp. Report 352. Transport and Road Research Laboratory, Berkshire.

Hauer, E. (1982). Traffic conflicts and exposure. *Accident Analysis & Prevention, 14*(50), 359–364.

Hauer, E., & Garder P. (1986). Research into validity of traffic conflicts technique. *Accident Analysis and Prevention, 18,* 471–481.

Hayward, J. C. (1971). *Near misses as a measure of safety at urban intersections.* M.S. Thesis, The Pennsylvania State University, Philadelphia, PA.

Hossain, M., & Muromachi, Y. (2011). Understanding crash mechanism and selecting appropriate interventions for real-time hazard mitigation on urban expressways. *Transportation Research Record, 2213*, 53–62.

Hourdos, N., Garg, V., Michalopoulos, G., & Davis, G. (2006). Real-time detection of crash-prone conditions at freeway high-crash locations. *Transportation Research Record, 1968*, 83–91.

Huang, F., Liu, P., Yu, H., & Wang, W. (2013). Identifying if VISSIM simulation model and SSAM provide reasonable estimates for field measured traffic conflicts at signalized intersections. *Accident Analysis and Prevention, 50,* 1014–1024.

Hunt, J. E., Vandervalk, A., & Snyder, D. (2011). *Roadway measurement system evaluation.* SHRP 2 Report S2-S03-RW-1. Transportation Research Board, Washington, D.C.

Hydén, C. (1987). *The development of a method for traffic safety evaluation: The Swedish traffic conflicts technique*. Lund: Lund Institute of Technology.

Jonasson, J. K., & Rootzén, H. (2014). Internal validation of near-crashes in naturalistic driving studies: A continuous and multivariate approach. *Accident Analysis and Prevention, 62*, 102–109.

Klebelsberg, D. (1964). Derzeitiger sand der verhaltensanalyse des kraftfahrens. Arbeit und leitsung. *Ablt. Arbeitswissenscaft Soziale Betriebspraxis, 18*, 33–37.

Laureshyn, A., Svensson, A., & Hydén, C. (2010). Evaluation of traffic safety, based on micro-level behavioural data: Theoretical framework and first implementation. *Accident Analysis and Prevention, 42*, 1637–1646.

Migletz, D. J., Glauz, W. D., & Bauer, K. M. (1985). *Relationship between traffic conflicts and accidents.* Report No. FHWA/RD-84/042. Federal Highway Administration, Richmond, VA.

Minderhoud, M. M., & Bovy, P. H. (2001). Extended time-to-collision measures for road traffic safety assessment. *Accident Analysis and Prevention, 33*, 89–97.

Nilsson, G. (2004). *Traffic safety dimensions and the power model to describe the effect of speed on safety.* Bulletin 221. Department of Technology and Society, Traffic Engineering, Lund Institute of Technology, Lund.

Oh, C., Oh, J., & Ritchie, S. (2005). Real-time hazardous traffic condition warning system: Framework and evaluation. *IEEE Transactions on Intelligent Transportation Systems, 6*(3), 265–272.

Ozbay, K., Yang, H., Bartin, B., & Mudigonda, S. (2008). Derivation and validation of new simulation-based surrogate safety measure. *Transportation Research Record: Journal of the Transportation Research Board, 2083,* 105–113.

Pande, A., & Abdel-Aty, M. (2006). Assessment of freeway traffic parameters leading to lane-change related collisions. *Accident Analysis and Prevention, 38*(5), 936–948.

Perkins, R. S., & Harris, J. I. (1968). Traffic conflict characteristics: Accident potential at intersections. *Highway Research Record, 224*, 35–43.

Ropeik, D. (2002). *Risk.* New York, NY: Houghton Mifflin Company.

Sacchi, E., Sayed, T., & deLeur, P. (2013). A comparison of collision-based and conflict-based safety evaluations: The case of right-turn smart channels. *Accident Analysis and Prevention, 59*, 260–266.

Saunier, N., & Sayed, T. (2008). Probabilistic framework for automated analysis of exposure to road collisions. *Transportation Research Record: Journal of the Transportation Research Board, 2083*, 96–104.

Saunier, N., Sayed, T., & Ismail, K. (2010). Large-scale automated analysis of vehicle interactions and collisions. *Transportation Research Record: Journal of the Transportation Research Board, 2147*, 42–50.

Sayed, T., Ismail, K., Zaki, M. H., & Autey, J. (2012). Feasibility of computer vision-based safety evaluations: Case study of a signalized right-turn safety treatment. *Transportation Research Record: Journal of the Transportation Research Board, 2280*, 18–27.

Sayed, T., Zaki, M. H., & Autey, J. (2013). Automated safety diagnosis of vehicle-bicycle interactions using computer vision analysis. *Safety Science, 59*, 163–172.

Sayed, T., & Zein, S. (1999). Traffic conflict standards for intersections. *Transportation Planning and Technology, 22,* 309–323.

Shinar, D. (1984). The traffic conflict technique: A subjective vs. objective approach. *Journal of Safety Research, 15,* 153–157.

Songchitruksa, P., & Tarko, A. (2006). The extreme value theory approach to safety estimation. *Accident Analysis and Prevention, 28*, 811–822.

Svensson, Å. (1998). *A method for analysing the traffic process in a safety perspective.* Dissertation, Lund University.

Svensson, Å., & Hydén, C. (2006). Estimating the severity of safety related behavior. *Accident Analysis and Prevention, 38*, 379–385.

Tarko, A. P. (2012). Use of crash surrogates and exceedance statistics to estimate road safety. *Accident Analysis and Prevention, 45*, 230–240.

Tarko, A. P., & Songchitruksa, P. (2005). Measuring roadway safety. In *Conference Proceedings: Road safety on four continents,* Swedish National Road and Transport Research Institute, Warsaw, Poland.

Ueyama, M. (1997). Effects of vehicular behavior at the pre-accident stage on decision-making at intersections without signals. In *Proceedings of the third international symposium on intersections without traffic signals,* Portland, OR, July 21–23, 303–309.

van der Horst, A. R. A. (1982). The analysis of traffic behaviour by video. In J. H. Kraay (Ed.), *Proceedings of the third international workshop on traffic conflicts technique* (pp. 26–39). SWOV Institute for Road Safety Research, Leidschendam: The Netherlands.

van der Horst, A. R. A. (1990). *A time-based analysis of road user behavior in normal and critical encounter*. Dissertation, Delft University of Technology.

van der Horst, A. R. A., & Kraay, J. H. (1986). The Dutch conflict observation technique 'DOCTOR'. In *Proceedings of the workshop traffic conflicts and other intermediate measures in safety evaluation*, Budapest, September 8–10. KTI Institute for Transport Sciences, Budapest, Hungary.

van der Parker, M. R., & Zegeer, C. V. (1989). *Traffic conflict techniques for safety and operations-observers manual.* Publication No. FHWA-IP-88-027. Federal Highway Administration, U.S. Department of Transportation, Washington, D.C.

Vogel, K. (2003). A comparison of headway and time to collision as safety indicators. *Accident Analysis and Prevention, 35,* 427–433.

Williams, M. J. (1981). Validity of the traffic conflict techniques. *Accident Analysis and Prevention, 13,* 133–145.

Wu K. F., & Jovanis P. P. (2012). Crashes and crash-surrogate events: Exploratory modeling with naturalistic driving data. *Accident Analysis and Prevention, 45*, 507–516.

Wu, K. F., & Jovanis P. P. (2013). Defining and screening crash surrogate events using naturalistic driving data. *Accident Analysis and Prevention, 61*, 10–22.

Xu, C., Tarko, A. P., Wang, W., & Liu, P. (2013). Predicting crash likelihood and severity on freeways with real-time loop detector data. *Accident Analysis and Prevention, 57*, 30–39.

Zegeer, C. V., & Deen, R. C. (1977). *Traffic conflicts as a diagnostic tool in highway safety,* Research Report 482. Division of Transportation, Commonwealth of Kentucky, Washington, D.C.

Zheng, L., Ismail, K., & Meng, X. H. (2014a). Freeway safety estimation using extreme value theory approaches: A comparative study. *Accident Analysis and Prevention, 62,* 32–41.

Zheng, L., Ismail, K., & Meng, X. H. (2014b). Shifted Gamma-Generalized Pareto Distribution model to map the safety continuum and estimate crashes. *Safety Science, 64,* 155–162.

Zheng, L., Ismail, K., & Meng, X. H. (2014c). Traffic conflict techniques for road safety analysis: Open questions and some insights. *Canadian Journal of Civil Engineering, 41*(7), 633–641.

METHODS FOR EVALUATING SAFETY IMPACTS OF COUNTERMEASURES

CHAPTER 18

BEFORE–AFTER EVALUATIONS

Bhagwant Persaud

ABSTRACT

Purpose – *The focus of this chapter is on state-of-the-art methodology for observational before–after evaluations. The purpose of such evaluations is to acquire information on the safety effects of site-specific treatments that could be used by road agencies for future interventions that may have an impact on safety.*

Methodology – *Information, illustrated by a detailed example, is presented mainly on what is still regarded as the gold standard – the Empirical Bayes (EB) method. The methodology accounts for observed changes in crash frequencies that may be due to regression-to-the-mean, changes in traffic volume and time trends. Two related approaches – a comparison group method and the Full Bayes approach, which have been legitimately used for evaluations under special circumstances – are discussed in brief.*

Findings – *The EB before–after methodology remains the gold standard but must be applied properly.*

Research implications – *Some issues, for example, the selection of reference groups, still require further research.*

Safe Mobility: Challenges, Methodology and Solutions
Transport and Sustainability, Volume 11, 409–424

ISSN: 2044-9941/doi:10.1108/S2044-994120180000011020

Practical implications – *Key issues in the application of the before–after evaluation methods in general are also discussed.*

Keywords Empirical Bayes; treatment evaluation; before–after studies

INTRODUCTION

This chapter is the first of two chapters that describe methods for evaluating the safety impacts of safety investments or countermeasures. This chapter is focused on before–after evaluations, which are widely considered to be the most reliable methodology for drawing definitive conclusions about the safety effects of treatments when designed and administered correctly. These evaluations rely on an ability to finance and conduct a research study. This chapter complements Chapter 19, which is focused on the use of meta-analysis for drawing definitive conclusions a about the safety effects of safety countermeasures – more appropriate when resources are not available for financing and conducting a local study.

The focus of this chapter is on state-of-the-art methodology for before–after evaluations to learn about the safety effectiveness of site-specific road safety interventions. This methodology is already well documented in key resources (see, e.g., AASHTO, 2010; Gross, Persaud, & Lyon, 2010; Hauer, 1997; Persaud & Lyon, 2007). The intent, in the main, is therefore to provide information that complements that in those resources, in many cases citing directly from them since the author of this chapter was often a key contributor. As is the case for these other resources, this chapter is written for those with experience and/or education in the theory and practice of road safety engineering, including basic analytical procedures and statistical concepts.

The focus is also on *observational* before–after evaluations, since the opportunity to do experimental studies is rare and, in any case, the methodology for such studies is well covered in the general statistical literature. As noted by Gross et al. (2010) in the 'CMF Development Guidebook', with observational studies, data are collected retrospectively by observing the performance of an existing road system where the treatment has already been implemented at some sites, usually not on the basis of a planned experiment but on engineering considerations, including safety. This presents challenges in doing before–after evaluations, and it is these challenges that are the focus of the chapter, which outlines methodologies to overcome them.

CONTEXT

Purpose of Before–After Evaluations

The purpose of before–after evaluations is to acquire information on the safety effects of treatments that could be used by road agencies for future interventions that may have an impact on safety. This information is usually presented in the form of crash modification factors (CMFs) that would be multiplied by the expected number of crashes without a planned intervention to estimate the number of crashes that would be expected following the intervention. Thus, a CMF of 0.8 applied to an expected number of crashes without treatment of 2.5 per year would indicate that there would be 2.0 crashes per year if the treatment were implemented, a reduction of 100(1 – 0.80) = 20%. CMFs may also be specified as estimates from a crash modification function (CMFunction) that relates the CMF to application circumstances, reflecting the reality that CMFs are likely to vary from application to application with factors such as site characteristics.

Before–After Evaluations in the Context of other Types of Observational Studies

A variety of study designs have been used to derive CMFs. By far the two most common ones are before–after studies and cross-sectional studies (Chapter 3, Volume 9b of the series). According to the CMF Development Guidebook (Gross et al., 2010), 'where there are sufficient applications of a specific countermeasure, the before–after design is preferred. Cross-sectional designs are particularly useful for estimating CMFs where there are insufficient instances, where a countermeasure was applied to conduct a before–after study (Gross et al., 2010, p. 28). For example, there may be few or no projects where horizontal curvature is flattened from 4° to 3°, so deriving a CMF for such a change from a before–after study would be impossible. Instead, cross-sectional data would be used to build a regression model that relates crashes to the degree of curvature, controlling for other traffic and geometric factors, and this model would be used to estimate the CMF. As discussed in Chapter 12, a key issue with the cross-sectional design is the difficulty of controlling for the confounding factors, such that CMFs from cross-sectional data tend to be larger (safety benefits smaller) than those derived from before–after evaluations.

THE GOLD STANDARD – EMPIRICAL BAYES BEFORE–AFTER EVALUATION

Overview

The objective of the empirical Bayes (EB) methodology, as it is for other less robust before–after evaluations, is to estimate the number of crashes (*B*) that would have occurred at an individual treated site in the after period had there been no intervention. The safety effect of the intervention is estimated by comparing the sum of the estimates of *B* over several treated sites with the actual number of crashes after treatment at these sites.

The methodology accounts for observed changes in crash frequencies that may be due to regression-to-the-mean (RTM). This phenomenon, which has been well documented in the literature (see e.g., Hauer, 1997), arises when sites are selected for treatment on the basis of a randomly high crash frequency, which 'regresses towards the mean' even if the sites were not treated. It is this mean in the absence of treatment that needs to be estimated and compared to the post-treatment frequency in a robust before–after evaluation. In providing a procedure for estimating this mean, the methodology also facilitates the accounting for changes in safety due to traffic volumes and time trends that are unrelated to the treatment.

Fundamental to the EB methodology is the use of safety performance functions (SPFs) that are applied to do the 'corrections' for RTM: changes in traffic volume and time trends. These SPFs are usually estimated with negative binomial regression that relates crashes to traffic volume and other characteristics of a group of untreated 'reference' sites that are similar to the treatment sites. Time trends due to changes in weather, crash reporting practices, so on, are accounted for through calibrated annual SPF multipliers.

Mathematics Essentials

In the EB approach the estimated change in safety at a treated site is

$$\Delta Safety = \lambda - \pi \qquad (1)$$

where λ = expected number of crashes that would have occurred in the after period without the treatment, and π = number of crashes recorded in the after period (note that subsets of the total after period can be evaluated separately to assess changes in safety effects over time).

In the EB procedure the SPF is used to first estimate the number of crashes that would be expected in each year of the before period at locations with traffic volumes and other characteristics similar to the one being analysed (i.e., reference sites). The expected number of crashes (m) before treatment is then estimated as a weighted average of the sum of the annual SPF estimates (P) and the count of crashes (x) in the before period at a treatment site. This estimate of m and its variance is

$$m = w(P) + (1 - w)(x) \tag{2}$$

$$\text{VAR}\{m\} = (1 - w)(m) \tag{3}$$

where w is the weight for the SPF crash prediction relative to the crash count and is derived from the SPF estimate and the over-dispersion parameter as

$$w = \frac{1}{1 + kP} \tag{4}$$

where k = negative binomial over-dispersion parameter estimated in the SPF calibration. (As seen from Eq. (3), the weight given to the SPF prediction P is relatively large compared to that given to the crash count if the over-dispersion parameter k is small, as would be expected of a more robust SPF and vice versa.)

A factor is then applied to m to account for the length of the after period and differences in traffic volumes between the before and after periods. This factor, denoted as r, is the sum of the annual SPF predictions for the after period divided by P, the sum of these predictions for the before period. The result, after applying this factor, is an estimate of λ. The procedure also produces an estimate of the variance of λ. Thus:

$$\lambda = (r)(m) \tag{5}$$

$$\text{VAR}\{\lambda\} = (r^2)[\text{VAR}\{m\}] \tag{6}$$

The estimate of λ is then summed over all treatment sites to obtain λ_{sum}, which is then compared with the count of crashes observed during the after period at those sites (π_{sum}). The variance of λ is also summed over all sites in the treatment group.

The Index of Effectiveness (θ) is estimated as

$$\theta = \frac{\pi_{sum}/\lambda_{sum}}{1 + \left(\text{Var}(\lambda_{sum})/\lambda_{sum}^2\right)} \tag{7}$$

The standard deviation of θ is given by

$$\mathrm{StDev}(\theta)=\sqrt{\frac{\theta^2\left(\frac{\mathrm{Var}(\pi_{\mathrm{sum}})}{\pi_{\mathrm{sum}}^2}+\frac{\mathrm{Var}(\lambda_{\mathrm{sum}})}{\lambda_{\mathrm{sum}}^2}\right)}{\left(1+\frac{\mathrm{Var}(\lambda_{\mathrm{sum}})}{\lambda_{\mathrm{sum}}^2}\right)^2}} \tag{8}$$

The per cent change in crashes is calculated as $100(1 - \theta)$; thus, a value of $\theta = 0.8$ with a standard deviation of 0.09 indicates a 20% reduction in crashes with a standard deviation of 8%.

Hauer (2011) argues that it is also important to estimate the standard deviation of the *distribution of the CMF* (denoted as σ) in order to explicitly account for the fact that a CMF is a random variable that does not necessarily have a unique value since it depends on many factors, including the variation in application circumstances of the treatment across sites. In the approach outlined in Hauer (2011), σ is estimated by considering the 'Law of Total Variance', as follows:

$$\sigma=\sqrt{V+\mathrm{Var}(\bar{\theta})} \tag{9}$$

where:

$$V=\begin{cases}\dfrac{\sum_{i=1}^{n}(\theta_i-\bar{\theta})^2}{n}-\dfrac{\sum_{i=1}^{n}\mathrm{Var}(\theta_i)}{n} & \textit{if positive}\\ 0 & \textit{otherwise}\end{cases}$$

Since $\mathrm{StdDev}(\theta)=\sqrt{\mathrm{Var}(\bar{\theta})}$ and $V \geq 0$, it follows that $\sigma \geq \mathrm{StDev}(\theta)$.

A Detailed Example

This example is for a single site; for a real study there would be several such sites with the same treatment, forming a composite entity. To get an aggregate effect, use Eqs. (7) and (8).

Example: A stop-controlled intersection converted to a roundabout in 2004 with $x = 34$ accidents in the period from 2000 before conversion and 14 accidents after, to the end of 2007.

The SPF is as follows, with the over-dispersion parameter k estimated as 0.25:

$$\mathbf{E}\{\tau_y\} \text{ (crashes / year)}$$
$$= \alpha \times (\text{major road AADT})^{0.256} \times (\text{minor road AADT})^{0.831}$$

The α parameter and the average annual daily traffic (AADT) varies from year to year, as shown in the shaded cells in Table 1. In this table the SPF is applied for each year, using the AADTs for that year. For 2004, the SPF predictions are for partial years in the before and after period in proportion to the number of months in that period.

From Eq. (4), $w = [1/(1 + kP)] = 0.1571$

From Eq. (2), $m = w(P) + (1 - w)(x) = 0.1571(21.4583) + (1 - 0.1571)(34) = 32.0295$

From Eq. (3), $\text{VAR}\{m\} = (1 - w)(m) = (1 - 0.1571)32.0295 = 26.997$

The adjustment for m to account for temporal differences, and differences in traffic volume and period length between the before and after periods, is simply the ratio of SPF estimates for the after and before periods:

$$r = E\{\tau\}_a / E\{\tau\}_b$$

From Eqs. (5) and (6):

$$\lambda(j) = (r)(m) = (16.1394 / 21.4583)\ (32.0295) = 24.090$$
$$\text{VAR}\{\lambda\} = r^2\ \text{VAR}\{m\} = (16.1394 / 21.4583)^2 (26.997) = 15.272$$

From Eq. (1), $\delta = \lambda - \pi = 24.090 - 14 = 10.090$

Applying Eqs. (7) and (8) to a single site, gives

$$\theta = \pi / \lambda\ [1 + \text{VAR}(\lambda)\} / (\lambda)^2] = 14 / 24.090\ [1 + 15.272 / 24.090^2] = 0.566$$

and

$$\begin{aligned} \text{StDev}\{\theta\} &= \theta\{(1/\pi) + [\text{VAR}\{\lambda\} / \lambda^2]\}^{0.5} / \{1 + [\text{VAR}\{\lambda\} / \lambda^2]\} \\ &= 0.596\ \{(1/14) + [15.272 / 24.090^2]\}^{0.5} / \\ &\quad \{1 + [15.272 / 24.090^2]\}^2 = 0.1723 \end{aligned}$$

VARIATION FROM THE GOLD STANDARD

Two methods that are related to, but vary in important ways from, the EB methodology have been legitimately used under special circumstances and are discussed in this section in brief.

Table 1. Input Data for the Before-After Study Example.

	2000	2001	2002	2003	Jan.–Aug. 2004	Nov.–Dec. 2004	2005	2006	2007
Major AADT	10,228	10,441	10,761	10,867	10,974	12,076	11,597	11,836	12,315
Minor AADT	4,503	4,597	4,738	4,785	4,832	5,317	5,106	5,211	5,422
Annual α	0.000383	0.000388	0.000392	0.000358	0.000391	0.000391	0.000389	0.000362	0.000367
Yearly SPF estimate $E\{\tau_y\}$	4.4238	4.5831	4.7846	4.4167	3.2501	0.9016	5.1505	4.9005	5.1868
Period SPF Estimate $E\{\tau\}=\Sigma[E\{\tau_y\}]$	Before Period Sum = $E\{\tau\}_b = 21.4583 = P$ (Eq. (2))					After Period Sum = $E\{\tau\}_a = 16.1394$			

Comparison Group Study

For the special circumstance where there is no RTM, the EB methodology and SPFs are not strictly required and a simpler comparison group (C-G) study may be undertaken. Since RTM results from a treatment site selection bias, it would not be an issue where this bias is non-existent; that is, where crash frequency is not considered in selecting a site for safety treatment, as would be the case where a change is implemented strictly for operational reasons. Another situation where there would be no site selection bias is where a blanket treatment is applied to all sites of relevance. As the CMF Guidebook (Gross et al., 2010) notes, 'in practice, except for blanket treatments, it is difficult to ascertain that there is no RTM and only a truly random selection of sites for treatment will ensure that there is no selection bias' (p. 14).

Where it is appropriate to do a C-G study it is still necessary to account for time trends in crashes and changes due to traffic volume. There are two ways of doing so. In the traditional, simpler method, an untreated C-G of sites similar to the treated ones needs to be identified to account for the changes in safety unrelated to the treatment. 'Similarity' requires that the sample mean and variance of the safety-related variable be as close as possible and, ideally, one-to-one matching on traffic volumes to account for changes in this variable – a task that could be challenging. The C-G data are then used to calculate a comparison ratio as the crash count in the after period divided by that in the before period. The crash count in the before period at a treatment site group, which is unadjusted for RTM, is then multiplied by this comparison ratio to estimate the expected crashes at the treatment group had no treatment been applied. This estimate is equivalent to λ in Eq. (5), with the before period count x used instead of m since there is no RTM adjustment as per Eq. (2). The estimate of λ is then compared to the observed crashes in the after period (π) at the treatment site group to estimate the safety effect of the treatment. Eqs. (7) and (8) can be adapted for this purpose, assuming a value for variance for the comparison ratio. More details are provided in Hauer (1997) and Gross et al. (2010).

Alternatively, Hauer (1997) proposes a hybrid with the EB method, whereby an SPF developed from a C-G that is not necessarily closely matched on traffic volumes is used to estimate time trend and traffic volume change effects, but not for an RTM correction as in the EB method. In this approach, the value of r in Eq. (5) is estimated as per the above example and is applied to the before period count, x, instead of m as before, since there is no RTM adjustment as per Eq. (2). More details on this approach are provided in Hauer (1997).

Full Bayes Approach

A Full Bayes (FB) approach has been proposed as an alternative and an improvement to the EB study (see, e.g., Park, Park, & Lomax, 2010; Pawlovich, Li, Carriquiry, & Welch, 2006). However, it is quite complex and not easily implementable in software, so has not typically been used for before–after evaluation studies. In addition, limited research (see, e.g., Persaud, Lan, Lyon, & Bhim, 2010) suggests that results of the FB approach are no more reliable than those of a properly done EB study with adequate sample sizes of reference and treatment sites. The CMF Development Guide (Gross et al., 2010) provides a good, concise discussion of the pros and cons of FB versus EB, so much of the text in the rest of this section is taken from that resource.

According to the CMF Development Guide (Gross et al., 2010), the FB approach is not a type of evaluation study per se, but rather is a modelling approach which can be used in the same way as the generalised linear modelling approach typically employed for SPFs in EB before–after studies.

In the FB approach to before–after studies a reference population is also used, similar to the EB method. However, instead of using a point estimate of the expected crash frequency (m in Eq. (2)) and its variance, a distribution of likely values is generated. This distribution of likely values is then combined with the observed crash frequency to estimate the long-term expected crash frequency in the absence of treatment (equivalent to λ in Eq. (5)). Through the use of a distribution, rather than a point estimate of the expected crash frequency, the variance of the long-term crash frequency and the variance of the estimated CMF can be calculated more precisely. Other identified advantages of the FB approach include the following:

i. The modelling process, in contrast to the conventional generalised linear modelling used for EB studies, allows for complex model forms to be specified, such as those that include both multiplicative and the additive terms that may be required for point hazards.
ii. The properties of FB models should also allow the estimation of valid models with smaller sample sizes. This may be particularly valuable for relatively rare crash types, such as those involving pedestrians, or for reference groups with limited sites.
iii. The FB approach has the ability to consider spatial correlation between sites, which, for before–after studies, will likely be an issue since both treated and comparison/reference sites tend to be geographically close to each other for pragmatic as well as theoretical reasons. For the latter consideration, the comparison/reference group selected should be as close in

proximity as possible to the treated sites since the traffic exposure and exposure effects are more likely to be similar than if they were farther away.

iv. This approach provides the opportunity to incorporate prior knowledge from previous research deemed to be reliable for the parameter estimates of the crash prediction model or for the estimated CMF. This will improve the estimate of the CMF in principle.

KEY ISSUES AFFECTING THE ROBUSTNESS OF BEFORE–AFTER EVALUATIONS

There are many issues in conducting before–after evaluations which, if not properly addressed, will affect the validity of the results. Issues, grouped into two main areas – the treatment sample and the reference/C-G sample – are addressed here.

Treatment Group Sample

The key issue in identifying a treatment group sample for an EB before–after safety evaluation is to ensure that there are enough data to detect, with appropriate statistical significance, the expected change in safety. Surprising as it may seem, there is currently no formal method for determining required sample sizes for the EB before–after approach. However, there is a method presented in Hauer (1997), which pertains to the C-G method and has been used by many analysts as an approximation in conservatively estimating the sample size required for an EB study. The issue is further addressed in Hauer (2008).

In Hauer's (1997) method, an index effectiveness θ, as defined in Eq. (7), is postulated based on various logical considerations, including a review of literature on evaluations of similar treatments. For a given size of treatment and C-Gs, in terms of number of crashes and lengths of before and after periods, the variance of θ may be estimated from Eq. (10):

$$\text{StDev}(\theta) = \{(\theta / r_d + \theta^2) / x_b + \theta^2[(1 / r_d + 1) / C_b + \text{VAR}(\omega)]\}^{0.5} \qquad (10)$$

where r_d is the ratio of the after and before period durations, and x_b and C_b are the number of crashes in the treatment and C-Gs respectively in the before period.

In the term $\omega = r_C/r_T$, $r_C = C_a/C_b$, is the ratio of crashes in the after and before periods in the C-G and r_T is the expected change in the treatment group due to non-treatment factors. ω is in fact a measure of the suitability of the comparison in matching the treatment group, and ideally should have a value of 1.0 with no variance. However, since r_T cannot easily be known, it is suggested that sensitivity to assuming values between 0.001 and 0.01 be examined.

The size of StDev(θ) relative to the postulated value of θ is then used to decide if the effect can be detected with a given level of significance for a given sample size. For example, for a treatment with a postulated value of $\theta = 0.6$, assuming an equal number of sites (20), years (3) and crashes ($C_b = 480$) in the treatment and C-Gs, and a 1.5 year after period ($r_d = 0.5$) and a conservative value of 0.01 for VAR$\{\omega\}$, gives StDev(θ) = 0.17. For this sample, the postulated effect, if realised, would be statistically significant at the 5% level since a value of 1.0 (implying no effect) is outside the 95% confidence interval (0.36–0.94).

Reference/Comparison Group Sample

This section addresses issues in identifying a suitable reference/C-G. Needless to say, none of the treatment site data, except in rare circumstances, should be used for this purpose. The challenge is usually not to find sufficient untreated sites but in establishing 'similarity' of the untreated sites to the treated sites as they were before the treatment. The two main characteristics that define similarity are discussed below, along with the very issue of defining sites that are strictly untreated.

Assessing Similarity in Trends

As documented by Persaud and Lyon (2007) and proposed by Hauer (1997), a test of comparability that evaluates a time series of crash counts should be applied to evaluate the suitability of the reference groups for similarity in time trends to the treatment group. If a potential reference group is a good one, then the yearly trends in crash frequencies track each other well over time. For this test, a sequence of sample odds ratios using 1 year of data for a hypothetical 'before' period and the following year as the 'after' data, starting with the first and second years and incrementally increasing by 1 year. From this set of odds ratios the sample mean and standard error are determined.

If this sample mean is not sufficiently close to 1.0 (as determined by, say, a 95% confidence interval), then the candidate reference group is unsuitable. An example based on data from one of the cities used to study the effects of red-light cameras (Persaud, Council, Lyon, Eccles, & Griffith, 2005) is cited, for which the mean and standard errors of the sequence of sample odds ratios for total crashes of all severities and types combined were estimated to be 1.045 and 0.150, respectively. A 95% confidence interval for the odds ratio was estimated to be between 0.751 and 1.339 (which includes a value of 1.0), suggesting that the candidate reference group is suitable in terms of time trend.

Assessing Similarity in Characteristics

In a controlled experiment, similarity between the treatment and non-treatment sites can be assured through random assignment of potential treatment sites to the two groups. However, in observational before–after studies, the focus of this chapter and the approach of choice for almost all safety evaluations, the closest that one can come to this desideratum would be to use as reference/comparison sites those sites that are identified for the treatment of interest but are yet to be treated for budgetary reasons. In practice, this possibility is quite rare, so analysts tend to resort to subjective judgement in assessing similarity (see Lord & Kuo, 2012 for additional discussion on this topic). In this case, the data assembled to calibrate the SPF used to account for non-treatment effects need to be such that the SPF will effectively control for all of the factors that are different between the treatment and reference groups. The more similar the sites are to the treated sites, the less onerous this requirement is. So the key is to properly match the treatment and non-treatment sites. While subjective judgement has been typically used, more formal means are now available and are being explored in new research. The most promising appears to be propensity score matching. Details of this procedure are beyond the scope of this chapter, but can be obtained from recent references (Aul & Davis, 2006; Park & Saccomanno, 2007; Sasidharan & Donnell, 2013). The ANCOVA has also been suggested, but this can still be biased, as discussed by Kuo and Lord (2013).

Separating Treatment and Reference/Comparison Sites – Recognising the Potential for Potential Reference Sites to be Affected by the Treatment

For some treatments, in particular those intended to achieve a general deterrence effect, a treatment may affect the most logical reference group.

As Persaud and Lyon (2006) note, red-light camera programs are a classical example, but there is evidence of this 'spill-over' effect for other measures, such as traffic calming, all-way stop installation and raised pavement markers. In such cases, the safety effect of the treatment sites is underestimated if a decrease in crashes in the reference group that may have 'spilled over' from the treatment group is applied to the treatment site data to correct for time trends unrelated to the treatment. In addition, not claiming this bonus benefit will further underestimate the overall treatment effect. The converse is true for those treatments, such as those designed to achieve speed reduction, that may result in a compensatory 'migration' effect at nearby untreated sites.

Persaud and Lyon (2006) cite an example of how the issue can be resolved based on the EB evaluation of red-light cameras by Persaud et al. (2005). Because of the possibility of spill-over effects to neighbouring signalised intersections used to develop SPFs for the reference sites, annual multipliers were estimated for the period *after* the first camera installation from the trend in annual multipliers of SPFs calibrated for a C-G of *unsignalised* intersections in the same city. The illustration in Table 2 is taken from Persaud and Lyon (2006), based on information for one of the cities that supplied data for the red-light camera evaluation study. Using the SPFs calibrated, using the reference data collected for unsignalised and signalised intersections, annual multipliers were estimated for each year. The year of the first camera installation was 1996. In case there was a spill-over effect at the non-treated signalised intersections, the annual multipliers estimated from the signalised reference group were applied only for the years 1992–1995.

Table 2. Development of Yearly Trend Factors when a Treatment May Affect Logical Reference Sites.

Reference Group	Unsignalised	Adjusted Unsignalised	Signalised	Adjusted Signalised
1992	1.34		1.61	
1993	1.20		1.52	
1994	1.17		1.44	
1995	0.99		1.15	
Avg. 1992–1995	1.18		1.43	
1996	0.84	0.71	0.89	1.02
1997	0.78	0.66	0.88	0.95
1998	0.93	0.79	0.94	1.13
1999	0.78	0.66	1.01	0.95
2000	1.23	1.05	1.09	1.50
2001	0.84	0.71	1.21	1.02
2002	1.96	0.82	1.09	1.17

Source: Persaud and Lyon (2006).

Yearly multipliers for 1996–2002 were then estimated by first determining, for both the unsignalised and signalised reference groups, the average yearly multiplier between 1992 and 1995. The unsignalised reference group yearly multipliers for the years 1996–2002 were divided by the average from 1992 to 1995. These results are shown in the column labelled 'Adjusted Unsignalised'. Yearly multipliers for the signalised reference group for 1996–2002 were then estimated by multiplying the average from 1992 to 1995 by the adjusted unsignalised factor for each year.

The data in Table 2 show that, with the exception of 2001, and to a lesser extent 1999, the yearly multipliers for the signalised reference group (the last column) have increased significantly with the adjustment, with the mean of these annual factors for the 1996–2002 period increasing from 1.02 to 1.11, indicating a possible spill-over effect, and further justifying the approach used.

CONCLUSIONS

The aim of the chapter was to provide information on state-of-the-art methodology for observational before–after evaluations to learn about the safety effectiveness of site-specific road safety interventions. This information complements that in extant resources. As is the case for these other resources, this chapter is written for those with experience and/or education in the theory and practice of road safety engineering, including basic analytical procedures and statistical concepts. Although a worked example is provided, the reader is encouraged to review the many recently published studies that have utilised the main methodology presented here.

REFERENCES

AASHTO. (2010). *Highway safety manual*. Washington, DC: American Association of State Highway and Transportation Officials.

Aul, N., & Davis, G. (2006). Use of propensity score matching method and hybrid Bayesian method to estimate crash modification factors of signal installation. *Transportation Research Record, 1950*, 17–23.

Gross, F., Persaud B., & Lyon, C. (2010). *A guide to developing quality crash modification factors*. FHWA-SA-10-032, Federal Highway Administration Report. US Department of Transportation.

Hauer, E. (1997). *Observational before–after studies in road safety: Estimating the effect of highway and traffic engineering measures on road safety*. Oxford: Pergamon Press, Elsevier Science Ltd.

Hauer, E. (2008). How many accidents are needed to show a difference? *Accident Analysis and Prevention*, *40*(4), 1634–1635.

Hauer, E. (2011). *Estimating the mean and variance of CMFs and how they can be changed by new research* (Appendix D of Task 1 report of NCHRP Project 17-48: Development of a Strategic National Highway Infrastructure Safety Research Agenda).

Kuo, P.-F., & Lord D. (2013). Accounting for site-selection bias in before-after studies for continuous distributions: Characteristics and application using speed data. *Transportation Research – Part A*, *49*, 256–269.

Lord, D., & Kuo, P.-F. (2012). Examining the effects of site selection criteria for evaluating the effectiveness of traffic safety improvement countermeasures. *Accident Analysis and Prevention*, *47*, 52–63.

Park E., Park J., & Lomax, T. (2010). A fully Bayesian multivariate approach to before–after safety evaluation. *Accident Analysis & Prevention*, *42*(4), 1118–1127.

Park, P., & Saccomanno, F. (2007). Reducing treatment selection bias for estimating treatment effects using propensity score method. *Journal of Transportation Engineering*, *133*, 112–118.

Pawlovich, M., Li, W., Carriquiry, A., & Welch, T. (2006). Iowa's experience with "road diet" measures: Impacts on crash frequencies and crash rates assessed following a Bayesian approach. *Transportation Research Record – Journal of the Transportation Research Board*, *1953*, 163–171.

Persaud, B., Lan, B., Lyon, C., & Bhim, R. (2010). Comparison of empirical Bayes and full Bayes approaches for before-after road safety evaluations. *Accident Analysis and Prevention*, *42*(1), 38–43.

Persaud, B., Council, F., Lyon, C., Eccles, K., & Griffith, M. (2005). Multi-jurisdictional safety evaluation of red light cameras. *Transportation Research Record: Journal of the Transportation Research Board*, *1922*, 29–37.

Persaud & Lyon (2006)

Persaud, B., & Lyon, C. (2007). Empirical Bayes before-after safety studies: Lessons learned from two decades of experience and future directions. *Accident Analysis and Prevention*, *39*(3), 546–555.

Sasidharan, L., & Donnell, E. (2013). Application of propensity scores and potential outcomes to estimate effectiveness of traffic safety countermeasures: Exploratory analysis using intersection lighting data. *Accident Analysis & Prevention*, *50*, 539–553.

CHAPTER 19

META-ANALYTIC METHODS

Rune Elvik

ABSTRACT

Purpose – *This chapter gives an overview of meta-analytic methods and illustrates the use of these methods for synthesising research findings. The advantages of performing a meta-analysis are described. Pitfalls in meta-analyses are also discussed. The chapter is intended to present the main elements of a meta-analysis and guide readers to literature presenting meta-analytic methods in greater detail.*

Methodology – *Key references in the meta-analysis literature are quoted and examples of meta-analyses are presented.*

Findings – *A meta-analysis is a useful tool for summarising knowledge in fields where a large number of studies have been reported. In addition to providing summary estimates of results, a meta-analysis can be applied to identify factors that produce systematic variation in study findings.*

Research implications – *Methods of meta-analyses keep developing to deal with complex data structures, thus extending the type of research findings that are amenable to meta-analyses.*

Practical implications – *Performing a meta-analysis saves labour by eliminating the need to read and digest a large number of studies in order to get*

Safe Mobility: Challenges, Methodology and Solutions
Transport and Sustainability, Volume 11, 425–447

ISSN: 2044-9941/doi:10.1108/S2044-994120180000011022

an overview of the current state-of-knowledge in a field. Moreover, a meta-analysis establishes a system for easily and quickly updating knowledge as new studies become available.

Keywords Meta-analysis; techniques; systematic review; models

INTRODUCTION

This chapter is the second of two chapters that describe methods for evaluating the safety impacts of safety investments, or countermeasures. This chapter is focussed on the use of meta-analysis for drawing definitive conclusions about the safety effects of safety countermeasures – most appropriate when resources are not available for financing and conducting a local study. Chapter 18, in contrast, is focussed on before–after evaluations, widely considered to be the most reliable methodology for drawing definitive conclusions about the safety effects of treatments when designed and administered correctly.

WHAT IS A META-ANALYSIS?

A meta-analysis is the statistical analysis of a set of numerical research results for the purpose of developing a weighted mean result and identifying sources of systematic variation in individual results (Borenstein, Hedges, Higgins, & Rothstein., 2009; Card, 2012; Cooper, 2010; Cooper, Hedges, & Valentine, 2009; Egger, Davey Smith, & Altman, 2001; Hedges & Olkin, 1985; Higgins & Green, 2008; Lipsey & Wilson, 2001; Littell, Corcoran, & Pillai, 2008; Lund, 2002; Petitti, 2000; Petticrew & Roberts, 2006; Schmidt & Hunter, 2015; Sutton, Abrams, Jones, Sheldon, & Song, 2000). There are several techniques for meta-analysis. The essential elements of the most commonly used of these are described. For detailed technical presentations the reader should consult the textbooks cited above.

Meta-analyses normally form part of systematic literature reviews. A systematic literature review is a review which is performed according to rules that are intended to make it replicable. A review is replicable if two or more researchers working independently and following the same rules achieve identical results, that is, identify the same studies dealing with a topic. Systematic literature reviews are described further in the section dealing with preparation for meta-analysis.

WHY PERFORM A META-ANALYSIS?

Traditionally, literature surveys were presented in a narrative format. The reviewer wrote a short description of each study and perhaps added an informal assessment of the study. In many fields of knowledge this format is no longer practical; there are simply too many studies to tell a short story about each of them. For example, Espey (1998) retrieved 101 studies containing a total of 277 estimates of the short-term price elasticity of the demand for motor fuel, and a total of 363 estimates of the long-term price elasticity. Lindhjem, Navrud, Biasque and Braathen (2012) synthesised evidence from 74 studies containing 856 estimates of the value of preventing a fatality. Elvik (2013a) identified 115 studies containing 526 estimates of the relationship between speed and road safety. When there are so many studies and results, it becomes impractical to use a narrative format to summarise knowledge. The first reason for performing a meta-analysis is therefore to help organise and gain an overview of a large body of literature.

The second reason for performing a meta-analysis is that a need is often felt for summarising or synthesising (the terms are used interchangeably in this chapter) the results of studies dealing with a certain topic. What do these studies typically find? Meta-analysis is a well-suited tool for answering this question.

A third reason for conducting a meta-analysis is that it reduces the element of subjectivity that may creep into narrative reviews. A meta-analysis forces the analyst to code objective characteristics of each study and apply statistical techniques to summarise the results of studies. These features of meta-analysis do not eliminate all choices made by the analyst, but it makes them more transparent.

A fourth, and very important, reason is to increase statistical power to detect effects. It is not unusual for the results of the majority of studies about a topic to fail to reach statistical significance at conventional levels. Fig. 1 illustrates this for studies that have evaluated the effect of using antidepressant drugs on accident risk (Elvik, 2013b). Fig. 1 presents 25-point estimates and their 95% confidence intervals. It is seen that 16 of the 25 estimates are not statistically significant. The summary estimates, shown at the bottom, are highly statistically significant.

A fifth reason for performing a meta-analysis is to explain why the results of different studies vary. In Fig. 1, studies span a period of 17 years and we might be interested in whether the results of recent studies differ from those of older studies. Another source of variation in study findings often attracting interest is differences in method; very often studies using different methods get different results.

Model	Study name	Subgroup within study	Statistics for each study				
			Odds ratio	Lower limit	Upper limit	Z-Value	p-Value
	Ray 1992	Injury	2.160	1.316	3.544	3.048	0.002
	Leveille 1994	Injury current user	2.300	1.101	4.805	2.218	0.027
	Leveille 1994	Injury past user	0.700	0.200	2.450	-0.558	0.577
	Currie 1995	Injury	2.660	0.311	22.747	0.893	0.372
	Barbone 1998	PDO anti	1.070	0.805	1.422	0.467	0.641
	Barbone 1998	PDO sero	0.890	0.534	1.482	-0.448	0.654
	Barbone 1998	Serious injury anti	1.200	0.493	2.919	0.402	0.688
	Barbone 1998	Serious injury sero	1.610	0.210	12.322	0.459	0.646
	Barbone 1998	Slight injury anti	0.590	0.309	1.126	-1.599	0.110
	Barbone 1998	Slight injury sero	0.600	0.203	1.775	-0.923	0.356
	Neutel 1998	Injury	1.000	0.500	2.000	0.000	1.000
	McGwin 2000	Mostly PDO	0.670	0.238	1.890	-0.757	0.449
	Wadsworth 2003	Injury	1.180	0.671	2.074	0.575	0.565
	Etminan 2004	Injury	1.800	1.000	3.240	1.960	0.050
	Lam 2005	Injury	1.750	0.600	5.107	1.024	0.306
	Wadsworth 2005	Injury	1.720	0.569	5.195	0.962	0.336
	Sagberg 2006	Mostly PDO	1.700	1.127	2.564	2.531	0.011
	Bramness 2008	Injury newer	1.600	1.503	1.703	14.720	0.000
	Bramness 2008	Injury sedating	1.400	1.201	1.632	4.307	0.000
	Hours 2008	Injury	3.630	1.340	10.948	2.506	0.012
	Rapoport 2008	Mostly PDO	1.820	1.558	2.127	7.537	0.000
	Vingilis 2008	Injury	2.550	1.742	3.734	4.812	0.000
	Gibson 2009	Injury group 1	1.020	0.893	1.164	0.293	0.770
	Gibson 2009	Injury group 2	0.950	0.815	1.107	-0.657	0.511
	Woratanarat 2009	Injury	4.000	0.250	63.975	0.980	0.327
Fixed			1.429	1.366	1.495	15.515	0.000
Random			1.359	1.164	1.586	3.892	0.000

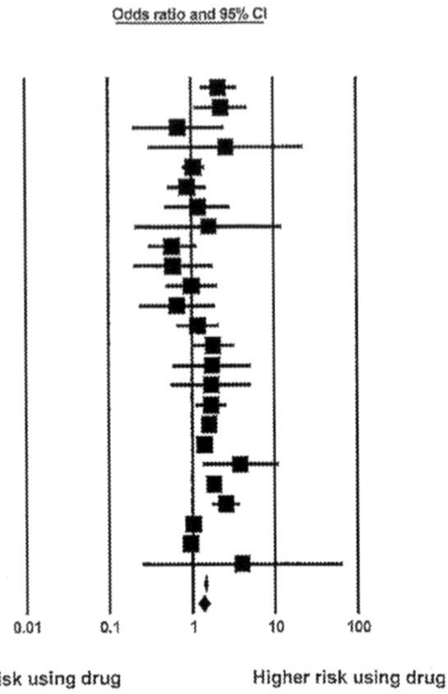

Fig. 1. Forest Plot of Studies of the Risk of Accident Involvement Associated with the Use of Antidepressant Drugs. *Source*: Elvik (2013b).

Finally, a meta-analysis establishes an easy system for updating knowledge. Studies included in a meta-analysis are normally entered into a spreadsheet and new studies can easily be added.

PREPARING FOR A META-ANALYSIS

Preparing for a meta-analysis includes formulating the research problem, conducting a systematic literature survey, defining study inclusion criteria, developing a coding form and study protocol, coding and extracting data from each study and assessing study quality.

Meta-analyses typically deal with studies that have evaluated the effects of something. Examples of questions that may serve as the starting point for a meta-analysis are:

- What are the effects on travel demand of changes in fuel prices?
- Are users of illicit and medicinal drugs at higher risk of accident involvement than other drivers?

- What is the shape and strength of the relationship between changes in the speed of traffic and changes in the number and severity of accidents?
- What are the effects on accidents of converting intersections to roundabouts?

Common to all these questions is that many studies are likely to have been undertaken to answer them. When there are many studies seeking to answer the same question the first condition for performing a meta-analysis is fulfilled.

To retrieve relevant studies a systematic literature survey should be performed. A systematic literature survey starts with a search of relevant bibliographic databases using key words that are likely to capture relevant studies. It is normally necessary to search more than one literature database; even then – and even if experienced librarians do the search – there is no guarantee of identifying all relevant studies (Lawrence, 2008). Searching literature databases should therefore be complemented by other types of literature searches. One type of search is the so-called ancestry method, which consists of examining the list of references of studies that have been retrieved. The ancestry method will usually identify older studies, some of which may not be included in literature databases. Another type of search is to contact prominent researchers in the field to ask them about ongoing or recently completed studies that have not yet been published. A third method is to visit research institutes' or government agencies' websites. These sites may contain 'grey' literature, that is, reports that have not been published in peer-reviewed journals.

To guide the literature review and screening of studies once they have been retrieved, study inclusion criteria should be defined. The primary inclusion criterion is relevance: the study should deal with the topic of the meta-analysis and not a different topic. Topics, however, can sometimes be very broadly defined. At the inception of the meta-analysis its pioneers, Smith and Glass (1977) published a meta-analysis of the effects of psychotherapy, prompting the famous psychologist H. J. Eysenck to label meta-analysis 'An exercise in mega-silliness' (Eysenck, 1978, p. 517). To Eysenck, psychotherapy was a much too broad topic – broad enough, he thought, to be meaningless.

The definition of the topic and the research problem should be as precise as possible. Once a set of relevant studies has been defined, other inclusion criteria may be considered. Perhaps the most common of these is study language. If a study is in a language, few people can understand that may be an argument for not including it. Ideally speaking, however, language restrictions should be avoided. Other criteria that have been discussed are restrictions on study age and study quality. For example, the Cochrane collaboration, which mostly conducts systematic reviews and meta-analyses of medical treatments (Higgins & Green, 2008), gives priority to randomised controlled trials and may omit non-randomised studies if a sufficient number of randomised controlled trials are available for meta-analysis. In most areas of transport research, however, there are few randomised controlled trials. Restrictions based on study quality will then, as a rule, be poorly defined. This does not mean that study quality does not vary or is not an issue merely that it is better dealt with within the meta-analysis than by excluding studies from the analysis. Once studies have been excluded, it is no longer possible to learn how they influence the results of the meta-analysis. Similar comments apply to restrictions on study age – these should also be avoided.

To be included in a meta-analysis using the inverse-variance technique, a study should provide two pieces of information:

1. One or more estimates of the result of interest.
2. The standard error of each estimate.

It will sometimes be the case that results are reported, but not their standard errors. In such cases it may still be possible to conduct a meta-analysis, although the inverse-variance technique cannot be applied. When reporting a meta-analysis a list of studies that were excluded should always be provided, stating why each study was excluded.

To extract data from each study, it is useful to develop a coding form. Saving the data extracted from each study on a spreadsheet is useful, as all but the most complex analyses can be done on a spreadsheet. Data normally extracted from each study include:

- Study identifier, usually by author and publication year.
- Year of publication of study.
- If available years data analysed in study refer to (this would normally be a few years before the publication year).
- Country or countries of origin of the study.
- Descriptors of study design.
- Descriptors of estimators of effect used in study.
- Descriptors of the intervention, risk factor or measure whose effects are studied.
- Potential moderator variables (a moderator variable is any variable influencing the size of an effect).
- Estimates of effect (*d*, *r*, odds ratio, etc., discussed in the next section).
- The standard error of each estimate of effect.

Finally, assessing study quality may be part of the preparation for a meta-analysis. There is wide agreement that some sort of study quality assessment should normally be part of the meta-analysis, but disagreement exists about how best to include a study quality assessment in a meta-analysis. More is covered on this topic later.

EXPLORATORY META-ANALYSIS

The purpose of an exploratory meta-analysis is to decide whether doing a full meta-analysis makes sense or not. An exploratory meta-analysis consists of: listing the type of study results found in the relevant studies; determining the need for converting these to a common metric; converting or transforming results to a common metric; applying graphical tools, in particular funnel plots to examine the distribution of results and deciding whether a formal synthesis of results makes sense.

There are many types of study results. The most common include:

- Correlation coefficients or regression coefficients showing the relationship between two variables.
- Numerical measurements and differences between such measurements, such as air temperature, precipitation, friction coefficient, etc.
- Mean values of variables or differences between mean values, for example scores on tests, the mean speed of traffic, etc.
- Proportions or percentages and differences between percentages, such as the modal split of travel in different regions or before and after an intervention.
- Diagnostic classifications, for example, rates of correct and false positives, positive predictive values, etc.
- Coefficients estimated in multivariate analyses.
- Estimators of effect, for instance, accident rates, simple odds, odds ratios, ratios of odds ratios or relative risks.

A body of literature may contain studies that present their results using more than one of these metrics or, even if the metrics are formally the same (e.g., only mean values), the definition of the variables may not be identical. Is it possible to perform a meta-analysis at all when studies report results in different metrics? The general answer is yes – provided the different metrics can be converted to a common metric. A large body of technical literature exists about how to do this; for reasons of space only a few illustrations are given here.

Differences in means are often converted to standardised mean differences (Borenstein et al., 2009, Eqs. (4.18) and (4.19)):

$$d = \frac{\overline{X}_1 - \overline{X}_2}{S_{\text{within}}}$$

in which $\overline{X}_1$ and $\overline{X}_2$ are the means in the two groups being compared and S_{within} is the pooled within-groups standard deviation:

$$S_{\text{within}} = \sqrt{\frac{(n_1 - 1)S_1^2 + (n_2 - 1)S_2^2}{n_1 + n_2 - 2}}$$

where n_1 and n_2 are the sample sizes in the two groups and S_1^2 and S_2^2 are the variances in the two groups.

Correlation coefficients have well-known statistical properties. The most comprehensive textbook explaining how to exploit these properties, for example, to adjust estimated correlations for attenuation due to measurement errors, is Schmidt and Hunter (2015). Borenstein et al. (2009) explain how to convert d (standardised mean difference) to log the odds ratio (the natural logarithm of the odds ratio) and vice versa, and how to convert d to r (the correlation coefficient) and vice versa.

Odds ratios are normally converted to natural logarithms, as odds ratios in natural units have a lognormal distribution. Odds ratios can easily be converted to relative risks and vice versa. Thus d, r, odds ratios and relative risks are all metrics that can be converted from one to the other. Whenever all results in a set of studies are in the same metric, meta-analysis is feasible.

How about situations when the different metrics cannot be converted to a single metric? Multiple estimators of effect tend to be used in road safety evaluation studies: accident rate ratio, simple odds (after/before), odds ratios, ratios of odds ratios, relative risk ratios and model coefficients. These estimators cannot be converted to the same metric. It may still be possible to perform a meta-analysis, but then results based on the different estimators of effect should form subgroups in the meta-analysis.

Several graphical tools have been developed to assist exploratory meta-analysis. The most widely used of these is the funnel plot (Light & Pillemer, 1984). A funnel plot is a scatter diagram of results, in which the results are plotted on the abscissa and an indicator of their precision on the ordinate. The data points then ought to distribute like a funnel turned upside down, that is, have small dispersion at the top of the diagram and be more widely

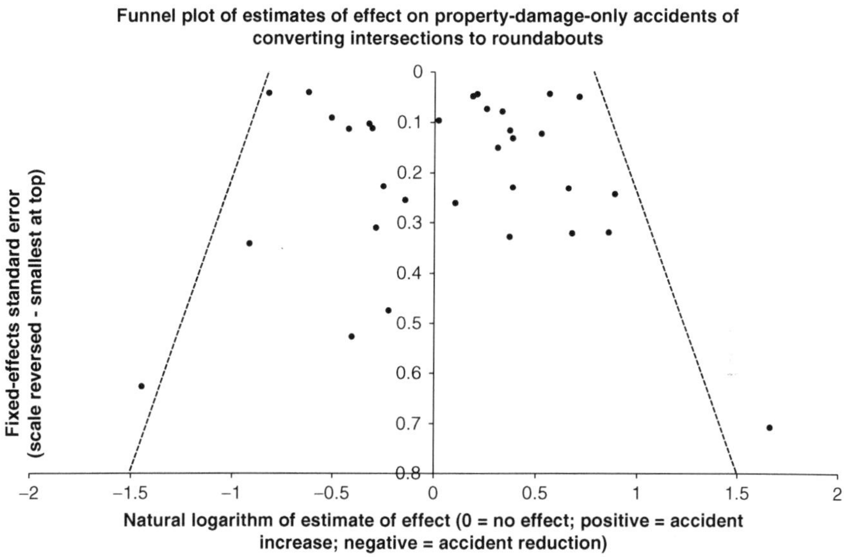

Fig. 2. Funnel Plot of Estimates of Effect on Property-Damage-Only Accidents of Converting Intersections to Roundabouts.

spread at the bottom. There are various options for the choice of scale on the axes of a funnel plot. Sterne and Egger (2001) discuss these options and recommend using the natural logarithm of the results for the abscissa and the standard error for the ordinate.

If the funnel plot does not clearly show the funnel shape, it may not be informative to perform a meta-analysis. A funnel plot that does not show the expected shape is shown in Fig. 2; instead it shows a wide dispersion of estimates of the effect on property-damage-only accidents of converting intersections to roundabouts.

As seen in Fig. 2, the dispersion of the data points is almost as wide at the top of the diagram as at the bottom. There is only a weak tendency for data points to converge to an average effect. In this case, most data points are located far from the summary estimate of effect (which is very close to zero). The summary estimate of effect would thus not be very informative about the typical result of a study. In general, it makes sense to perform a meta-analysis for the purpose of estimating a weighted summary result when the funnel plot:

- Clearly shows the shape of a funnel turned upside down.
- Is symmetrical around the weight mean estimate – asymmetry indicates publication bias (see the section on sensitivity analysis).

- Is unimodal (has only one peak).
- Does not contain clearly outlying data points (see the section on sensitivity analysis).

If one or more of these points is not satisfied one may still opt to perform a meta-analysis. In that case, however, extra care has to be taken in performing a sensitivity analysis of the meta-analysis (see the following section).

MAIN ANALYSIS

Historically speaking, three techniques have been developed for performing a meta-analysis (Rosenthal, 1991). Today, only one of these is widely used.

The oldest and simplest technique for summarising a set of results is the vote count. Results are divided into, for example, those that show a favourable effect of a treatment and those that show an adverse effect. The number of results in each group is counted and the largest group declared the winner. Obviously, this is very crude and says nothing about the size or strength of the effect.

The next technique that was applied was to combine *p*-values. As noted earlier, one of the arguments for performing a meta-analysis is that it increases statistical power to detect effects. Combined *p*-values can show that there is a significant relationship between variables when all results are combined, even if none of the individual estimates reach statistical significance. However, combining *p*-values only shows the statistical significance of a relationship, not the magnitude of an effect.

In most meta-analyses we are interested in learning about the magnitude of an effect of a treatment. This is by far the most common problem motivating meta-analysis. The weighted mean estimate of an effect based on g independent is defined as follows:

$$\text{Summary mean} = \overline{Y} = \frac{\sum_{i=1}^{g} Y_i \cdot W_i}{\sum_{I=1}^{g} W_i}$$

Here $\overline{Y}$ is the estimate of the weighted summary mean, based on g individual estimates, each of which is assigned a statistical weight as follows:

$$\text{Statistical weight} = W_i = \frac{1}{\text{SE}_i^2}$$

Statistical weights that are proportional to the inverse value of the sampling variance of each estimate (which is the square of the standard error of the estimate) are efficient. They minimise the variance of the summary estimate; that is, produce a summary estimate, which is more precise than a summary estimate based on any other definition of the statistical weights. The notation is general; that is, it applies to all metrics discussed above, such as *d*, *r*, log odds ratio or log risk ratio.

There are two models for an inverse-variance meta-analysis: the fixed-effects model and the random-effects model. The fixed-effects model is based on the assumption that the variation in individual results consists of sampling variance only; that is, it is fully explained in terms of the sampling variance of each result. The random-effects model is based on the assumption that there is a systematic between-study variation in results; that is, variation greater than the sampling variance accounts for. To determine whether there is systematic between-study variation in results, a statistical test is performed using the following test-statistic:

$$Q = \sum_{i=1}^{g} W_i \cdot Y_i^2 - \frac{\left(\sum_{i=1}^{g} W_i \cdot Y_i\right)^2}{\sum_{i=1}^{g} W_i}$$

Q is an estimate of variance. It is chi-square distributed with $g - 1$ degrees of freedom. The statistical weights used in a random-effects model are modified by adding a variance component, tau-squared, which is estimated as follows:

$$\tau^2 = \frac{Q - \text{df}}{C}$$

C is estimated as follows:

$$C = \sum W_i - \frac{\sum W_i^2}{\sum W_i}$$

The modified statistical weight for each study becomes:

$$W_i = \frac{1}{\text{SE}_i^2 + \tau^2}$$

A random-effects model should be used when tau-squared is positive. If negative, it is set to zero. Please note that while the sampling variance (SE^2)

varies from study to study, the between-study variance parameter (τ^2) is a constant. It therefore flattens the statistical weights, which can vary considerably less from study to study in the random-effects model than in the fixed-effects model. There may be interest in knowing how much sampling variance and between-study variance contributes to the overall variance in results in a set of studies. The following statistic indicates this:

$$I^2 = \left(\frac{Q - \mathrm{df}}{Q}\right) \cdot 100$$

The I^2 statistic indicates, as a percentage (hence the multiplication by 100), the share of total variance which is between-study variance. In social science applications of meta-analysis, this share will typically be around 70–90%.

The main part of a meta-analysis usually starts by conducting a fixed-effects and random-effects analysis of the entire data set. Following this, three types of further analysis may be performed. The first is subgroup analysis. It is important to decide on the groups to be formed before analysis; otherwise there is a risk of data mining and spurious findings. In applications to road safety evaluation studies, there is often an interest in whether the effects of a risk factor or a safety treatment vary according to accident severity. Subgroups can then be formed according to accident severity, for example fatal accidents, injury accidents and property-damage-only accidents.

Table 1 shows a meta-analysis for two levels of accident severity for the 25 results regarding antidepressant drugs shown in Fig. 1. There were 20 results for injury accidents and five for property-damage-only accidents. The summary estimates of risk do not vary much according to accident severity. It is seen that the confidence intervals are wider in the random-effects models than in the fixed-effects models.

How much wider the random-effects confidence intervals are than the fixed-effects confidence intervals depends on the value of τ^2. The larger the value of τ^2, the wider the random-effects confidence intervals. This can be verified by comparing the width of the confidence intervals in Table 1 for the three values of τ^2: 0.069, 0.077 and 0.104.

A second type of analysis that can be informative as part of the main analysis is a cumulative meta-analysis. A cumulative meta-analysis summarises results chronologically, starting with the oldest, then the second oldest and so on until all results are included. This way, the history of the research is reconstructed and one may see if results have changed over time. Fig. 3 shows a cumulative meta-analysis of the 25 estimates of the effect on the risk of accident involvement of using antidepressant drugs.

Table 1. Summary Estimates of Risk Associated with the Use of Antidepressant Drugs.

Group	Number of Results	Model	Best Estimate	Lower 95 % Limit	Upper 95 % Limit	*Q*-Statistic	*I*-Squared	Tau-Squared
All studies	25	Fixed	1.429	1.366	1.495	114.856	79.1	0.069
	25	Random	1.359	1.164	1.586			
Injury accidents	20	Fixed	1.413	1.346	1.483	95.355	80.1	0.077
	20	Random	1.383	1.148	1.667			
PDO accidents	5	Fixed	1.542	1.361	1.747	17.862	77.6	0.104
	5	Random	1.275	0.903	1.800			
Odds ratio	20	Fixed	1.547	1.391	1.719	42.505	58.0	0.101
	20	Random	1.381	1.111	1.717			
Relative risk	5	Fixed	1.404	1.336	1.476	67.039	94.0	0.074
	5	Random	1.308	1.012	1.691			

Source: Elvik (2013b).
Note: PDO, property-damage-only accidents.

Study name	Subgroup within study	Cumulative statistics: Point	Lower limit	Upper limit	Z-Value	p-Value
Ray 1992	Injury	2.160	1.316	3.544	3.048	0.002
Leveille 1994	Injury current user	2.203	1.460	3.322	3.766	0.000
Leveille 1994	Injury past user	1.970	1.333	2.912	3.404	0.001
Currie 1995	Injury	1.990	1.355	2.921	3.509	0.000
Barbone 1998	PDO anti	1.332	1.060	1.674	2.462	0.014
Barbone 1998	PDO sero	1.246	1.011	1.534	2.064	0.039
Barbone 1998	Serious injury anti	1.243	1.015	1.523	2.101	0.036
Barbone 1998	Serious injury sero	1.246	1.018	1.525	2.136	0.033
Barbone 1998	Slight injury anti	1.166	0.962	1.414	1.562	0.118
Barbone 1998	Slight injury sero	1.143	0.945	1.382	1.377	0 169
Neutel 1998	Injury	1.132	0.943	1.360	1.328	0.184
McGwin 2000	Mostly PDO	1.114	0.930	1.334	1.176	0.240
Wadsworth 2003	Injury	1.120	0.943	1.330	1.295	0.195
Etminan 2004	Injury	1.163	0.986	1.371	1.793	0.073
Lam 2005	Injury	1.174	0.997	1.382	1.928	0.054
Wadsworth 2005	Injury	1.183	1.007	1.390	2.048	0.041
Sagberg 2006	Mostly PDO	1.242	1.069	1.443	2.831	0.005
Bramness 2008	Injury newer	1.541	1.455	1.633	14.675	0.000
Bramness 2008	Injury sedating	1.523	1.443	1.607	15.251	0.000
Hours 2008	Injury	1.526	1.446	1.611	15.359	0.000
Rapoport 2008	Mostly PDO	1.556	1.478	1.637	16.981	0.000
Vingilis 2008	Injury	1.569	1.492	1.650	17.469	0.000
Gibson 2009	Injury group 1	1.486	1.417	1.557	16.426	0.000
Gibson 2009	Injury group 2	1.429	1.366	1.495	15.501	0.000
Woratanarat 2009	Injury	1.429	1.366	1.495	15.515	0.000
		1.429	1.366	1.495	15.515	0.000

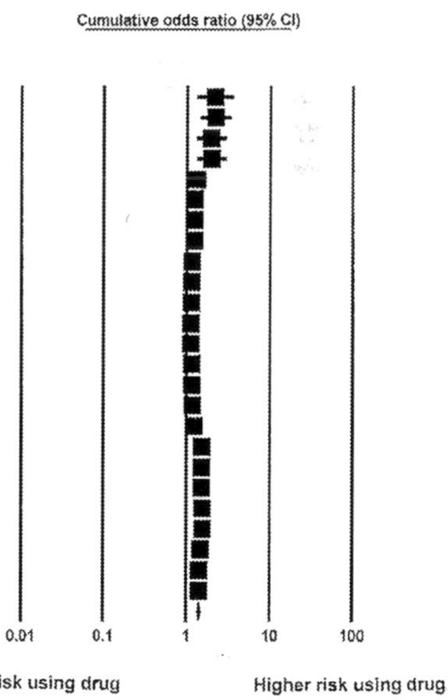

Fig. 3. Cumulative Meta-Analysis of Risk of Accident Involvement Associated with the Use of Antidepressant Drugs. *Source*: Elvik (2013b).

As seen in Fig. 3, the summary estimate of risk has changed over time. Until 1995 studies found that using antidepressants doubled the risk of accident involvement. Then a series of studies followed indicating a lower risk, until summary risk reached a value of 1.11 in 2000 and was no longer statistically significant. Subsequently, the summary estimate of risk has, in general, crept upward and ended at the value of 1.43 in 2009. The trend over time does not show a consistent direction, and the summary estimate of risk has always remained above 1.0.

A third type of analysis is meta-regression. Meta-regression is suitable for studying the influence of continuous moderator variables on the summary estimate. Fig. 4 shows the relationship between the summary estimate of the risk associated with using antidepressant drugs and the number of potentially confounding variables controlled for in the primary studies.

Although the best estimates of the coefficients indicate a negative relationship, only the intercept is statistically significant. The slope is, however, more likely to be negative than positive. The analysis therefore indicates that there is a tendency for studies controlling for many potentially confounding variables to find a smaller increase in accident risk associated with using antidepressant drugs than studies controlling for few or no confounding variables.

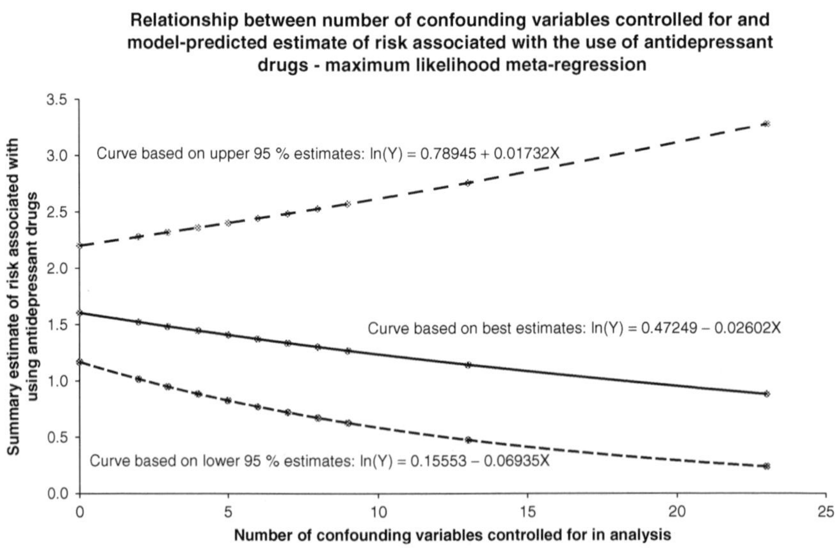

Fig. 4. Relationship between Number of Confounding Variables Controlled for and Model-Predicted Estimate of Risk Associated with the Use of Antidepressant Drugs. *Source*: Elvik (2013b).

SENSITIVITY ANALYSIS

In any meta-analysis a number of analytic choices are made that can influence the results. In addition, characteristics of the data may influence the results. A sensitivity analysis should therefore always be part of a meta-analysis (Elvik, 2005). It should, as a minimum, address the possible presence of publication bias, outlying data points, study quality assessment and, if relevant, an estimator of effect.

Publication bias (Rothstein, Sutton, & Borenstein, 2005) denotes the tendency not to publish studies where the findings are either not statistically significant or unexpected and therefore regarded as difficult to interpret. In road safety studies one would typically expect studies showing that safety treatments are ineffective or have adverse effects to be suppressed. There are many techniques for detecting publication bias based on funnel plots; all of them rest on the assumption that the funnel plot should be symmetric if there is no publication bias.

A simple technique that both detects and adjusts for publication bias is the trim-and-fill technique (Duval, 2005; Duval & Tweedie, 2000a, 2000b). It trims away data points in a funnel plot until the plot becomes symmetric around the weighted (trimmed) mean. Then, the mirror image of the trimmed data points is filled in, symmetrically around the trimmed mean. Fig. 5 shows an example of the results of a trim-and-fill analysis. It refers to studies that have assessed the relationship between the use of benzodiazepines and accident involvement.

There were 51 data points, to which 26 were added in the trim-and-fill analysis. In the fixed-effects analysis, 34 data points were added. In general, the trim-and-fill method adjusts less for publication bias in a random-effects model than in a fixed-effects model (Peters, Sutton, Jones, Abrams, & Rushton, 2007). The new data points generated by trim-and-fill are fictitious and serve only to illustrate the possible magnitude of the bias. These data points should therefore never be included in the main meta-analysis but only be part of a sensitivity analysis.

An outlying data point is any data point that has a decisive influence on the summary estimate. To test for the presence of outlying data points the meta-analysis is redone omitting one data point at a time and estimating a weighted mean based on the remaining $g-1$ data points. If all estimates developed this way lie within the confidence intervals of the summary estimate based on all data points, there are no outlying data points. If one or more of the summary estimates based on $g-1$ data points differ significantly from the overall summary estimate, the data points associated with these differences are outlying.

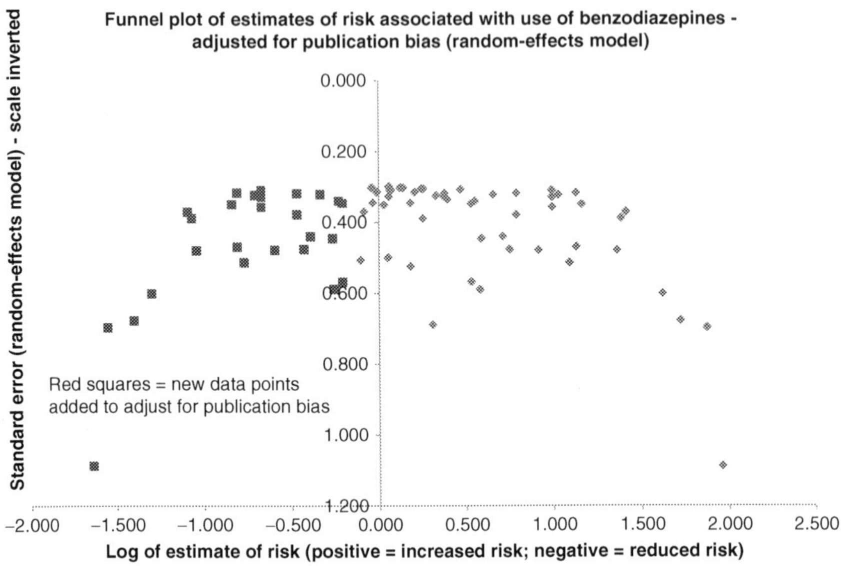

Fig. 5. Funnel Plot of Estimates of Risk Associated with the Use of Benzodiazepines – Adjusted for Publication Bias. *Source*: Elvik (2013b).

Fig. 6 shows the results of an analysis probing for outlying data points in studies of the association between using antidepressant drugs and the risk of accident involvement.

The results in Fig. 6 are remarkably stable, but there is one outlying data point, the first result in the 'Bramness 2008' study. The third item considered in sensitivity analysis is study quality. For the studies of the association between accident involvement rate and the use of drugs (Elvik, 2013b), a quantitative quality score was developed, ranging from 0 to 1. Most studies scored between 0.25 and 0.65 on this scale. Fig. 7 shows the results of a maximum likelihood meta-regression analysis of the relationship between quality score and estimate of the risk of accident involvement when using antidepressant drugs.

As illustrated in Fig. 7, a clear relationship between study quality and estimate of risk was found. Both the intercept and slope were statistically highly significant. The higher the quality of a study, the lower the estimate of risk. Quality score explained 46% of the systematic between-study variation in estimates of risk.

Finally, a comparison was made between two estimators of risk used in the studies dealing with antidepressant drugs and traffic risk. Most studies

Study name	Subgroup within study	Point	Lower limit	Upper limit	Z-Value	p-Value
		Statistics with study removed				
Ray 1992	Injury	1.424	1.361	1.490	15.301	0.000
Leveille 1994	Injury current user	1.427	1.364	1.493	15.409	0.000
Leveille 1994	Injury past user	1.431	1.367	1.497	15.546	0.000
Currie 1995	Injury	1.429	1.366	1.495	15.500	0.000
Barbone 1998	PDO anti	1.440	1.376	1.507	15.640	0.000
Barbone 1998	PDO sero	1.435	1.371	1.501	15.616	0.000
Barbone 1998	Serious injury anti	1.430	1.367	1.496	15.515	0.000
Barbone 1998	Serious injury sero	1.429	1.366	1.495	15.509	0.000
Barbone 1998	Slight injury anti	1.435	1.372	1.502	15.665	0.000
Barbone 1998	Slight injury sero	1.431	1.368	1.498	15.567	0.000
Neutel 1998	Injury	1.431	1.368	1.498	15.548	0.000
McGwin 2000	Mostly PDO	1.431	1.368	1.497	15.563	0.000
Wadsworth 2003	Injury	1.431	1.368	1.497	15.519	0.000
Etminan 2004	Injury	1.427	1.364	1.493	15.410	0.000
Lam 2005	Injury	1.429	1.366	1.495	15.486	0.000
Wadsworth 2005	Injury	1.429	1.368	1.495	15.489	0.000
Sagberg 2006	Mostly PDO	1.426	1.363	1.492	15.330	0.000
Bramness 2008	Injury newer	1.265	1.185	1.350	7.075	0.000
Bramness 2008	Injury sedating	1.432	1.366	1.501	14.908	0.000
Hours 2008	Injury	1.427	1.364	1.493	15.422	0.000
Rapoport 2008	Mostly PDO	1.398	1.334	1.465	13.929	0.000
Vingilis 2008	Injury	1.418	1.355	1.483	15.052	0.000
Gibson 2009	Injury group 1	1.494	1.424	1.567	16.396	0.000
Gibson 2009	Injury group 2	1.486	1.417	1.558	16.440	0.000
Woratanarat 2009	Injury	1.429	1.366	1.495	15.501	0.000
		1.429	1.366	1.495	15.515	0.000

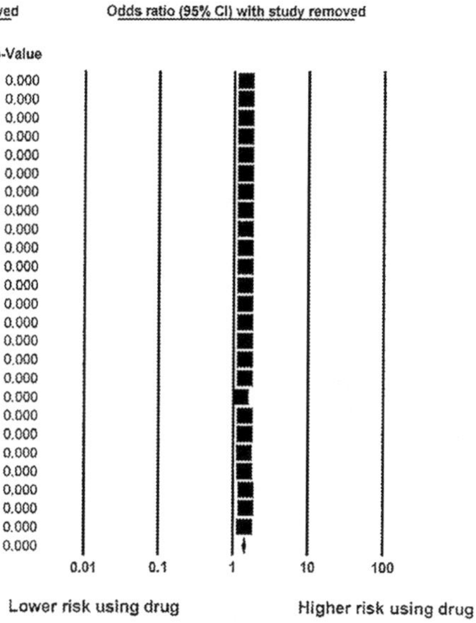

Fig. 6. Sensitivity Analysis to the Presence of Outlying Data Points in Studies of the Risk Associated with the Use of Antidepressant Drugs. *Source*: Elvik (2013b).

use the odds ratio of accident involvement as an estimator of risk. However, a few studies have used relative risk. When a risk factor is rare, odds ratios and relative risk are close. It was judged that using antidepressant drugs is sufficiently rare among drivers that only a negligible error would be made by treating odds ratios and relative risks as equivalent estimators. The lower part of Table 1 tests whether this is the case or not.

In general, one would expect estimates of risk-based relative risks to be slightly lower than those based on odds ratios. Table 1 confirms that this is the case. The difference in summary estimates is statistically significant at the 5% level in the fixed-effects model, but not in the random-effects model.

CONTROVERSIES, PITFALLS, RESEARCH NEEDS

There are a number of controversies in meta-analysis, some pitfalls and some needs for further research. This section briefly discusses some of these issues.

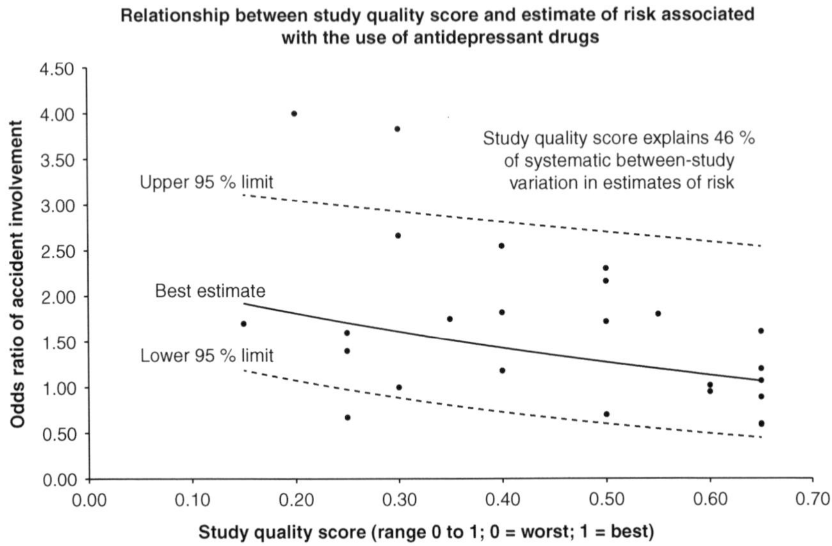

Fig. 7. Relationship Between Study Quality Score and Estimate of Risk Associated with the Use of Antidepressant Drugs. *Source*: Elvik (2013b).

One of the oldest controversies in meta-analysis concerns how much heterogeneity can be accepted before a meta-analysis becomes meaningless, or at least uninformative. The most restrictive point of view, probably held by few today, is that a meta-analysis should only be performed when there is no heterogeneity in results, that is, only when the *Q*-statistic is not statistically significant. This would seriously limit the conditions under which meta-analysis can be performed. A more common approach is therefore to try to explain heterogeneity by means of meta-regression or subgroup analysis. It is, however, possible to imagine circumstances when unexplained heterogeneity would be regarded as problematic. If, for example, a drug is highly effective in 75% of patients but harmful for the remaining 25%, and there is no way of predicting which group a patient belongs to, one can understand why doctors would hesitate to use the drug. There will be heterogeneity in virtually all data sets containing a sizable number of studies (more than 25). This need not be a problem if the distribution of results is 'well-behaved' (symmetric, unimodal, with the weighted mean located in the middle of the data points) and some sources of heterogeneity can be identified.

A very large amount of heterogeneity, combined with highly variable study quality, can make meta-analysis challenging. Fig. 8 presents a case in point.

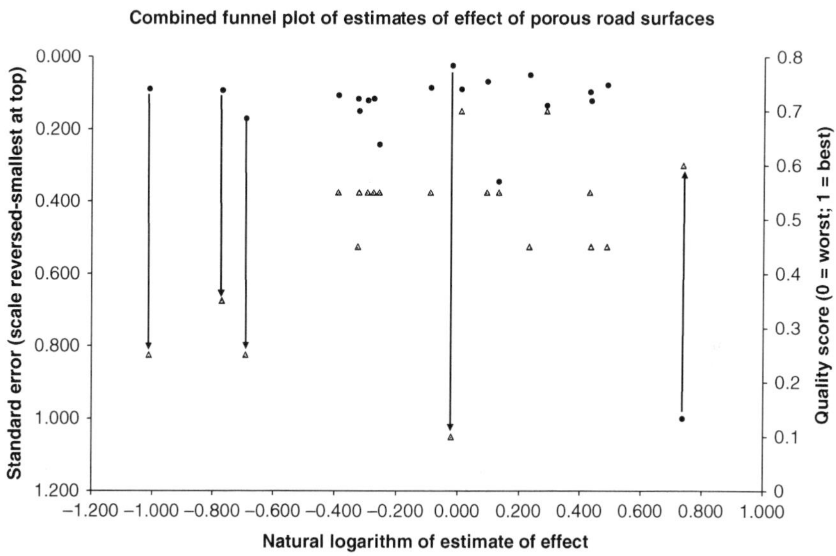

Fig. 8. Combined Funnel Plot of Estimates of Effect of Porous Road Surfaces. *Source*: Elvik and Greibe (2005).

It presents 20 estimates of the effect on accidents of porous road surfaces, along with the quality scores assigned to these estimates.

Several observations can be made. First, there is enormous dispersion in estimates of effect; stated as percentages, they range from an accident reduction of 64% to an increase of 109%. Second, there is no tendency for the more precise estimates to be less dispersed than the less precise estimates. Third, some of the more precise estimates score very low for study quality, as indicated by the arrows pointing down in Fig. 8. Fourth, the least precise estimate of all scores is fairly high for study quality (arrow pointing up in Fig. 8). This is a mess. Elvik and Greibe (2005) nevertheless made a heroic attempt to summarise the evidence by means of meta-analysis. One could argue that the analysis made little sense; at the very least it was inconclusive, as noted by the authors.

Another long-lasting controversy in meta-analysis concerns study quality assessment. Greenland (1994) condemns it in colourful terms; others (Cooper, 2010) defend it and explain in detail how to assess study quality. Assessing study quality is certainly not an exact science and any numerical scale intended to indicate study quality will to some extent be arbitrary and subjective (i.e., reflect the points of view held by the researchers who developed the scale). Yet, as long as the assessment or the scale is explained, others can decide whether or not they think the assessment or scale makes sense. In some cases, even a crude assessment of study quality may help prevent researchers from drawing misleading conclusions. Thus, Elvik (1999), in a meta-analysis of studies that evaluated the effects on accidents of studded tires, formed three crudely defined groups of studies by study quality. Estimates of effects were found to vary enormously between the groups; had no categories for study quality been formed, and all studies pooled, the summary estimate of effect would have been highly misleading.

A third topic which has so far not been entirely resolved is when an analysis of regression coefficients, fitted by means of multivariate statistical analyses, makes sense. Card (2012) argues for a restrictive point of view, stating that a formal synthesis of regression coefficients only makes sense if the models are identical in all respects; that is, they include exactly the same variables and are specified in exactly the same way. The practice of meta-analysis of regression coefficients does not follow this strict guideline. One only has to open any journal that publishes meta-analyses to find examples of such analyses, which are obviously based on models that were not identical in all respects, as demanded by Card. Thus, Brons, Pels, Nijkamp and Rietveld (2002) present a meta-analysis of price elasticities of demand for passenger air travel. The studies they summarise were not identical in all aspects. They include both cross-sectional and time-series studies, which can be very different from a statistical point of view and do not necessarily produce identical results.

It is nevertheless clear that Card (2012) has a point. Coefficients estimated in different models can differ simply because the models are different, not because the underlying relationship they are intended to capture differs. In some cases it is possible to test the stability of regression coefficients in different model specifications. Elvik and Bjørnskau (2017) performed such a test. Finding that coefficients were stable in different model specifications they proceeded to formally synthesise regression coefficients estimated in different studies.

A fourth topic where different approaches have been taken is the treatment of statistical dependence between multiple estimates of effect in the same study. There are at least five approaches to the treatment of multiple, statistically dependent results from the same study. The first approach is to include only a single estimate from each study in the meta-analysis. A choice must then be made, either by the analyst or by the authors of the study, about which estimate to include. The second approach is to formally synthesise multiple findings of a single study into a single result. This approach is somewhat self-contradictory, since the dependence between study results is regarded as a problem when synthesising the results. The third approach, explained by

Gleser and Olkin (2009), is to statistically model the dependence between results and perform a meta-regression, which adjusts for the degree of correlation between the individual results of a study. Such an analysis is complex and few examples of it are found in the literature. A fourth approach is illustrated by Elvik, Christensen and Amundsen (2004). They compared the relative contributions of (random) sampling variation and systematic variation to the total variance in studies providing 1, 2, 3, so on estimates of effect. The idea was that if there is a dependency between the results of a single study, the relative contribution of (pseudo) between-studies (between-estimates) variation should be smaller than in a set of studies each providing only a single estimate of effect. They did not find such a tendency. Finally, a fifth approach (Nelson & Kennedy, 2009) is to down-weight each estimate in a study giving N estimates by $1/N$.

A fifth topic deserving a short discussion is whether it is defensible to perform a meta-analysis when none of the formal methods, such as the inverse-variance method, can be applied. Thus, in the analysis reported by Elvik and Bjørnskau (2017), the inverse-variance method could only be applied to 21 of 30 estimates for motor vehicle volume, 11 of 16 estimates for pedestrian volume and 11 of 14 estimates of cycle volume. However, all estimates could be included in simpler analyses. Table 2 compares the results.

As can be seen from Table 2, the results are very similar. At least in this case, therefore, strict adherence to rigorous methods did not make a large difference for the results of the analysis.

Table 2. Summary Estimates of Coefficients for Traffic Volume Based on Different Methods of Analysis.

	Mean Values of Coefficients (95% Confidence Limits in Parentheses)		
Method of synthesis	Motor vehicle volume	Cyclist volume	Pedestrian Volume
Simple mean (not weighted)	0.540 (0.452; 0.628)	0.431 (0.339; 0.523)	0.518 (0.446; 0.590)
Mean weighted by number of accidents	0.473 (0.462; 0.484)	0.519 (0.503; 0.535)	0.567 (0.550; 0.584)
Fixed-effects inverse-variance meta-analysis mean	0.498 (0.452; 0.544)	0.516 (0.473; 0.560)	0.517 (0.493; 0.543)
Random-effects inverse-variance-meta-analysis mean	0.489 (0.415; 0.563)	0.432 (0.333; 0.530)	0.510 (0.409; 0.611)

Source: Elvik and Bjørnskau (2017).

REFERENCES

Borenstein, M., Hedges, L. V., Higgins, J. T., & Rothstein, H. (2009). *Introduction to meta-analysis.* New York, NY: John Wiley and Sons.

Brons, M., Pels, E., Nijkamp, P., & Rietveld, P. (2002). Price elasticities of demand for passenger air travel: A meta-analysis. *Journal of Air Transport Management, 8*, 165–175.

Card, N. A. (2012). *Applied meta-analysis for social science research.* New York, NY: The Guilford Press.

Cooper, H. M. (2010). *Research synthesis and meta-analysis. A step-by-step approach* (4th ed.). Thousand Oaks, CA: Sage Publications.

Cooper, H., Hedges, L. V., & Valentine, J. C. (Eds.). (2009). *The handbook of research synthesis and meta-analysis* (2nd ed.). New York, NY: Russell Sage Foundation.

Duval, S. (2005). The trim and fill method. In H. Rothstein, A. J. Sutton, & M. Borenstein (Eds.), *Publication bias in meta-analysis: Prevention, assessment and adjustments* (pp. 127–144). Chichester: John Wiley and Sons.

Duval, S., & Tweedie, R. (2000a). Trim and fill: A simple funnel plot based method of testing and adjusting for publication bias in meta-analysis. *Journal of the American Statistical Association, 95*, 89–98.

Duval, S., & Tweedie, R. (2000b). A non-parametric trim and fill method of assessing publication bias in meta-analysis. *Biometrics, 56*, 455–463.

Egger, M., Davey Smith, G., & Altman, D. G. (2001). *Systematic reviews in health care. Meta-analysis in context.* London: BMJ Publishing Group.

Elvik, R. (1999). The effects on accidents of studded tires and laws banning their use: A meta-analysis of evaluation studies. *Accident Analysis and Prevention, 31*, 125–134.

Elvik, R. (2005). Can we trust the results of meta-analyses? A systematic approach to sensitivity analysis in meta-analyses. *Transportation Research Record, 1908*, 221–229.

Elvik, R. (2013a). A re-parameterisation of the power model of the relationship between the speed of traffic and the number of accidents and accident victims. *Accident Analysis and Prevention, 50*, 854–860.

Elvik, R. (2013b). Risk of road accident associated with the use of drugs: A systematic review and meta-analysis of evidence from epidemiological studies. *Accident Analysis and Prevention, 60*, 254–267.

Elvik, R., & Bjørnskau, T. (2017). Safety-in-numbers: A systematic review and meta-analysis of evidence. *Safety Science, 92*, 274–282.

Elvik, R., Christensen, P., & Amundsen, A. (2004). *Speed and road accidents. An evaluation of the power model.* Report 740. Oslo: Institute of Transport Economics.

Elvik, R., & Greibe, P. (2005). Road safety effects of porous asphalt: A systematic review of evaluation studies. *Accident Analysis and Prevention, 37*, 515–522.

Espey, M. (1998). Gasoline demand revisited: An international meta-analysis of elasticities. *Energy Economics, 20*, 273–295.

Eysenck, H. J. (1978). An exercise in mega-silliness. *American Psychologist, 33*, 517.

Gleser, L. J., & Olkin, I. (2009). Stochastically dependent effect sizes. In H. Cooper, L. V. Hedges, & J. C. Valentine (Eds.), *The handbook of research synthesis and meta-analysis* (2nd ed., Chap. 19, pp. 357–376). New York, NY: Russell Sage Foundation.

Greenland, S. (1994). Invited commentary: A critical look at some popular meta-analytic methods. *American Journal of Epidemiology, 140*, 290–296, 300–302.

Hedges, L. V., & Olkin, I. (1985). *Statistical methods for meta-analysis.* San Diego, CA: Academic Press.

Higgins, J. P. T., & Green, S. (2008). *Cochrane handbook for systematic reviews of interventions*. Chichester: John Wiley and Sons.

Lawrence, D. W. (2008). *The information-seeking behaviors of professionals and information sources in the field of injury prevention and safety promotion*. Ph.D. thesis, Karolinska Institutet, Stockholm.

Light, R. J., & Pillemer, D. B. (1984) *Summing up. The science of reviewing research*. Cambridge, MA: Harvard University Press.

Lindhjem, H., Navrud, S., Biasque, V., & Braathen. N. A. (2012). *Mortality risk valuation in environment, health and transport policies*. Paris: OECD Publishing.

Lipsey, M. W., & Wilson, D. B. (2001). *Practical meta-analysis*. Thousand Oaks, CA: Sage Publications.

Littell, J. H., Corcoran, J., & Pillai, V. (2008). *Systematic reviews and meta-analysis*. Oxford: Oxford University Press.

Lund, T. (Red). (2002). *Innføring i forskningsmetodologi*. Oslo: Unipub.

Nelson, J. P., & Kennedy, P. E. (2009). The use (and abuse) of meta-analysis in environmental and natural resource economics: An assessment. *Environmental and Resource Economics*, *42*, 345–377.

Peters, J. L., Sutton, A. J., Jones, D. R., Abrams, K. R., & Rushton, L. (2007). Performance of the trim and fill method in the presence of publication bias and between-study heterogeneity. *Statistics in Medicine*, *26*, 4544–4562.

Petitti, D. B. (2000). *Meta-analysis, decision analysis and cost-effectiveness analysis* (2nd ed.). New York, NY: Oxford University Press.

Petticrew, M., & Roberts, H. (2006). *Systematic reviews in the social sciences. A practical guide*. Oxford: Blackwell Publishing.

Rosenthal, R. (1991). *Meta-analytic procedures for social research. Applied social research methods series* (Vol. 6). Newbury Park, CA: Sage Publications.

Rothstein, H., Sutton, A. J., & Borenstein, M. (2005). *Publication bias in meta-analysis. Prevention, assessment and adjustments*. Chichester: John Wiley and Sons.

Schmidt, F. L., & Hunter, J. E. (2015). *Methods of meta-analysis. Correcting error and bias in research findings* (3rd ed.). Thousand Oaks, CA: Sage Publications.

Smith, M. L., & Glass, G. V. (1977). Meta-analysis of psychotherapy outcome studies. *American Psychologist*, *32*, 752–760.

Sterne, J. A., & Egger, M. (2001). Funnel plots for detecting bias in meta-analysis: Guidelines on choice of axis. *Journal of Clinical Epidemiology*, *54*, 1046–1055.

Sutton, A. J., Abrams, K. R., Jones, D. R., Sheldon, T. A., & Song, F. (2000). *Methods for meta-analysis in medical research*. New York, NY: John Wiley and Sons.

SUMMARY AND CONCLUSIONS

CHAPTER 20

CONCLUSIONS AND FUTURE DIRECTIONS

Dominique Lord and Simon Washington

ABSTRACT

Purpose – *This chapter highlights main thrusts discussed in the book in its entirety.*

Methodology – *This chapter reviews the content of the book, drawing together the challenges safety analysts and other important road safety stakeholders confront while trying to understand and reduce transport system risks.*

Findings – *The chapter describes the challenges the profession is confronting. These challenges include the difficulty in analysing crashes and drawing reliable conclusions when crashes are rare, the complexities involved with piecing together and identifying actual causes of crashes, how driver behaviour is a dominant influence on crash risk, how a diverse mix of road users is challenging to manage, that estimating the safety benefits of treatments is fraught with complexity, how surrogate measures of safety – despite the analytical research needed on their linkages with crashes – can be used for proactively improving design and operational decisions, and how real-time data collected on monitored highways, which need the development of solid theoretical underpinnings moving forward – can be used reliably for*

Safe Mobility: Challenges, Methodology and Solutions
Transport and Sustainability, Volume 11, 451–458

ISSN: 2044-9941/doi:10.1108/S2044-994120180000011023

estimating crash risks and manage safety. The chapter briefly summarises how these challenges may be addressed moving forward.

The chapter also identifies future research opportunities that can be pursued to further improve safe mobility. These opportunities include the use of alternate methodologies for 'measuring safety' such as advanced vision recognition techniques, the use of surrogate measures of safety such as time to collision, driverless and connected vehicles, naturalistic driving experiments and data, the reshaping of urban cities, and the rapid growth and motorisation of developing countries.

Practical implications – *By fostering a better understanding of the challenges the profession is facing, and prompting a dialogue on how they can be overcome, will lead to a reduction in the number and severity of crashes of global transport networks, making them more sustainable. The chapter shows how some of these challenges can be tackled.*

Originality/value of chapter – *This chapter draws from all the chapters of this book that describes various challenges road safety professionals currently face, and directs readers to different parts of the book for more in-depth treatment on how these challenges can be met. The chapter also describes research opportunities where further safety gains can be obtained moving forward.*

Keywords: Road crashes; road safety; societal risk; risk management; transport engineering; driver behaviour; pedestrians; transport network management; bicyclists

CHALLENGES CONFRONTING THE PROFESSION

While operating and managing transport systems safely is sufficiently difficult, conducting rigorous scientific studies to quantify and understand crash risk in the transport network is fraught with challenges. The following partial list identifies many of the ongoing challenges and how these challenges were addressed in this book:

1. Road crashes are relatively rare events with respect to how much time is spent driving, giving rise to analysis challenges. Mannering (Chapter 12) discusses how many of the insights are provided through analysis of cross-section data – most suited for analysing specific types of transport network

locations. Quddus (Chapter 13) describes how crashes over time can be analysed, better suited to understanding how seasonal effects such as weather, vacation travel and safety programs and policies can be assessed. Lord and Geedipally (Chapter 14) discuss the particular challenges associated with the rarity of crashes, as often there is limited evidence available to make defensible conclusions of the impacts of investments on safety. The profession will continue to explore ways to improve the methods used to examine crash data and yield as yet unknown insights as to causes of and remedies for motor vehicle crashes.
2. It is difficult to piece together and identify the actual causes of crashes. While a known major cause of crashes is human error, the design of transport network locations also contributes in significant ways to crash risk. Quantifying the relative contributions to crash risks remains an ongoing challenge. The crash risks arising from engineering design are dependent on the type of facility being considered. Brijs and Pirdavani (Chapter 5) discuss the known engineering design effects of urban and suburban arterials, while La Torre (Chapter 6) examines those related to high speed, controlled access facilities. Intersections are relatively complex driving environments and place large cognitive loads onto drivers. What is known about rural and urban intersection safety is discussed by Tay in Chapter 7, while roundabouts are covered by Montella in Chapter 8.
3. Road crashes are predominately caused by driver errors. Sorting out the contribution of driver behaviour to road crashes is a challenging business. Whether the behaviour issues are related to driver training and licensing (Bates, Filtness and Watson; Chapter 2), aggressive driving and speeding (Kockelman and Ma; Chapter 3), or distracted driving (Cunningham and Regan; Chapter 4) – the challenges with identifying the causes and subsequently selecting appropriate mitigations is fraught with difficulty. As a main source of knowledge on driver behaviour related risk is through simulator-based studies, one area of fruitful research will be on quantifying, or calibrating the linkage between simulator-derived and actual crash risk – if it is possible. Moreover, identifying effective strategies for mitigating the crash risks posed by driver behaviour related issues will remain an important area of research.
4. A mixture of road users presents challenges to maintaining road safety. When users with differing levels of protection and travel speeds share the same, adjacent parts of the same or proximal facilities, then crash risks tend to increase and crashes also tend to be more severe. These risks are described for pedestrians by Gårder in Chapter 10 and cyclists by Haworth and Fuller in Chapter 11. Providing safely for multiple road users will

remain a challenge moving into the future, especially as the world population becomes increasingly urbanised, increasing the mix of users on many of our roads.

5. Studies aimed at assessing the impact of a particular treatment on road safety – such as installation of guardrail, pedestrian crossings or hazard warning signs – rely predominately on before–after studies, which are threatened by a lack of experimental control on the myriad of factors that can affect road crashes. Methodologically the before–after study – presented in Chapter 18 by Persaud – remains the most reliable and time-tested approach for reaching reliable conclusions about the safety effects of countermeasures – be they road features, policies or operational enhancements. When designed and administered with care, before–after studies are widely believed to be superior to alternative approaches for assessing road safety. Meta-analytical approaches – discussed in Chapter 19 (Elvik) – are extremely useful and reliable for assessing the prior evidence on specific safety investments and are particularly useful for informing decision – making when a local or regional cannot be conducted. In contrast, simulator studies that can be controlled experimentally to a significant extent relative to observational studies are much better suited at isolating the contributors to crashes and for assessing relative risks but are unreliable for measuring absolute levels of risk (see Chapter 4). Research will continue into making each of these methods for assessing road safety more reliable and overcoming their individual weaknesses.
6. Crash surrogate measures that quantify crash 'pre-cursor' events such as near-misses are extremely promising tools for evaluating road safety. Tarko (Chapter 17) describes this growing field of research, which will support a transition of the safety profession away from reactive and towards proactive methods for assessing transport network safety. New developments in image processing combined with advances in computing power make this area of research and proactive safety management extremely promising.
7. Real-time crash prediction, with the aim of taking a proactive and immediate response to crash mitigation, will require the establishment of reliable, theoretical relationships between crash risk and microscopic traffic conditions. Abdel-Aty, Shi, Pande and Yu (Chapter 9) describe this area of research, which to date is dominated by empirical studies aimed at establishing linkages between traffic operations and crash risk. As sensors and near real-time analytics capabilities support the development of this work, this focussed research area will remain both a challenge and an opportunity for improving road safety going forward, and for linking with on-road messaging to road users about impending road safety hazards.

SIGNIFICANT RESEARCH OPPORTUNITIES MOVING FORWARD

Improving transport system safety from a transport manager's perspective and personal safety from a traveller's perspective are challenges facing developing and developed countries alike. While the challenges with delivering safe transport systems are numerous – as described in the previous section – there are also numerous opportunities confronting the safety profession. If implemented deliberately and effectively, new programs, technologies and policies are likely to have significant positive impacts on the safety of future mobility. The remainder of this section describes several of the most important opportunities that are likely to have significant impacts on the safety and sustainability of future mobility.

Surrogate Measures of Road Safety

Road safety managers prefer to rely on an evidence-based analysis of road safety prior to investing valuable public funds to improve road safety. Governments in general require empirically based methods and procedures for justifying transport network safety-related expenditures (see Chapters 12–16). These requirements have in large part driven the dramatic development of reactive safety methods for managing transport system safety the profession has seen over the past three decades or so. These requirements have placed governments and road-safety managers in an extremely undesirable predicament – requiring the accumulation of road crashes in order to interpret them, and subsequently make investments to eliminate them.

The rapid development of computer vision capability over the past decade, along with advances in computer processing speed, has enabled the development of prototype vision processing software that can capture, record, quantify and tabulate surrogate measure of road safety at selected transport system locations. Metrics like time-to-collision (see Chapter 17) – which captures the time until two transport system users (e.g. vehicle and pedestrian) – will have a direct impact unless current speeds and/or trajectories are changed, are now automatically collected by such systems for every possible interaction pair at a particular location (e.g. intersection). Processing a video recording of a busy signalised intersection, for example, which could have more than 15,000 vehicles per hour passing through it, would have every single vehicle tracked along with its time to potential collision with other vehicles, pedestrians and cyclists. The detailed metrics arising from such automated observation – which can be done over a relatively short period of time (e.g. 2 days) – can reveal significant insights into the operational performance of the intersection.

While the advancement offered through video processing is transformational, the remaining challenge and opportunity facing the profession are in the establishment of reliable relationships between these surrogate measures and actual crashes. For example, is the risk of collision for two vehicles that will collide in 1 second unless diverted the same for head-on and angle crashes? What about the crash risk between vehicles and pedestrians or cyclists, or even between pedestrians and cyclists? It remains the opportunity in future research to describe and quantify these relationships, so we can then better interpret the results of surrogate measure of road safety and make appropriate subsequent investments.

Driverless Vehicles

The current design and operation of transport facilities when occupied by driverless vehicles will fundamentally change – with these changes having profound impacts on safety. The changes are likely to shift where risk is occurring on transport networks. In-vehicle crash risk is likely to be reduced significantly during automated driving; however, transitions from manual and automated driving may pose new risks. Interactions between pedestrians and cyclists – who will not be automated, are likely to present unique challenges with respect to crash risk.

The impact on risks associated with vehicle failure, both in relation to the vehicle itself (say the automated braking not working or a flat tire, etc.) and the roadway communications system (e.g., traffic signal talking to vehicle) are also largely unknown. If and when an automated system component fails, what are the risks, especially with regard to crash severity? Will these risks be pre-programmed? If so, how will those programs be written to minimise societal damage and risk?

It is too early to know exactly how transport system crash risks will be shifted as a result of computers delivering the driving task – but we do anticipate that shifts will occur. The extent to which road-safety professionals currently involved in road safety assessments will engage in assessing the new risks in the new environment. It is quite likely that entirely new methodologies for assessing risks will be needed to support the new paradigm.

Naturalistic Driving Experiments and Data

With the rapid advancement in recording technology, sensors and computing power, safety analysts have now collected and begun to access a wide range of on-road or 'naturalistic' data that can be used to gain an improved understanding of factors that increase or reduce crash risk. The advantage of these naturalistic data is that we do not have to rely on $P(B|C)$ to make statements about $P(C|B)$, where P is *probability*, C is *crash* and B is *driver behaviour*. In other words, crash data that form the basis of most crash studies to date are based on data that provide the probability of an event having occurred (e.g., talking to a passenger, driving on a wet road at high speed), conditional on a crash having occurred, whereas analysts we would like to know the probability of a crash, given that certain events have occurred – the exact opposite conditional probability.

Naturalistic data provide an opportunity to obtain the correct conditional probability needed to make objective statements about road risk as they relate to driver behaviours. These data are collected as part of naturalistic driving studies, using instrumented vehicles equipped with video cameras and various kinematic sensors and radars. It is not uncommon to see more than 1,000 terabytes of collected data from naturalistic driving studies.

Analyses of these data has started yielding new insights regarding the interaction between the driver, vehicle and the environment. However, given the size of the data assembled and the newness of these datasets, only a fraction of the data has been analysed to date. Consequently, new and innovative statistical methods are needed to efficiently analyse such large datatsets. Mannering (Chapter 12) and Lord and Geedipally (Chapter 14) have provided some new avenues for undertaking these analyses, but additional tools will be needed that are specifically tailored for these datatsets. The new tools needed may include data mining techniques, Bayesian neural networks and machine learning techniques.

Reshaping of Cities

Major cities around the world are becoming unsustainable with respect to traffic congestion, pollution, social isolation and the toll attributed to road safety. These externalities are impinging upon economic productivity, social welfare and wellbeing and the public health burden. Injuries and deaths arising from motor vehicle crashes are in most countries increasing in risk relative to other societal injury and death risks such as falls, cancer, wars and other chronic diseases. For these reasons and others, urban and transport planners are looking for alternative ways to configure the way people live and interact within cities.

The world is starting to see dramatic responses to these problems in mostly urban areas. Some city governments are looking to or have already restricted vehicular access to the heart of cities. City governments are looking to progressive land uses, economic and social policies to influence mobility demand, particularly for single occupant vehicle use. As these policies, investments and programs are trialled around the world, we are likely to see increase in active travel, with greater challenges in providing safe environments for vehicles, public transport and active travellers to co-mingle (see Chapters 10 and 11 for a discussion of these current challenges).

Future safety research is likely to focus on safety impacts of these initiatives. As many of these programs and policies have region-wide effects, the methodological techniques needed to examine these programs will also need to be developed and improved.

Rapid Growth in Mobility in Developing Countries

Rapid motorisation in developing countries presents significant challenges in maintaining safe mobility systems. While many countries view increased personal mobility via access to personal vehicles as proportional to increases to quality of life for the local population, this motorisation often coincides with increased collisions between active travellers and vehicles – which tend to be severe – along with other externalities as well. Compounding these increased potentially lethal interactions, often the design standards in developing countries cannot keep up with the increased demand for vehicular travel. Moreover, and perhaps even more challenging, is that the social behaviour of drivers in many developing countries is inconsistent with the laws the govern driving in those same countries, with effective police enforcement and feedback mechanisms to improve these social norms lacking or largely ineffective.

Hindering effective safety analysis in developing countries is a general lack of data, which in contrast to developed countries, is routinely available. Thus, the fundamental notion of evidence-based safety programming and investment is difficult to implement in developing countries. Another argument, then, for implementation of proactive surrogate safety analysis methodologies (see Chapter 17) is to rapidly assist the improvement of safety in developing countries. In addition, the cumulative evidence about road safety is more reliable than relying on individual studies – in which case meta-analysis can be extremely useful (see Chapter 19) for guiding safety investments in developing countries.

INDEX

Note: Page numbers followed by "*f*" and "*t*" refer to figures and tables.